GRADUATE STUDIES IN MATHEMATICS 171

Nonlinear Elliptic Equations of the Second Order

Qing Han

American Mathematical Society
Providence, Rhode Island

2010 *Mathematics Subject Classification.* Primary 35J60, 35J25, 35J93, 35J96.

For additional information and updates on this book, visit
www.ams.org/bookpages/gsm-171

Library of Congress Cataloging-in-Publication Data

Names: Han, Qing, 1964–
Title: Nonlinear elliptic equations of the second order / Qing Han.
Description: Providence, Rhode Island : American Mathematical Society, [2016] | Series: Graduate studies in mathematics; volume 171 | Includes bibliographical references and index.
Identifiers: LCCN 2015043419 | ISBN 9781470426071 (alk. paper)
Subjects: LCSH: Differential equations, Elliptic. | Differential equations, Nonlinear. | AMS: Partial differential equations – Elliptic equations and systems – Nonlinear elliptic equations. msc | Partial differential equations – Elliptic equations and systems – Boundary value problems for second-order elliptic equations. msc | Partial differential equations – Elliptic equations and systems – Quasilinear elliptic equations with mean curvature operator. msc | Partial differential equations – Elliptic equations and systems – Elliptic Monge-Ampère equations. msc
Classification: LCC QA377 .H31825 2016 | DDC 515/.3533–dc23 LC record available at http://lccn.loc.gov/2015043419

∞ The paper used in this book is acid-free and falls within the guidelines established to ensure permanence and durability.
Visit the AMS home page at `http://www.ams.org/`

10 9 8 7 6 5 4 3 2 1   21 20 19 18 17 16

To Yansu, Raymond, and Tommy

Contents

Preface

The theory of nonlinear elliptic partial differential equations of the second order has flourished in the past half-century. The pioneering work of de Giorgi in 1957 opened the door to the study of general quasilinear elliptic differential equations. Since then, the nonlinear elliptic differential equation has become a diverse subject and has found applications in science and engineering. In mathematics, the development of elliptic differential equations has influenced the development of the Riemannian geometry and complex geometry. Meanwhile, the study of elliptic differential equations in a geometric setting has provided interesting new questions with fresh insights to old problems.

This book is written for those who have completed their study of the linear elliptic differential equations and intend to explore the fascinating field of nonlinear elliptic differential equations. It covers two classes of nonlinear elliptic differential equations, quasilinear and fully nonlinear, and focuses on two important nonlinear elliptic differential equations closely related to geometry, the mean curvature equation and the Monge-Ampère equation.

This book presents a detailed discussion of the Dirichlet problems for quasilinear and fully nonlinear elliptic differential equations of the second order: quasilinear uniformly elliptic equations in arbitrary domains, mean curvature equations in domains with nonnegative boundary mean curvature, fully nonlinear uniformly elliptic equations in arbitrary domains, and Monge-Ampère equations in uniformly convex domains. Global solutions of these equations are also characterized. The choice of topics is influenced by my personal taste. Some topics may be viewed by others as too advanced for a graduate textbook. Among those topics are the curvature estimates for minimal surface equations, the complex Monge-Ampère equation, and the generalized solutions of the (real) Monge-Ampère equations. Inclusion of these topics reflects their importance and their connections to many of the most active current research areas.

There is an inevitable overlap with the successful monograph by Gilbarg and Trudinger. This book, designed as a textbook, is more focused on basic materials and techniques. Many results in this book are presented in special forms. For example, the quasilinear and fully nonlinear uniformly elliptic differential equations studied in this book are not in their most general form. The study of these equations serves as a prerequisite to the study of the mean curvature equation and the Monge-Ampère equation, respectively. More notably, our discussion of the Monge-Ampère equations is confined to the *pure* Monge-Ampère equations, instead of the Monge-Ampère type equations.

This book is based on one-semester courses I taught at Peking University in the spring of 2011 and at the University of Notre Dame in the fall of 2011. Part of it was presented in the Special Lecture Series at Peking University in the summer of 2007, in the Summer School in Mathematics at the University of Science and Technology of China in the summer of 2008, and in a graduate course at Beijing International Center of Mathematical Research in the spring of 2010.

During the writing of the book, I benefitted greatly from comments and suggestions of many friends, colleagues, and students in my classes. Chuanqiang Chen, Xumin Jiang, Weiming Shen, and Yue Wang read the manuscript at various stages. Chuanqiang Chen and Jingang Xiong helped write Chapter 8. Bo Guan, Marcus Khuri, Xinan Ma, and Yu Yuan provided valuable suggestions on the arrangement of the book.

It is with pleasure that I record here my gratitude to my thesis advisor, Fanghua Lin, who guided me into the fascinating world of elliptic differential equations more than twenty years ago.

I am grateful to Arlene O'Sean, my editor at the American Mathematical Society, for reading the manuscript and guiding the effort to turn it into a book. Last but not least, I thank Sergei Gelfand at the AMS for his help in bringing the book to press.

The research related to this book was partially supported by grants from the National Science Foundation.

Qing Han

Introduction

The primary goal of this book is to study nonlinear elliptic differential equations of the second order, with a focus on quasilinear and fully nonlinear elliptic differential equations. Chapter 1 is a brief review of linear elliptic differential equations. Then in Part 1 and Part 2, we study quasilinear elliptic differential equations and fully nonlinear elliptic differential equations, respectively.

In Chapter 1, we review briefly three basic topics in the theory of linear elliptic equations: the maximum principle, Krylov-Safonov's Harnack inequality, and the Schauder theory. These topics form the foundation for further studies of nonlinear elliptic differential equations.

Part 1 is devoted to quasilinear elliptic differential equations and consists of three chapters.

In Chapter 2, we discuss quasilinear uniformly elliptic equations. We derive various a priori estimates for their solutions, the estimates of the L^∞-norms of solutions and their first derivatives by the maximum principle, and the estimates of the Hölder semi-norms of the first derivatives by Krylov-Safonov's Harnack inequality. As a consequence of these estimates, we solve the Dirichlet boundary-value problem by the method of continuity.

In Chapter 3, we discuss equations of the prescribed mean curvatures, or the mean curvature equations. We derive various a priori estimates for their solutions, in particular, the boundary gradient estimates, the global gradient estimates, and the interior gradient estimates. As a consequence, we solve the Dirichlet boundary-value problem by the method of continuity. Difficulties in studying the mean curvature equations are due to a lack of the uniform ellipticity. The structure of the equation plays an important role.

In Chapter 4, we discuss minimal surface equations. Needless to say, the minimal surface equation is a special class of the mean curvature equations; namely, the mean curvature vanishes identically. It might appear that this chapter should be included in the previous one. However, there is a reason for an independent chapter. Results in this chapter are proved by analysis "upon surfaces". In other words, we treat minimal surfaces as submanifolds in the ambient Euclidean spaces and write equations on these submanifolds. In this chapter, we will derive an improved interior gradient estimate and an interior curvature estimate for solutions of the minimal surface equation.

Part 2 is devoted to fully nonlinear elliptic differential equations and consists of four chapters.

In Chapter 5, we discuss fully nonlinear uniformly elliptic equations. We derive various a priori estimates for their solutions, the estimates of the L^∞-norms of solutions and their first and second derivatives by the maximum principle, and the estimates of the Hölder semi-norms of the second derivatives by Krylov-Safonov's Harnack inequality. As a consequence of these estimates, we solve the Dirichlet boundary-value problem by the method of continuity.

In Chapter 6, we discuss Monge-Ampère equations. We derive various a priori estimates for their solutions, in particular, the boundary Hessian estimates, the global Hessian estimates, and the interior Hessian estimates. As a consequence, we solve the Dirichlet boundary-value problem by the method of continuity. Difficulties in studying the Monge-Ampère equations are due to a lack of the uniform ellipticity. The structure of the equation plays an important role.

In Chapter 7, we extend results in the previous chapter to the complex case and discuss complex Monge-Ampère equations.

In Chapter 8, we discuss generalized solutions of (real) Monge-Ampère equations. Such solutions are defined only for convex functions, which are not assumed to be C^2 to begin with. We prove various regularity results under appropriate assumptions on the corresponding Monge-Ampère measures. In particular, we prove the strict convexity and the interior $C^{1,\alpha}$-regularity for solutions if the Monge-Ampère measures satisfy a doubling condition. We also derive the optimal interior $C^{2,\alpha}$-regularity for solutions under the condition that the Monge-Ampère measures are induced by positive Hölder continuous functions. The discussion is based on the level set approach.

Concerning the arrangement of this book, Part 1 is not a prerequisite for Part 2. Those who are interested only in fully nonlinear elliptic equations can skip Part 1 entirely.

We now list some basic notations to be used in this book.

We denote by x points in $\mathbb{R}^n$ and write $x = (x_1, \ldots, x_n)$ in terms of its coordinates. For any $x \in \mathbb{R}^n$, we denote by $|x|$ the standard *Euclidean norm*, unless otherwise stated. Namely, for any $x = (x_1, \ldots, x_n)$, we have

$$|x| = \left(\sum_{i=1}^{n} x_i^2\right)^{\frac{1}{2}}.$$

Sometimes, we need to distinguish one particular direction and write points in $\mathbb{R}^n$ as (x', x_n) for $x' = (x_1, \ldots, x_{n-1}) \in \mathbb{R}^{n-1}$. We also denote by $\mathbb{R}^n_+$ the upper half-space; i.e., $\mathbb{R}^n = \{x \in \mathbb{R}^n : x_n > 0\}$.

Let Ω be a domain in $\mathbb{R}^n$, that is, an open and connected subset in $\mathbb{R}^n$. We denote by $L^\infty(\Omega)$ the collection of all bounded functions in Ω and define the L^∞*-norm* by

$$|u|_{L^\infty(\Omega)} = \sup_{\Omega} |u|.$$

We denote by $C(\Omega)$ the collection of all continuous functions in Ω, by $C^m(\Omega)$ the collection of all functions with continuous derivatives up to order m, for any integer $m \geq 1$, and by $C^\infty(\Omega)$ the collection of all functions with continuous derivatives of arbitrary order. For any $u \in C^m(\Omega)$, we denote by $\nabla^m u$ the collection of all partial derivatives of u of order m. For $m = 1$ and $m = 2$, we usually write $\nabla^m u$ in special forms. For first-order derivatives, we write ∇u as a vector of the form

$$\nabla u = (\partial_1 u, \ldots, \partial_n u).$$

This is the *gradient vector* of u. For second-order derivatives, we write $\nabla^2 u$ in the matrix form

$$\nabla^2 u = \begin{pmatrix} \partial_{11}u & \partial_{12}u & \cdots & \partial_{1n}u \\ \partial_{21}u & \partial_{2n}u & \cdots & \partial_{2n}u \\ \vdots & \vdots & \ddots & \vdots \\ \partial_{n1}u & \partial_{n2}u & \cdots & \partial_{nn}u \end{pmatrix}.$$

This is a symmetric matrix, called the *Hessian matrix* of u. For derivatives of order higher than two, we need to use multi-indices. A multi-index $\beta \in \mathbb{Z}^n_+$ is given by $\beta = (\beta_1, \ldots, \beta_n)$ with nonnegative integers $\beta_1, \ldots, \beta_n$. We write

$$|\beta| = \sum_{i=1}^{n} \beta_i.$$

The partial derivative $\partial^\beta u$ is defined by

$$\partial^\beta u = \partial_1^{\beta_1} \cdots \partial_n^{\beta_n} u,$$

and its order is $|\beta|$. For any positive integer m, we define

$$|\nabla^m u| = \left(\sum_{|\beta|=m} |\partial^\beta u|^2 \right)^{\frac{1}{2}},$$

and the C^m-*norm* by

$$|u|_{C^m(\Omega)} = \sum_{k=0}^{m} |\nabla^k u|_{L^\infty(\Omega)}.$$

For a constant $\alpha \in (0,1)$, we denote by $C^\alpha(\Omega)$ the collection of all Hölder continuous functions in Ω with the Hölder exponent α and by $C^{m,\alpha}(\Omega)$ the collection of all functions in $C^m(\Omega)$ whose derivatives of order m are Hölder continuous in Ω with the Hölder exponent α. We define the *Hölder semi-norm* by

$$[u]_{C^\alpha(\Omega)} = \sup_{\substack{x,y\in\Omega \\ x\neq y}} \frac{|u(x) - u(y)|}{|x-y|^\alpha},$$

and the C^α-*norm* by

$$|u|_{C^\alpha(\Omega)} = |u|_{L^\infty(\Omega)} + [u]_{C^\alpha(\Omega)}.$$

For any positive integer m and a contant $\alpha \in (0,1)$, we also define the $C^{m,\alpha}$-*norm* by

$$|u|_{C^{m,\alpha}(\Omega)} = |u|_{C^m(\Omega)} + \sum_{|\beta|=m} [\nabla^\beta u]_{C^\alpha(\Omega)}.$$

Accordingly, we can define $C(\bar{\Omega})$, $C^\alpha(\bar{\Omega})$, $C^m(\bar{\Omega})$, $C^{m,\alpha}(\bar{\Omega})$, and $C^\infty(\bar{\Omega})$ if $\partial\Omega$ is appropriately regular and define $[\cdot]_{C^\alpha(\bar{\Omega})}$, $|\cdot|_{C^\alpha(\bar{\Omega})}$, $|\cdot|_{C^m(\bar{\Omega})}$, and $|\cdot|_{C^{m,\alpha}(\bar{\Omega})}$ similarly.

We adopt the summation convention on repeated indices throughout the book. The general form of the *linear equations* of the second order is given by

$$a_{ij}(x)\partial_{ij}u + b_i(x)\partial_i u + c(x)u = f(x) \quad \text{in } \Omega,$$

where a_{ij}, b_i, c, and f are given functions in Ω. Very often, we write derivatives as $u_i = \partial_i u$ and $u_{ij} = \partial_{ij} u$ for brevity. In this way, we can express linear equations in the following form:

$$a_{ij}u_{ij} + b_i u_i + cu = f \quad \text{in } \Omega.$$

Subscripts here have different meanings for coefficients and solutions. Similarly, the general forms of the *quasilinear equations* and the *fully nonlinear equations* of the second order are given, respectively, by

$$a_{ij}(x, u, \nabla u)u_{ij} = f(x, u, \nabla u) \quad \text{in } \Omega$$

and

$$F(x, u, \nabla u, \nabla^2 u) = 0 \quad \text{in } \Omega.$$

A significant portion of the book is devoted to the derivation of a priori estimates, where certain norms of solutions are bounded by a positive constant C depending only on a set of known quantities. In a given context, the same letter C will be used to denote different constants depending on the same set of quantities.

Chapter 1

Linear Elliptic Equations

In this chapter, we review briefly three basic topics in the theory of linear elliptic equations: the maximum principle, Krylov-Safonov's Harnack inequality, and the Schauder theory.

In Section 1.1, we review Hopf's maximum principle. The maximum principle is an important method to study elliptic differential equations of the second order. In this section, we review the weak maximum principle and the strong maximum principle and derive several forms of a priori estimates of solutions.

In Section 1.2, we review Krylov-Safonov's Harnack inequality. The Harnack inequality is an important result in the theory of elliptic differential equations of the second order and plays a fundamental role in the study of nonlinear elliptic differential equations.

In Section 1.3, we review the Schauder theory for uniformly elliptic linear equations. Three main topics are a priori estimates in Hölder norms, the regularity of arbitrary solutions, and the solvability of the Dirichlet problem. Among these topics, a priori estimates are the most fundamental and form the basis for the existence and the regularity of solutions. We will review both the interior Schauder theory and the global Schauder theory.

These three sections play different roles in the rest of the book. In the study of quasilinear elliptic equations in Part 1, the maximum principle will be used to derive estimates of derivatives up to the first order, the Harnack inequality will be used to derive estimates of the Hölder semi-norms of derivatives of the first order, and the Schauder theory will be used to solve the linearized equations. In the study of fully nonlinear elliptic equations in Part 2, the maximum principle will be used to derive estimates of derivatives up to the second order, the Harnack inequality will be used to derive estimates of the Hölder semi-norms of derivatives of the second order, and the Schauder theory will be used to solve the linearized equations.

It is not our intention to present a complete review of the linear theory. Notably missing from this short review are the $W^{2,p}$-theory for linear equations of the nondivergence form and the H^k-theory and the de Giorgi-Moser theory for linear equations of the divergence form. Refer to Chapters 2–9 of [**59**] for a complete account of the linear theory.

1.1. The Maximum Principle

The maximum principle is an important method to study elliptic differential equations of the second order. In this section, we review the weak maximum principle and the strong maximum principle and derive several forms of a priori estimates of solutions. Refer to Chapter 3 of [**59**] for details.

Throughout this section, we let Ω be a bounded domain in $\mathbb{R}^n$ and let a_{ij}, b_i, and c be bounded and continuous functions in Ω, with $a_{ij} = a_{ji}$. We consider the operator L given by

$$Lu = a_{ij}\partial_{ij}u + b_i\partial_i u + cu \quad \text{in } \Omega, \tag{1.1.1}$$

for any $u \in C^2(\Omega)$. The operator L is always assumed to be *strictly elliptic* in Ω; namely, for any $x \in \Omega$ and $\xi \in \mathbb{R}^n$,

$$a_{ij}(x)\xi_i\xi_j \geq \lambda|\xi|^2, \tag{1.1.2}$$

for some positive constant λ. For later reference, L is called *uniformly elliptic* if, for any $x \in \Omega$ and $\xi \in \mathbb{R}^n$,

$$\lambda|\xi|^2 \leq a_{ij}(x)\xi_i\xi_j \leq \Lambda|\xi|^2, \tag{1.1.3}$$

for some positive constants λ and Λ, which are usually called the *ellipticity constants*.

1.1.1. The Weak Maximum Principle.

In this subsection, we review the weak maximum principle and its corollaries. We first introduce subsolutions and supersolutions.

Definition 1.1.1. For some $f \in C(\Omega)$, a $C^2(\Omega)$-function u is called a *subsolution* (or *supersolution*) of $Lw = f$ if $Lu \geq f$ (or $Lu \leq f$) in Ω.

If $a_{ij} = \delta_{ij}$, $b_i = c = 0$, and $f = 0$, subsolutions (or supersolutions) are subharmonic (or superharmonic).

Now we prove the weak maximum principle for subsolutions. Recall that u^+ is the nonnegative part of u, defined by $u^+ = \max\{0, u\}$.

Theorem 1.1.2. *Let Ω be a bounded domain in $\mathbb{R}^n$ and L be given by* (1.1.1), *for some a_{ij}, b_i, $c \in L^\infty(\Omega) \cap C(\Omega)$ satisfying $c \leq 0$ in Ω and* (1.1.2). *Suppose that $u \in C(\bar{\Omega}) \cap C^2(\Omega)$ satisfies $Lu \geq 0$ in Ω. Then, u attains on $\partial\Omega$ its nonnegative maximum in $\bar{\Omega}$; i.e.,*

$$\max_{\bar{\Omega}} u \leq \max_{\partial\Omega} u^+.$$

Proof. We first consider the special case $Lu > 0$ in Ω. If u has a local nonnegative maximum at a point x_0 in Ω, then $u(x_0) \geq 0$, $\nabla u(x_0) = 0$, and the Hessian matrix $\big(\nabla^2 u(x_0)\big)$ is negative semi-definite. By (1.1.2), the matrix $\big(a_{ij}(x_0)\big)$ is positive definite. Then,

$$Lu(x_0) = (a_{ij}\partial_{ij}u + b_i\partial_i u + cu)(x_0) \leq 0.$$

This leads to a contradiction. Hence, the nonnegative maximum of u in $\bar{\Omega}$ is attained only on $\partial\Omega$.

Now we consider the general case $Lu \geq 0$ in Ω. For any $\varepsilon > 0$, consider

$$w(x) = u(x) + \varepsilon e^{\mu x_1},$$

where μ is a positive constant to be determined. Then,

$$Lw = Lu + \varepsilon e^{\mu x_1}(a_{11}\mu^2 + b_1\mu + c).$$

Since b_1 and c are bounded and $a_{11} \geq \lambda > 0$ in Ω, by choosing $\mu > 0$ large enough, we get

$$a_{11}\mu^2 + b_1\mu + c > 0 \quad \text{in } \Omega.$$

This implies $Lw > 0$ in Ω. By the special case we just discussed, w attains its nonnegative maximum only on $\partial\Omega$ and hence,

$$\max_{\bar{\Omega}} w \leq \max_{\partial\Omega} w^+.$$

Then,

$$\max_{\bar{\Omega}} u \leq \max_{\bar{\Omega}} w \leq \max_{\partial\Omega} w^+ \leq \max_{\partial\Omega} u^+ + \varepsilon \max_{x\in\partial\Omega} e^{\mu x_1}.$$

We have the desired result by letting $\varepsilon \to 0$ and using the fact that $\partial\Omega \subset \bar{\Omega}$. □

If $c \equiv 0$ in Ω, we can draw conclusions about the maximum of u rather than its nonnegative maximum. A similar remark holds for the strong maximum principle.

A continuous function in $\bar{\Omega}$ always attains its maximum in $\bar{\Omega}$. Theorem 1.1.2 asserts that any subsolution continuous up to the boundary attains its maximum on the boundary $\partial\Omega$, but possibly also in Ω. Theorem 1.1.2 is referred to as the weak maximum principle. A stronger version asserts that subsolutions attain their maximum only on the boundary, unless they are constant.

As a simple consequence of Theorem 1.1.2, we have the following result.

Corollary 1.1.3. *Let Ω be a bounded domain in $\mathbb{R}^n$ and L be given by (1.1.1), for some a_{ij}, b_i, $c \in L^\infty(\Omega) \cap C(\Omega)$ satisfying $c \leq 0$ in Ω and (1.1.2). Suppose that $u \in C(\bar{\Omega}) \cap C^2(\Omega)$ satisfies $Lu \geq 0$ in Ω and $u \leq 0$ on $\partial\Omega$. Then, $u \leq 0$ in Ω.*

More generally, we have the following comparison principle.

Corollary 1.1.4. *Let Ω be a bounded domain in $\mathbb{R}^n$ and L be given by (1.1.1), for some a_{ij}, b_i, $c \in L^\infty(\Omega) \cap C(\Omega)$ satisfying $c \leq 0$ in Ω and (1.1.2). Suppose that $u, v \in C(\bar{\Omega}) \cap C^2(\Omega)$ satisfy $Lu \geq Lv$ in Ω and $u \leq v$ on $\partial\Omega$. Then, $u \leq v$ in Ω.*

The comparison principle provides a reason that functions u satisfying $Lu \geq f$ are called subsolutions. They are less than a solution v of $Lv = f$ with the same boundary value.

In the following, we simply say *by the maximum principle* when we apply Theorem 1.1.2, Corollary 1.1.3, or Corollary 1.1.4.

A consequence of the maximum principle is the uniqueness of solutions of Dirichlet problems.

Corollary 1.1.5. *Let Ω be a bounded domain in $\mathbb{R}^n$ and L be given by (1.1.1), for some a_{ij}, b_i, $c \in L^\infty(\Omega) \cap C(\Omega)$ satisfying $c \leq 0$ in Ω and (1.1.2). Then for any $f \in C(\Omega)$ and $\varphi \in C(\partial\Omega)$, there exists at most one solution $u \in C(\bar{\Omega}) \cap C^2(\Omega)$ of*

$$\begin{aligned} Lu &= f \quad \text{in } \Omega, \\ u &= \varphi \quad \text{on } \partial\Omega. \end{aligned}$$

1.1.2. The Strong Maximum Principle. The weak maximum principle asserts that subsolutions of linear elliptic equations attain their nonnegative maximum on the boundary under suitable conditions. In fact, these subsolutions can attain their nonnegative maximum only on the boundary, unless they are constant. This is the strong maximum principle.

For any C^1-function u in $\bar{\Omega}$ that attains its maximum on $\partial\Omega$, say at $x_0 \in \partial\Omega$, we have $\frac{\partial u}{\partial \nu}(x_0) \geq 0$, where ν is the exterior unit normal to Ω at x_0. The Hopf lemma asserts that this normal derivative is in fact positive if u is a subsolution in Ω.

Theorem 1.1.6. *Let B be an open ball in $\mathbb{R}^n$ with $x_0 \in \partial B$ and L be given by (1.1.1), for some a_{ij}, b_i, $c \in L^\infty(B) \cap C(B)$ satisfying $c \leq 0$ in B and*

(1.1.2). *Suppose that* $u \in C^1(\bar{B}) \cap C^2(B)$ *satisfies* $Lu \geq 0$ *in* B, $u(x) < u(x_0)$ *for any* $x \in B$, *and* $u(x_0) \geq 0$. *Then,*

$$\frac{\partial u}{\partial \nu}(x_0) > 0,$$

where ν *is the exterior unit normal to* B *at* x_0.

Proof. Without loss of generality, we assume $B = B_R$ for some $R > 0$. By the continuity of u up to ∂B_R, we have, for any $x \in \bar{B}_R$,

$$u(x) \leq u(x_0).$$

For positive constants μ and ε to be determined, we set

$$w(x) = e^{-\mu|x|^2} - e^{-\mu R^2}$$

and

$$v(x) = u(x) - u(x_0) + \varepsilon w(x).$$

We consider w and v in $D = B_R \setminus \bar{B}_{R/2}$.

A direct calculation yields

$$\begin{aligned} Lw &= e^{-\mu|x|^2}\left\{4\mu^2 a_{ij}x_ix_j - 2\mu a_{ij}\delta_{ij} - 2\mu b_ix_i + c\right\} - ce^{-\mu R^2} \\ &\geq e^{-\mu|x|^2}\left\{4\mu^2 a_{ij}x_ix_j - 2\mu\big(a_{ij}\delta_{ij} + b_ix_i\big) + c\right\}, \end{aligned}$$

where we used $c \leq 0$ in B_R. By the strict ellipticity (1.1.2), we have

$$a_{ij}(x)x_ix_j \geq \lambda|x|^2 \geq \frac{1}{4}\lambda R^2 \quad \text{in } D.$$

Hence,

$$Lw \geq e^{-\mu|x|^2}\left\{\mu^2\lambda R^2 - 2\mu\big(a_{ij}\delta_{ij} + b_ix_i\big) + c\right\} \geq 0 \quad \text{in } D$$

if we choose μ sufficiently large. By $c \leq 0$ and $u(x_0) \geq 0$, we obtain, for any $\varepsilon > 0$,

$$Lv = Lu + \varepsilon Lw - cu(x_0) \geq 0 \quad \text{in } D.$$

Next, we discuss v on ∂D in two cases. First, on $\partial B_{R/2}$, we have $u - u(x_0) < 0$, and hence $u - u(x_0) < -\varepsilon$ for some $\varepsilon > 0$, by the continuity of u on $\partial B_{R/2}$. Note that $w < 1$ on $\partial B_{R/2}$. Then for such an ε, we obtain $v < 0$ on $\partial B_{R/2}$. Second, on ∂B_R, we have $w = 0$ and $u \leq u(x_0)$. Hence, $v \leq 0$ on ∂B_R and $v(x_0) = 0$. Therefore, $v \leq 0$ on ∂D.

In conclusion, $Lv \geq 0$ in D and $v \leq 0$ on ∂D. By the maximum principle, we have

$$v \leq 0 \quad \text{in } D.$$

In view of $v(x_0) = 0$, v attains at x_0 its maximum in $\bar{D}$. Hence, we obtain

$$\frac{\partial v}{\partial \nu}(x_0) \geq 0,$$

and then

$$\frac{\partial u}{\partial \nu}(x_0) \geq -\varepsilon \frac{\partial w}{\partial \nu}(x_0) = 2\varepsilon \mu R e^{-\mu R^2} > 0.$$

This is the desired result. □

Theorem 1.1.6 still holds if we substitute for B any bounded C^1-domain which satisfies an *interior sphere condition* at $x_0 \in \partial\Omega$, namely, if there exists a ball $B \subset \Omega$ with $x_0 \in \partial B$. This is because such a ball B is tangent to $\partial\Omega$ at x_0. We note that the interior sphere condition always holds for C^2-domains.

Now, we are ready to prove the strong maximum principle due to Hopf [**84**].

Theorem 1.1.7. *Let Ω be a bounded domain in $\mathbb{R}^n$ and L be given by* (1.1.1), *for some a_{ij}, b_i, $c \in C(\Omega)$ satisfying $c \leq 0$ in Ω and* (1.1.2). *Suppose that $u \in C(\bar{\Omega}) \cap C^2(\Omega)$ satisfies $Lu \geq 0$ in Ω. Then, u attains only on $\partial\Omega$ its nonnegative maximum in $\bar{\Omega}$ unless u is constant.*

Proof. Let M be the nonnegative maximum of u in $\bar{\Omega}$ and set

$$D = \{x \in \Omega : \ u(x) = M\}.$$

We prove either $D = \emptyset$ or $D = \Omega$ by contradiction. Suppose D is a nonempty proper subset of Ω. It follows from the continuity of u that D is relatively closed in Ω. Then, $\Omega \setminus D$ is open and we can find an open ball $B \subset \Omega \setminus D$ such that $\partial B \cap D \neq \emptyset$. In fact, we may choose a point $x_* \in \Omega \setminus D$ with $\text{dist}(x_*, D) < \text{dist}(x_*, \partial\Omega)$ and then take the ball centered at x_* with the radius $\text{dist}(x_*, D)$. Suppose $x_0 \in \partial B \cap D$. Obviously, we have

$$Lu \geq 0 \quad \text{in } B$$

and

$$u(x) < u(x_0) \quad \text{for any } x \in B \text{ and } u(x_0) = M \geq 0.$$

By Theorem 1.1.6, we have

$$\frac{\partial u}{\partial \nu}(x_0) > 0,$$

where ν is the exterior unit normal to B at x_0. On the other hand, x_0 is an interior maximum point of u in Ω. This implies $\nabla u(x_0) = 0$, which leads to a contradiction. Therefore, either $D = \emptyset$ or $D = \Omega$. In the first case, u attains only on $\partial\Omega$ its nonnegative maximum in $\bar{\Omega}$, while in the second case, u is constant in Ω. □

The following result improves Corollary 1.1.3.

Corollary 1.1.8. *Let Ω be a bounded domain in $\mathbb{R}^n$ and L be given by* (1.1.1), *for some a_{ij}, b_i, $c \in C(\Omega)$ satisfying $c \le 0$ in Ω and* (1.1.2). *Suppose that $u \in C(\bar{\Omega}) \cap C^2(\Omega)$ satisfies $Lu \ge 0$ in Ω and $u \le 0$ on $\partial\Omega$. Then, either $u < 0$ in Ω or $u \equiv 0$ in Ω.*

1.1.3. A Priori Estimates. As Corollary 1.1.5 shows, an important application of the maximum principle is to prove the uniqueness of solutions of boundary-value problems. Equally or more important is to derive a priori estimates. In derivations of a priori estimates, it is essential to construct auxiliary functions. We will provide proofs of all results in this subsection, as constructing auxiliary functions is an important technique we will develop in this book. We point out that we need only the weak maximum principle in this subsection.

We first derive an estimate for subsolutions. In the next result, we write $u^+ = \max\{u, 0\}$ as before and $f^- = \max\{-f, 0\}$.

Theorem 1.1.9. *Let Ω be a bounded domain in $\mathbb{R}^n$ and L be given by* (1.1.1), *for some a_{ij}, b_i, $c \in L^\infty(\Omega) \cap C(\Omega)$ satisfying $c \le 0$ in Ω and* (1.1.2). *Suppose that $u \in C(\bar{\Omega}) \cap C^2(\Omega)$ satisfies*

$$Lu \ge f \quad \text{in } \Omega,$$

for some $f \in L^\infty(\Omega) \cap C(\Omega)$. Then,

$$\sup_\Omega u \le \max_{\partial\Omega} u^+ + Cd^2 \sup_\Omega f^-,$$

where $d = \mathrm{diam}(\Omega)$ and C is a positive constant depending only on n, λ, and $d|b_i|_{L^\infty(\Omega)}$, for $i = 1, \dots, n$.

Proof. Set

$$F = \sup_\Omega f^-, \quad \Phi = \max_{\partial\Omega} u^+.$$

Then, $Lu \ge -F$ in Ω and $u \le \Phi$ on $\partial\Omega$. Without loss of generality, we assume $\Omega \subset \{0 < x_1 < d\}$ for some constant $d > 0$. For some constant $\mu > 0$ to be chosen later, set

$$v = \Phi + d^2 \left(e^\mu - e^{\frac{\mu x_1}{d}}\right) F.$$

We note that $v \ge \Phi$ in $\bar{\Omega}$. Next, by a straightforward calculation and $c \le 0$ in Ω, we have

$$\begin{aligned} Lv &= -(a_{11}\mu^2 + db_1\mu)Fe^{\frac{\mu x_1}{d}} + c\Phi + cd^2\left(e^\mu - e^{\frac{\mu x_1}{d}}\right)F \\ &\le -(a_{11}\mu^2 + db_1\mu)F. \end{aligned}$$

Note that $a_{11} \geq \lambda$ in Ω by the strict ellipticity (1.1.2). We choose μ large so that

$$a_{11}\mu^2 + db_1\mu \geq 1 \quad \text{in } \Omega.$$

Then, $Lv \leq -F$ in Ω. Therefore,

$$\begin{aligned} Lu &\geq Lv \quad \text{in } \Omega, \\ u &\leq v \quad \text{on } \partial\Omega. \end{aligned}$$

By the maximum principle, we obtain

$$u \leq v \quad \text{in } \Omega,$$

and hence, for any $x \in \Omega$,

$$u(x) \leq \Phi + d^2 \left(e^{\mu} - e^{\frac{\mu x_1}{d}} \right) F.$$

This yields the desired result. □

The function v in the proof above is what we called an auxiliary function. In fact, auxiliary functions were already used in the proof of Theorem 1.1.6.

If L in Theorem 1.1.9 does not involve the lower-order terms, i.e., $L = a_{ij}\partial_{ij}$, we may take

$$v = \Phi + \frac{1}{2\lambda} F(d^2 - x_1^2).$$

In this case, a simple calculation yields

$$Lv = -\frac{1}{\lambda} a_{11} F \leq -F,$$

where we used the strict ellipticity (1.1.2).

By replacing u with $-u$, Theorem 1.1.9 extends to supersolutions and solutions of the equation $Lu = f$.

Theorem 1.1.10. *Let Ω be a bounded domain in $\mathbb{R}^n$ and L be given by (1.1.1), for some a_{ij}, b_i, $c \in L^\infty(\Omega) \cap C(\Omega)$ satisfying $c \leq 0$ in Ω and (1.1.2). Suppose that u is a $C(\bar{\Omega}) \cap C^2(\Omega)$-solution of*

$$\begin{aligned} Lu &= f \quad \text{in } \Omega, \\ u &= \varphi \quad \text{on } \partial\Omega, \end{aligned}$$

for some $f \in L^\infty(\Omega) \cap C(\Omega)$ and $\varphi \in C(\partial\Omega)$. Then,

$$\sup_\Omega |u| \leq \max_{\partial\Omega} |\varphi| + C \sup_\Omega |f|,$$

where C is a positive constant depending only on n, λ, diam(Ω)*, and the sup-norms of b_i.*

Next, we construct barrier functions for a large class of domains Ω and discuss global properties of solutions. The geometry of $\partial\Omega$ plays an important role. We consider the case where Ω satisfies an *exterior sphere condition* at $x_0 \in \partial\Omega$ in the sense that there exists a ball $B_R(y_0)$ such that

$$\Omega \cap B_R(y_0) = \emptyset, \quad \bar{\Omega} \cap \bar{B}_R(y_0) = \{x_0\}.$$

Lemma 1.1.11. *Let Ω be a bounded domain in $\mathbb{R}^n$ satisfying an exterior sphere condition at $x_0 \in \partial\Omega$ and L be given by (1.1.1), for some a_{ij}, b_i, $c \in L^\infty(\Omega) \cap C(\Omega)$ satisfying $c \le 0$ in Ω and (1.1.3). Then, there exists a function $w_{x_0} \in C(\bar{\Omega}) \cap C^2(\Omega)$ such that*

$$Lw_{x_0} \le -1 \quad \text{in } \Omega$$

and, for any $x \in \bar{\Omega} \setminus \{x_0\}$,

$$w_{x_0}(x_0) = 0, \quad w_{x_0}(x) > 0,$$

where w_{x_0} depends only on n, λ, Λ, the L^∞-norms of b_i, $\operatorname{diam}(\Omega)$, *and R in the exterior sphere condition.*

Proof. Set $D = \operatorname{diam}(\Omega)$. For the given $x_0 \in \partial\Omega$, consider an exterior ball $B_R(y)$ with $\bar{B}_R(y) \cap \bar{\Omega} = \{x_0\}$. Let $d(x)$ be the distance from x to $\partial B_R(y)$; i.e.,

$$d(x) = |x - y| - R.$$

Then, for any $x \in \Omega$,

$$0 < d(x) < D.$$

Consider a C^2-function ψ defined in $[0, D)$, with $\psi(0) = 0$ and $\psi > 0$ in $(0, D)$. Set

$$w = \psi(d) \quad \text{in } \Omega.$$

We now calculate Lw. A direct calculation yields

$$\partial_i d(x) = \frac{x_i - y_i}{|x - y|},$$

$$\partial_{ij} d(x) = \frac{\delta_{ij}}{|x - y|} - \frac{(x_i - y_i)(x_j - y_j)}{|x - y|^3}.$$

Hence, $|\nabla d| = 1$, $a_{ij}\partial_i d \partial_j d \ge \lambda$, and

$$\begin{aligned} a_{ij}\partial_{ij} d &= \frac{1}{|x - y|} a_{ij}\delta_{ij} - \frac{1}{|x - y|} a_{ij}\partial_i d \partial_j d \\ &\le \frac{n\Lambda}{|x - y|} - \frac{\lambda}{|x - y|} = \frac{n\Lambda - \lambda}{|x - y|} \le \frac{n\Lambda - \lambda}{R}. \end{aligned}$$

Next,

$$\partial_i w = \psi' \partial_i d, \quad \partial_{ij} w = \psi'' \partial_i d \partial_j d + \psi' \partial_{ij} d.$$

Then,

$$Lw = \psi'' a_{ij}\partial_i d \partial_j d + \psi'(a_{ij}\partial_{ij} d + b_i \partial_i d) + c\psi.$$

We now require $\psi' > 0$ and $\psi'' < 0$. Hence,

$$Lw \le \lambda\psi'' + \left(\frac{n\Lambda - \lambda}{R} + b_0\right)\psi',$$

where

$$b_0 = \sup_\Omega \left(\sum_{i=1}^n b_i^2\right)^{\frac{1}{2}}.$$

We write this as

$$Lw \le \lambda\left(\psi'' + a\psi' + b\right) - 1,$$

where

$$a = \frac{n\Lambda - \lambda}{\lambda R} + \frac{b_0}{\lambda}, \quad b = \frac{1}{\lambda}.$$

We need to find a function ψ in $[0, D)$ such that

$$\psi'' + a\psi' + b = 0 \quad \text{in } (0, D),$$
$$\psi'' < 0, \ \psi' > 0 \text{ in } (0, D), \text{ and } \psi(0) = 0.$$

First, general solutions of the ordinary differential equation above are given by

$$\psi(d) = -\frac{b}{a}d + C_1 - \frac{C_2}{a}e^{-ad},$$

for some constants C_1 and C_2. For $\psi(0) = 0$, we need $C_1 = C_2/a$. Hence we have, for some constant C,

$$\psi(d) = -\frac{b}{a}d + \frac{C}{a}(1 - e^{-ad}),$$

which implies

$$\psi'(d) = Ce^{-ad} - \frac{b}{a} = e^{-ad}\left(C - \frac{b}{a}e^{ad}\right),$$
$$\psi''(d) = -Cae^{-ad}.$$

In order to have $\psi' > 0$ in $(0, D)$, we need $C \ge \frac{b}{a}e^{aD}$. Then, $\psi' > 0$ in $(0, D)$, and hence $\psi > \psi(0) = 0$ in $(0, D)$. Therefore, we take

$$\psi(d) = -\frac{b}{a}d + \frac{b}{a^2}e^{aD}(1 - e^{-ad})$$
$$= \frac{b}{a}\left\{\frac{1}{a}e^{aD}(1 - e^{-ad}) - d\right\}.$$

Such a ψ satisfies all the requirements we imposed. □

Now we estimate the modulus of continuity of solutions with the help of barrier functions constructed in Lemma 1.1.11.

Theorem 1.1.12. *Let Ω be a bounded domain in $\mathbb{R}^n$ satisfying an exterior sphere condition at $x_0 \in \partial\Omega$ and L be given by (1.1.1), for some a_{ij}, b_i, $c \in L^\infty(\Omega) \cap C(\Omega)$ satisfying (1.1.3). Suppose that u is a $C(\bar{\Omega}) \cap C^2(\Omega)$-solution of*

$$\begin{aligned} Lu &= f \quad \text{in } \Omega, \\ u &= \varphi \quad \text{on } \partial\Omega, \end{aligned}$$

for some $f \in L^\infty(\Omega) \cap C(\Omega)$ and $\varphi \in C(\partial\Omega)$. Then, for any $x \in \Omega$,

$$|u(x) - u(x_0)| \le \omega(|x - x_0|),$$

where ω is a nondecreasing continuous function in $(0, D)$, with $D = \operatorname{diam}(\Omega)$ and $\lim_{r\to 0} \omega(r) = 0$, depending only on n, λ, Λ, the L^∞-norms of b_i and c, $\operatorname{diam}(\Omega)$, R in the exterior sphere condition, $\sup_\Omega |u|$, $\max_{\partial\Omega} |\varphi|$, $\sup_\Omega |f|$, and the modulus of continuity of φ on $\partial\Omega$.

Proof. Set

$$L_0 = a_{ij}\partial_{ij} + b_i\partial_i.$$

Then, $L_0 u = f - cu$ in Ω. Let $w = w_{x_0}$ be the function in Lemma 1.1.11 for L_0, i.e.,

$$L_0 w \le -1 \quad \text{in } \Omega,$$

and, for any $x \in \partial\Omega \setminus \{x_0\}$,

$$w(x_0) = 0, \quad w(x) > 0.$$

We set

$$F = \sup_\Omega |f - cu|, \quad \Phi = \max_{\partial\Omega} |\varphi|.$$

Then,

$$L_0(\pm u) \ge -F \quad \text{in } \Omega.$$

Let ε be an arbitrary positive constant. By the continuity of φ at x_0, there exists a positive constant δ such that, for any $x \in \partial\Omega \cap B_\delta(x_0)$,

$$|\varphi(x) - \varphi(x_0)| \le \varepsilon.$$

We then choose K sufficiently large so that $K \ge F$ and

$$Kw \ge 2\Phi \quad \text{on } \partial\Omega \setminus B_\delta(x_0).$$

We point out that K depends on ε through the positive lower bound of w on $\partial\Omega \setminus B_\delta(x_0)$. Then,

$$L_0(Kw) \le -F \quad \text{in } \Omega,$$

and

$$|\varphi - \varphi(x_0)| \le \varepsilon + Kw \quad \text{on } \partial\Omega.$$

Therefore,

$$L_0\big(\pm(u-\varphi(x_0))\big) \geq L_0(\varepsilon + Kw) \quad \text{in } \Omega,$$
$$\pm\big(u-\varphi(x_0)\big) \leq \varepsilon + Kw \quad \text{on } \partial\Omega.$$

By the maximum principle, we have $\pm\big(u-\varphi(x_0)\big) \leq \varepsilon + Kw$ in Ω, and hence

$$|u-\varphi(x_0)| \leq \varepsilon + Kw \quad \text{in } \Omega.$$

Note that the second term in the right-hand side converges to zero as $x \to x_0$. Then, there exists a positive constant $\delta' < \delta$ such that

$$|u-\varphi(x_0)| \leq 2\varepsilon \quad \text{in } \Omega \cap B_{\delta'}(x_0).$$

This yields the desired result. □

Remark 1.1.13. It is clear from the proof that Theorem 1.1.12 is a local result. If we assume that φ is continuous at $x_0 \in \partial\Omega$ and, in addition, u is bounded in a neighborhood of x_0, then we can estimate the modulus of continuity of u at x_0.

By a similar method as in the proof of Theorem 1.1.12, we can derive a boundary gradient estimate.

Theorem 1.1.14. *Let Ω be a bounded domain in $\mathbb{R}^n$ satisfying an exterior sphere condition at $x_0 \in \partial\Omega$ and L be given by* (1.1.1), *for some a_{ij}, b_i, $c \in L^\infty(\Omega) \cap C(\Omega)$ satisfying* (1.1.3). *Suppose that u is a $C(\bar{\Omega}) \cap C^2(\Omega)$-solution of*

$$Lu = f \quad \text{in } \Omega,$$
$$u = \varphi \quad \text{on } \partial\Omega,$$

for some $f \in L^\infty(\Omega) \cap C(\Omega)$ and $\varphi \in C^2(\bar{\Omega})$. Then, for any $x \in \Omega$,

$$|u(x) - u(x_0)| \leq C\Big(\sup_\Omega |u| + |\varphi|_{C^2(\bar{\Omega})} + \sup_\Omega |f|\Big)|x - x_0|,$$

where C is a positive constant depending only on n, λ, Λ, the L^∞-norms of b_i and c, diam(Ω), *and R in the exterior sphere condition.*

If $\partial\Omega$ is C^1 at x_0 and u is C^1 at x_0, taking the normal derivative at x_0, we obtain

$$\left|\frac{\partial u}{\partial \nu}(x_0)\right| \leq C\Big\{\sup_\Omega |u| + |\varphi|_{C^2(\bar{\Omega})} + \sup_\Omega |f|\Big\}.$$

Proof. Set

$$L_0 = a_{ij}\partial_{ij} + b_i\partial_i.$$

Then, $L_0 u = f - cu$ in Ω. By setting $v = u - \varphi$, we have

$$\begin{aligned} L_0 v &= f - cu - L_0\varphi \quad \text{in } \Omega, \\ v &= 0 \quad \text{on } \partial\Omega. \end{aligned}$$

Next, we set

$$F = \sup_{\Omega} |f - cu - L_0\varphi|.$$

Then,

$$\begin{aligned} L_0(\pm v) &\geq -F \quad \text{in } \Omega, \\ \pm v &= 0 \quad \text{on } \partial\Omega. \end{aligned}$$

Let $w = w_{x_0}$ be the function in Lemma 1.1.11 for L_0, i.e.,

$$L_0 w \leq -1 \quad \text{in } \Omega,$$

and, for any $x \in \partial\Omega \setminus \{x_0\}$,

$$w(x_0) = 0, \quad w(x) > 0.$$

Then,

$$\begin{aligned} L_0(\pm v) &\geq L_0(Fw) \quad \text{in } \Omega, \\ \pm v &\leq Fw \quad \text{on } \partial\Omega. \end{aligned}$$

By the maximum principle, we have $\pm v \leq Fw$ in Ω, and hence

$$|v| \leq Fw \quad \text{in } \Omega.$$

With $v = u - \varphi$ and $u(x_0) = \varphi(x_0)$, we obtain

$$|u - u(x_0)| \leq |u - \varphi| + |\varphi - \varphi(x_0)| \leq Fw + |\varphi - \varphi(x_0)|.$$

This implies the desired result. □

A remark similar to Remark 1.1.13 holds for Theorem 1.1.14.

It does not seem optimal to require φ to be C^2 in Theorem 1.1.14. It is natural to ask whether the result holds if $\varphi \in C^1(\partial\Omega)$. However, C^1-boundary values in general do not yield boundary gradient estimates, even for harmonic functions in balls. It is a good exercise to derive the best modulus of continuity for harmonic functions in balls with C^1-boundary values.

To end this section, we derive an estimate of the Hölder semi-norms near the boundary.

Theorem 1.1.15. *Let $\alpha \in (0,1)$ be a constant, Ω be a bounded domain in $\mathbb{R}^n$ satisfying an exterior sphere condition at $x_0 \in \partial\Omega$, and L be given by (1.1.1), for some a_{ij}, b_i, $c \in L^\infty(\Omega) \cap C(\Omega)$ satisfying (1.1.3). Suppose that u is a $C(\bar{\Omega}) \cap C^2(\Omega)$-solution of*

$$\begin{aligned} Lu &= f \quad \text{in } \Omega, \\ u &= \varphi \quad \text{on } \partial\Omega, \end{aligned}$$

for some $f \in L^\infty(\Omega) \cap C(\Omega)$ and $\varphi \in C(\partial\Omega)$ satisfying, for any $x \in \partial\Omega$,

$$|\varphi(x) - \varphi(x_0)| \leq \Phi_\alpha |x - x_0|^\alpha,$$

for some positive constant Φ_α. Then, for any $x \in \Omega$,

$$|u(x) - u(x_0)| \leq C\Big(\sup_\Omega |u| + \Phi_\alpha + \sup_\Omega |f| \Big) |x - x_0|^{\frac{\alpha}{1+\alpha}},$$

where C is a positive constant depending only on n, α, λ, Λ, the L^∞-norms of b_i and c, $\operatorname{diam}(\Omega)$, *and R in the exterior sphere condition.*

Proof. Set

$$L_0 = a_{ij}\partial_{ij} + b_i\partial_i.$$

Then, $L_0 u = f - cu$ in Ω. Next, we set

$$M = \sup_\Omega |u|, \quad F = \sup_\Omega |f - cu|.$$

Hence,

$$L_0\big(\pm (u - u(x_0))\big) \geq -F \quad \text{in } \Omega.$$

Now, we take a ball $B_R(y_0) \subset \mathbb{R}^n$ such that $\bar{\Omega} \cap \bar{B}_R(y_0) = \{x_0\}$. Then, for any $r \in (0, R]$, we take $y \in \mathbb{R}^n$ such that $\bar{\Omega} \cap \bar{B}_r(y) = \{x_0\}$. Note that

$$r < |x - y| < 2r \quad \text{for any } x \in \Omega \cap B_{2r}(y).$$

For any $x \in \partial\Omega \cap B_{2r}(y)$, we have

$$|u(x) - u(x_0)| \leq \Phi_\alpha |x - x_0|^\alpha \leq (3r)^\alpha \Phi_\alpha,$$

where we used the triangle inequality $|x - x_0| \leq |x - y| + |y - x_0| \leq 3r$. On $\Omega \cap \partial B_{2r}(y)$, we simply have

$$|u - u(x_0)| \leq 2M.$$

Therefore,

$$\begin{aligned} L_0\big(\pm (u - u(x_0))\big) &\geq -F \quad \text{in } \Omega \cap B_{2r}(y), \\ \pm\big(u - u(x_0)\big) &\leq (3r)^\alpha \Phi_\alpha \quad \text{on } \partial\Omega \cap B_{2r}(y), \\ \pm\big(u - u(x_0)\big) &\leq 2M \quad \text{on } \Omega \cap \partial B_{2r}(y). \end{aligned}$$

It is worth pointing out that r will vary later on.

Set, for some positive constant μ to be determined,

$$v(x) = 1 - \frac{r^\mu}{|x-y|^\mu}.$$

Then, $0 < v < 1$ in $\Omega \cap B_{2r}(y)$. A straightforward calculation yields, for any $i, j = 1, \ldots, n$,

$$\begin{aligned}\partial_i v &= \mu r^\mu |x-y|^{-\mu-2}(x_i - y_i),\\ \partial_{ij} v &= \mu r^\mu |x-y|^{-\mu-2}\left(-(\mu+2)\frac{(x_i-y_i)(x_j-y_j)}{|x-y|^2} + \delta_{ij}\right),\end{aligned}$$

and hence,

$$\begin{aligned}L_0 v = \mu r^\mu |x-y|^{-\mu-2}\Big(&-(\mu+2)\frac{a_{ij}(x_i-y_i)(x_j-y_j)}{|x-y|^2}\\ &+ \delta_{ij}a_{ij} + b_i(x_i-y_i)\Big).\end{aligned}$$

By the uniform ellipticity (1.1.3), we get

$$L_0 v \leq \frac{\mu r^\mu}{|x-y|^{\mu+2}}\big(-(\mu+2)\lambda + n\Lambda + 2rb_0\big) \quad \text{in } \Omega \cap B_{2r}(y),$$

where

$$b_0 = \sup_\Omega \left(\sum_{i=1}^n b_i^2\right)^{\frac{1}{2}}.$$

By choosing $\mu \geq 1$ sufficiently large, depending only on n, λ, Λ, b_0, and R, we have

$$L_0 v \leq -\frac{r^\mu}{|x-y|^{\mu+2}} \leq -\frac{1}{2^{\mu+2}r^2} \leq -\frac{1}{2^{\mu+2}R^2} \quad \text{in } \Omega \cap B_{2r}(y).$$

For some positive constant A to be determined, we set

$$w = Av + (3r)^\alpha \Phi_\alpha.$$

First, we have

$$L_0 w = AL_0 v \leq -\frac{A}{2^{\mu+2}R^2} \quad \text{in } \Omega \cap B_{2r}(y).$$

To have $L_0\big(\pm(u - u(x_0))\big) \geq L_0 w$ in $\Omega \cap B_{2r}(y)$, we require $A \geq 2^{\mu+2}R^2F$. Next, it is obvious that $\pm(u - u(x_0)) \leq w$ on $\partial\Omega \cap B_{2r}(y)$ for any $A > 0$ since $v \geq 0$ there. Last, for any $x \in \Omega \cap \partial B_{2r}(y)$, we have

$$v(x) = 1 - \frac{r^\mu}{|x-y|^\mu} = 1 - \frac{1}{2^\mu} \geq \frac{1}{2},$$

as long as $\mu \geq 1$. Hence, to have $\pm(u - u(x_0)) \leq w$ on $\Omega \cap \partial B_{2r}(y)$, we require $A \geq 4M$. Therefore, we take

$$A = 2^{\mu+2}R^2F + 4M.$$

Then,

$$\begin{aligned} L_0\big(\pm(u-u(x_0))\big) &\geq L_0 w \quad \text{in } \Omega\cap B_{2r}(y),\\ \pm\big(u-u(x_0)\big) &\leq w \quad \text{on } \partial(\Omega\cap B_{2r}(y)). \end{aligned}$$

By the maximum principle, we obtain $\pm\big(u-u(x_0)\big)\leq w$ in $\Omega\cap B_{2r}(y)$, and hence

$$|u-u(x_0)|\leq w \quad \text{in } \Omega\cap B_{2r}(y).$$

This implies, for any $x\in\Omega\cap B_{2r}(y)$,

$$|u(x)-u(x_0)|\leq A\left(1-\frac{r^\mu}{|x-y|^\mu}\right)+3r^\alpha\Phi_\alpha.$$

By the inequality $1-t^\mu\leq\mu(1-t)$ for any $t\in(0,1)$, we obtain, for any $x\in\Omega\cap B_{2r}(y)$,

$$\begin{aligned} 1-\frac{r^\mu}{|x-y|^\mu} &\leq \mu\left(1-\frac{r}{|x-y|}\right)\leq\frac{\mu(|x-y|-r)}{|x-y|}\\ &\leq\frac{\mu|x-x_0|}{|x-y|}\leq\frac{\mu|x-x_0|}{r}, \end{aligned}$$

where we used the triangle inequality and $|x_0-y|=r$. Therefore, for any $x\in\Omega\cap B_{2r}(y)$,

$$|u(x)-u(x_0)|\leq\mu A\frac{|x-x_0|}{r}+3r^\alpha\Phi_\alpha.$$

Now, we fix an $x\in\Omega$ and set $R_0=\min\{1,R\}$. If $|x-x_0|<R_0^{1+\alpha}$, take r such that $|x-x_0|=r^{1+\alpha}$. Obviously, $r<R_0$. With the ball $B_r(y)$ chosen as above with $\bar\Omega\cap\bar B_r(y)=\{x_0\}$, we have

$$|x-y|\leq|x-x_0|+|x_0-y|\leq r^{1+\alpha}+r<2r.$$

Therefore, $x\in\Omega\cap B_{2r}(y)$ and hence

$$|u(x)-u(x_0)|\leq(\mu A+3\Phi_\alpha)r^\alpha=(\mu A+3\Phi_\alpha)|x-x_0|^{\frac{\alpha}{1+\alpha}}.$$

If $|x-x_0|\geq R_0^{1+\alpha}$, then

$$|u(x)-u(x_0)|\leq 2M\leq\frac{2M}{R_0^\alpha}|x-x_0|^{\frac{\alpha}{1+\alpha}}.$$

Hence, we have the desired estimate. □

The power $\alpha/(1+\alpha)$ does not seem optimal. However, the present form is sufficient for applications later on.

1.2. Krylov-Safonov's Harnack Inequality

The Harnack inequality is an important result in the theory of elliptic differential equations of the second order and plays a fundamental rolc in the study of nonlinear elliptic differential equations. In this section, we review the Harnack inequality due to Krylov and Safonov. Refer to Chapter 9 of [**59**] for details.

1.2.1. Aleksandrov's Maximum Principle. We first prove a maximum principle due to Aleksandrov, which yields an estimate of solutions in terms of integral norms of nonhomogeneous terms. To do this, we need to introduce the concept of contact sets.

For any $u \in C(\Omega)$, we define

$$\begin{aligned}\Gamma^+ = \big\{y \in \Omega :\ & u(x) \leq u(y) + p\cdot(x-y)\\ & \text{for any } x\in\Omega \text{ and some } p = p(y)\in\mathbb{R}^n\big\}.\end{aligned}$$

The set Γ^+ is called the *upper contact set* of u. Clearly, u is concave if and only if $\Gamma^+ = \Omega$. If $u \in C^1(\Omega)$, then $p(y) = \nabla u(y)$ for $y \in \Gamma^+$. In this case, any support hyperplane must be a tangent plane to the graph, and

$$\Gamma^+ = \{y \in \Omega :\ u(x) \leq u(y) + \nabla u(y)\cdot(x-y) \ \text{ for any } x \in \Omega\}.$$

If $u \in C^2(\Omega)$, the Hessian matrix $\nabla^2 u = (\partial_{ij} u)$ is negative semi-definite in Γ^+. Lower contact sets can be defined similarly.

We first prove a result on the maximum of positive parts of C^2-functions.

Lemma 1.2.1. *Let Ω be a bounded domain in $\mathbb{R}^n$ and $g \in L^1_{\mathrm{loc}}(\mathbb{R}^n)$ be nonnegative. Then, for any $u \in C(\bar\Omega)\cap C^2(\Omega)$,*

$$\int_{B_{M/d}} g(p)\,dp \leq \int_{\Gamma^+} g(\nabla u)|\det \nabla^2 u|\,dx,$$

where Γ^+ is the upper contact set of u, $M = \sup_\Omega u - \max_{\partial\Omega} u^+$, and $d = \mathrm{diam}(\Omega)$.

Proof. Without loss of generality, we assume $u \leq 0$ on $\partial\Omega$. Set $\Omega^+ = \{u > 0\}$. By the area-formula for ∇u in $\Gamma^+\cap\Omega^+ \subset \Omega$, we have

$$\text{(1)}\qquad \int_{\nabla u(\Gamma^+\cap\Omega^+)} g(p)\,dp \leq \int_{\Gamma^+\cap\Omega^+} g(\nabla u)|\det(\nabla^2 u)|\,dx,$$

where $|\det(\nabla^2 u)|$ is the Jacobian of the map $\nabla u : \Omega \to \mathbb{R}^n$. In fact, we may consider $\chi_\varepsilon = \nabla u - \varepsilon\mathrm{Id} : \Omega \to \mathbb{R}^n$, where Id is the identity map in $\mathbb{R}^n$. Then, $\nabla\chi_\varepsilon = \nabla^2 u - \varepsilon I$, which is negative definite in Γ^+ and hence injective. By a change of variables, we have

$$\int_{\chi_\varepsilon(\Gamma^+\cap\Omega^+)} g(p)\,dp = \int_{\Gamma^+\cap\Omega^+} g(\chi_\varepsilon)|\det(\nabla^2 u - \varepsilon I)|\,dx.$$

This implies (1) if we let $\varepsilon \to 0$.

We assume $M = \sup_\Omega u > 0$. Now, we claim

$$B_{M/d} \subset \nabla u(\Gamma^+ \cap \Omega^+). \tag{2}$$

In other words, for any $a \in \mathbb{R}^n$ with $|a| < M/d$, there exists an $x \in \Gamma^+ \cap \Omega^+$ such that $a = \nabla u(x)$.

We assume u attains its maximum M at $0 \in \Omega$; i.e.,

$$u(0) = M.$$

For any $a \in \mathbb{R}^n$ with $|a| < M/d$, consider

$$u_a(x) = u(x) - a \cdot x.$$

First, we have $u_a(0) = M$ and, for any $x \in \partial\Omega$,

$$u_a(x) \leq -a \cdot x < M.$$

Hence, u_a attains its maximum at some $x_a \in \Omega$, with $u_a(x_a) \geq M$. Then,

$$u(x_a) = u_a(x_a) + a \cdot x_a \geq M + a \cdot x_a > 0,$$

and, for any $x \in \Omega$,

$$u(x) = u_a(x) + a \cdot x \leq u_a(x_a) + a \cdot x = u(x_a) + a \cdot (x - x_a).$$

Hence, $x_a \in \Gamma^+ \cap \Omega^+$ and $a = \nabla u(x_a)$.

By combining (1) and (2), we have

$$\int_{B_{M/d}} g(p)\,dp \leq \int_{\Gamma^+\cap\Omega^+} g(\nabla u)|\det(\nabla^2 u)|\,dx.$$

This is the desired result. □

In fact, the proof above can be interpreted geometrically. For any $a \in \mathbb{R}^n \setminus \{0\}$ with $|a| < M/d$, we consider the affine function $l(x) = M - a \cdot x$; the graph of l has the slope a and the value M at the origin. Note that l assumes positive values on the boundary by the choice of a. By moving the graph of l upward vertically, we assume it touches the graph of u at x_a for the last time. In other words, if we move the graph of l further upward, then it will never touch the graph of u. Then, $x_a \in \Gamma^+$.

We note that the integral domain Γ^+ in Lemma 1.2.1 can be replaced by

$$\Gamma^+ \cap \left\{ x \in \Omega : \ u(x) > \max_{\partial\Omega} u^+ \right\}.$$

This is clear from the proof.

A special case of Lemma 1.2.1 is given by the following result.

Corollary 1.2.2. *Let Ω be a bounded domain in $\mathbb{R}^n$. Then, for any $u \in C(\bar{\Omega}) \cap C^2(\Omega)$,*

$$\sup_{\Omega} u \leq \max_{\partial\Omega} u^+ + \frac{d}{\sqrt[n]{\omega_n}} \left(\int_{\Gamma^+} |\det \nabla^2 u| \, dx \right)^{\frac{1}{n}},$$

where Γ^+ is the upper contact set of u, $d = \mathrm{diam}(\Omega)$, and ω_n is the volume of the unit ball B_1 in $\mathbb{R}^n$.

Proof. The desired result follows from Lemma 1.2.1 by taking $g = 1$. □

In Corollary 1.2.2, we estimated the sup-norm of a function. We now improve such an estimate for values at any point in terms of its distance to the boundary. For simplicity, we consider only concave functions in convex domains.

Lemma 1.2.3. *Let Ω be a bounded convex domain in $\mathbb{R}^n$ and u be a concave $C(\bar{\Omega}) \cap C^2(\Omega)$-function with $u = 0$ on $\partial\Omega$. Then, for any $x_0 \in \Omega$,*

$$\big(u(x_0)\big)^n \leq C(\mathrm{diam}(\Omega))^{n-1} \, \mathrm{dist}(x_0, \partial\Omega) \int_{\Omega} |\det \nabla^2 u| \, dx,$$

where C is a positive constant depending only on n.

Proof. By the concavity, we have $u \geq 0$ in Ω. We assume $u > 0$ in Ω. By taking $g = 1$ in (1) in the proof of Lemma 1.2.1, we have

$$|\nabla u(\Omega)| \leq \int_{\Omega} |\det(\nabla^2 u)| \, dx. \tag{1}$$

For any given $x_0 \in \Omega$, let v be the concave function whose graph is the cone with vertex $(x_0, u(x_0))$ and base $\partial\Omega$, with $v = 0$ on $\partial\Omega$. In fact, v can be defined in the following way. For any $\widetilde{x} \in \partial\Omega$ and $t \in [0, 1]$, define

$$v\big(tx_0 + (1-t)\widetilde{x}\big) = tu(x_0).$$

In particular, $v(x_0) = u(x_0)$. By the concavity of u and $u = v = 0$ on $\partial\Omega$, we have $v \leq u$ in Ω. Set

$$\mathcal{S} = \{p \in \mathbb{R}^n : v(x) \leq v(x_0) + p \cdot (x - x_0) \text{ in } \Omega\}.$$

In other words, $\mathcal{S}$ is the collection of slopes of the supporting planes of the graph of v at $(x_0, v(x_0))$. As in the proof of Lemma 1.2.1, we can prove

$$\mathcal{S} \subset \nabla u(\Omega). \tag{2}$$

Set $D = \mathrm{diam}(\Omega)$ and $d = \mathrm{dist}(x_0, \partial\Omega)$. We claim

$$\left(\frac{v(x_0)}{D}\right)^{n-1} \cdot \frac{v(x_0)}{d} \leq C|\mathcal{S}|, \tag{3}$$

where C is a positive constant depending only on n. Then, the desired result follows from (1), (2), and (3).

We now prove (3). First, by the definition of v, we have, for any $p \in B_{v(x_0)/D}$ and any $x \in \Omega$,

$$v(x) \le v(x_0) + p \cdot (x - x_0).$$

Hence,

$$B_{v(x_0)/D} \subset \mathcal{S}. \tag{4}$$

Next, we note that

$$\text{there exists a } p_0 \in \mathcal{S} \text{ such that } |p_0| = v(x_0)/d. \tag{5}$$

To see this, we take a $\widetilde{x} \in \partial\Omega$ with $|\widetilde{x} - x_0| = d$ and a supporting plane H of Ω at $\widetilde{x}$ in $\mathbb{R}^n$. The existence of such an H follows from the convexity of Ω. The hyperplane in $\mathbb{R}^{n+1}$ generated by H and the point $(x_0, v(x_0))$ is a supporting plane of the graph of v with the slope p_0. This verifies (5). Last, by the definition of $\mathcal{S}$, we note that $\mathcal{S}$ is a convex set. By combining (4), (5), and $|p_0| \ge v(x_0)/D$, we conclude that $\mathcal{S}$ contains the convex hull of $B_{v(x_0)/D}$ and p_0. A geometric argument yields

$$C\left(\frac{v(x_0)}{D}\right)^{n-1} \cdot |p_0| \le |\mathcal{S}|.$$

This proves (3). □

Next, we relate $\det \nabla^2 u$ in Lemma 1.2.1 to subsolutions of linear differential equations.

Lemma 1.2.4. *Let Ω be a bounded domain in $\mathbb{R}^n$ and $g \in L^1_{\text{loc}}(\mathbb{R}^n)$ be nonnegative. Assume $a_{ij} \in C(\Omega)$, with $a_{ij} = a_{ji}$, such that $\big(a_{ij}(x)\big)$ is positive definite for any $x \in \Omega$, with $D^* = \sqrt[n]{\det(a_{ij})}$. Suppose that $u \in C(\bar{\Omega}) \cap C^2(\Omega)$ satisfies*

$$a_{ij}\partial_{ij}u \ge f \quad \text{in } \Omega,$$

for some $f \in C(\Omega)$ with $f^-/D^* \in L^n(\Omega)$. Then,

$$\int_{B_{M/d}} g(p)\,dp \le \int_{\Gamma^+} g(\nabla u)\left(\frac{f^-}{nD^*}\right)^n,$$

where Γ^+ is the upper contact set of u, $M = \sup_\Omega u - \max_{\partial\Omega} u^+$, and $d = \operatorname{diam}(\Omega)$.

Proof. Since the matrix $A = (a_{ij})$ is positive definite, we have

$$\det(A) \cdot \det(-\nabla^2 u) \le \left(\frac{-a_{ij}\partial_{ij}u}{n}\right)^n \quad \text{on } \Gamma^+.$$

Then, the desired result follows from Lemma 1.2.1. □

Now, we are ready to prove the maximum principle due to Alexandrov **[1]**, **[2]**, which yields an estimate of the L^∞-norms of solutions in terms of L^n-norms of nonhomogeneous terms. This is different from the maximum principle we discussed in Section 1.1, where we estimated the L^∞-norms of solutions in terms of L^∞-norms of nonhomogeneous terms.

Theorem 1.2.5. *Let Ω be a bounded domain in $\mathbb{R}^n$. Assume $a_{ij} \in C(\Omega)$, with $a_{ij} = a_{ji}$, such that $\big(a_{ij}(x)\big)$ is positive definite for any $x \in \Omega$, with $D^* = \sqrt[n]{\det(a_{ij})}$. Suppose that $u \in C(\bar{\Omega}) \cap C^2(\Omega)$ satisfies*

$$a_{ij}\partial_{ij}u \geq f \quad in\ \Omega,$$

for some $f \in C(\Omega)$ with $f^-/D^* \in L^n(\Omega)$. Then,

$$\sup_{\Omega} u \leq \max_{\partial\Omega} u^+ + Cd\left\|\frac{f^-}{D^*}\right\|_{L^n(\Gamma^+)},$$

where Γ^+ is the upper contact set of u, $d = \operatorname{diam}(\Omega)$, and C is a positive constant depending only on n.

Proof. We simply take $g = 1$ in Lemma 1.2.4. □

Remark 1.2.6. (i) The integral domain Γ^+ in Theorem 1.2.5 can be replaced by

$$\Gamma^+ \cap \left\{x \in \Omega : \ u(x) > \max_{\partial\Omega} u^+\right\}.$$

(ii) The uniform ellipticity is not needed in Theorem 1.2.5. This is important in applications to certain nonlinear elliptic equations, including the mean curvature equation and the Monge-Ampère equation.

1.2.2. The Harnack Inequality. In this subsection, we derive the Harnack inequality for positive solutions of uniformly elliptic differential equations and discuss its corollaries. We will focus on a priori estimates instead of the regularity. Our assumptions are more than what we need.

We let Ω be a bounded domain in $\mathbb{R}^n$ and let a_{ij} be continuous functions in Ω, with $a_{ij} = a_{ji}$. We consider the linear differential operator L given by

$$Lu = a_{ij}\partial_{ij}u \quad \text{in } \Omega, \tag{1.2.1}$$

for any $u \in C^2(\Omega)$. The operator L is always assumed to be *uniformly elliptic* in Ω; namely, for any $x \in \Omega$ and $\xi \in \mathbb{R}^n$,

$$\lambda|\xi|^2 \leq a_{ij}(x)\xi_i\xi_j \leq \Lambda|\xi|^2, \tag{1.2.2}$$

for some positive constants λ and Λ, which are usually called the *ellipticity constants.* This means that all eigenvalues of the matrix $\big(a_{ij}(x)\big)$ are between λ and Λ. We also set $D = \det(a_{ij})$ and $D^* = \sqrt[n]{D}$. Hence, D^* is the geometric mean of the eigenvalues of (a_{ij}). The uniform ellipticity (1.2.2) implies

$$\lambda \leq D^* \leq \Lambda. \tag{1.2.3}$$

First, we derive an upper bound for subsolutions.

Theorem 1.2.7. *Let B_R be a ball in $\mathbb{R}^n$ and L be given by (1.2.1) in B_R, for some $a_{ij} \in C(B_R)$ satisfying (1.2.2). Suppose that $u \in L^\infty(B_R) \cap C^2(B_R)$ satisfies $Lu \geq f$ in B_R, for some $f \in L^n(B_R) \cap C(B_R)$. Then, for any $p > 0$,*

$$\sup_{B_{R/2}} u \leq C\left\{R^{-\frac{n}{p}}\|u^+\|_{L^p(B_R)} + R\|f^-\|_{L^n(B_R)}\right\},$$

where C is a positive constant depending only on n, p, λ, and Λ.

Proof. Without loss of generality, we consider $R = 1$. We first introduce a cutoff function. For any constant $\beta \geq 1$, we define

$$\eta(x) = (1 - |x|^2)^\beta \quad \text{in } B_1.$$

By simple differentiations, we obtain

$$\partial_i \eta(x) = -2\beta x_i (1 - |x|^2)^{\beta - 1}$$

and

$$\partial_{ij}\eta(x) = -2\beta\delta_{ij}(1 - |x|^2)^{\beta-1} + 4\beta(\beta - 1)x_i x_j (1 - |x|^2)^{\beta-2}.$$

Hence,

$$|\nabla \eta(x)| = 2\beta|x|(1 - |x|^2)^{\beta-1} \leq 2\beta\eta^{1-\frac{1}{\beta}},$$

and

$$L\eta = 4\beta(\beta - 1)a_{ij}x_i x_j(1 - |x|^2)^{\beta-2} - 2\beta a_{ij}\delta_{ij}(1 - |x|^2)^{\beta-1}.$$

Setting $w = \eta u$, we have

$$Lw = \eta Lu + uL\eta + 2a_{ij}\partial_i\eta\partial_j u.$$

Consider the upper contact set Γ^+ of w in B_1. We clearly have $w > 0$ in Γ^+ since $w = 0$ on ∂B_1. Hence, for any $x \in \Gamma^+$,

$$|\nabla w(x)| \leq \frac{w(x)}{1 - |x|} \leq \frac{2w(x)}{1 - |x|^2},$$

which implies

$$|\nabla u(x)| = \frac{1}{\eta}|\nabla w - u\nabla\eta| \leq \frac{1}{\eta}\left(\frac{2w}{1 - |x|^2} + u|\nabla\eta|\right) \leq 2(1 + \beta)\eta^{-1-\frac{1}{\beta}}w.$$

Then in Γ^+, we have

$$|a_{ij}\partial_i\eta\partial_j u| \leq \Lambda|\nabla u||\nabla\eta| \leq 4\beta(1 + \beta)\Lambda\eta^{-\frac{2}{\beta}}w.$$

Note that we also have $u > 0$ in Γ^+. Hence,

$$\begin{aligned}
uL\eta &= -2\beta u a_{ij}\delta_{ij}(1-|x|^2)^{\beta-1} + 4\beta(\beta-1)u a_{ij}x_i x_j(1-|x|^2)^{\beta-2} \\
&\geq -2\beta u a_{ij}\delta_{ij}(1-|x|^2)^{\beta-1} \\
&\geq -2\beta n\Lambda u\eta^{1-\frac{1}{\beta}} = -2n\Lambda\beta w\eta^{-\frac{1}{\beta}} \\
&\geq -2n\Lambda\beta w\eta^{-\frac{2}{\beta}}.
\end{aligned}$$

Therefore, we obtain

$$Lw \geq -C\eta^{-\frac{2}{\beta}}w - \eta f^- \geq -\big(C\eta^{-\frac{2}{\beta}}w + f^-\big) \quad \text{on } \Gamma^+,$$

where C is a positive constant depending only on n, β, λ, and Λ. For $\beta \geq 2$, by applying Theorem 1.2.5, we obtain

$$\sup_{B_1} w^+ \leq C\left\{\|\eta^{-\frac{2}{\beta}}w^+\|_{L^n(B_1)} + \|f^-\|_{L^n(B_1)}\right\}.$$

Now writing $w^+ = (w^+)^{1-\frac{2}{\beta}}(w^+)^{\frac{2}{\beta}}$, we have

$$\sup_{B_1} w^+ \leq C\left\{\left(\sup_{B_1} w^+\right)^{1-\frac{2}{\beta}} \|(u^+)^{\frac{2}{\beta}}\|_{L^n(B_1)} + \|f^-\|_{L^n(B_1)}\right\}.$$

By choosing $\beta > 2$ and applying Young's inequality, we get

$$\begin{aligned}
\sup_{B_1} w^+ &\leq C\left\{\|(u^+)^{\frac{2}{\beta}}\|^{\frac{\beta}{2}}_{L^n(B_1)} + \|f^-\|_{L^n(B_1)}\right\} \\
&= C\left\{\|u^+\|_{L^{\frac{2n}{\beta}}(B_1)} + \|f^-\|_{L^n(B_1)}\right\}.
\end{aligned}$$

For $\beta > 2$, we have $2n/\beta \in (0, n)$. Hence, for any $p \in (0, n)$,

$$\sup_{B_1} w^+ \leq C\left\{\|u^+\|_{L^p(B_1)} + \|f^-\|_{L^n(B_1)}\right\}.$$

We apply the Hölder inequality to get the desired result for $p \geq n$. □

By replacing u with $-u$, Theorem 1.2.7 extends to supersolutions and solutions of the equation $Lu = f$.

Corollary 1.2.8. *Let B_R be a ball in $\mathbb{R}^n$ and L be given by (1.2.1) in B_R, for some $a_{ij} \in C(B_R)$ satisfying (1.2.2). Suppose that u is an $L^\infty(B_R) \cap C^2(B_R)$-solution of*

$$Lu = f \quad \text{in } B_R,$$

for some $f \in L^n(B_R) \cap C(B_R)$. Then, for any $p > 0$,

$$|u|_{L^\infty(B_{R/2})} \leq C\left\{R^{-\frac{n}{p}}\|u\|_{L^p(B_R)} + R\|f\|_{L^n(B_R)}\right\},$$

where C is a positive constant depending only on n, p, λ, and Λ.

Now, we turn our attention to supersolutions. We first introduce the Calderon-Zygmund decomposition. For any $x \in \mathbb{R}^n$ and $r > 0$, we define

$$Q_r(x) = \left\{ y \in \mathbb{R}^n : |y_i - x_i| \leq \frac{r}{2} \ \text{ for any } i = 1, \ldots, n \right\}.$$

For $x = 0$, we simply write Q_r instead.

Take the unit cube Q_1. Cut it equally into 2^n cubes, which we take as the first generation. Do the same cutting for these small cubes to get the second generation. Continue this process. These cubes (from all generations) are called *dyadic cubes.* Any cube Q in the $(k + 1)$th-generation arises from some cube $\widetilde{Q}$ in the kth-generation, which is called the *predecessor* of Q.

Lemma 1.2.9. *Suppose that f is a nonnegative $L^1(Q_1)$-function and that δ is a constant such that*

$$\frac{1}{|Q_1|} \int_{Q_1} f\, dx < \delta.$$

Then, there exists a sequence of (nonoverlapping) dyadic cubes $\{Q^j\}$ in Q_1 such that

$$f \leq \delta \quad \text{a.e. in} \ \ Q_1 \setminus \bigcup_j Q^j$$

and

$$\delta \leq \frac{1}{|Q^j|} \int_{Q^j} f dx < 2^n \delta.$$

Proof. Cut Q_1 into 2^n dyadic cubes and keep the cube Q if

$$\frac{1}{|Q|} \int_Q f\, dx \geq \delta. \tag{1}$$

Continue cutting for others, and always keep the cube Q if (1) is satisfied and cut the rest. Let $\{Q^j\}$ be the cubes we have kept during this infinite process. We need only verify

$$f \leq \delta \ \text{ a.e. in } \ Q_1 \setminus \bigcup_j Q^j. \tag{2}$$

For any $x \in Q_1 \setminus \bigcup_j Q^j$, from the way we collect $\{Q^j\}$, there exists a sequence of cubes $Q^{j'}$ containing x such that

$$\frac{1}{|Q^{j'}|} \int_{Q^{j'}} f\, dx < \delta$$

and

$$\operatorname{diam}(Q^{j'}) \to 0 \quad \text{as } j' \to \infty.$$

Then, (2) follows from the Lebesgue density theorem. □

Next, we prove a consequence of the Calderon-Zygmund decomposition, which will be needed in the proof of the weak Harnack inequality.

Lemma 1.2.10. *Suppose that A and B are measurable sets such that $A \subset B \subset Q_1$ and*

(1) $|A| < \delta$ *for some* $\delta \in (0,1)$;

(2) *for any dyadic cube* Q, $|A \cap Q| \geq \delta|Q|$ *implies* $\widetilde{Q} \subset B$ *for the predecessor* $\widetilde{Q}$ *of* Q.

Then,

$$|A| \leq \delta|B|.$$

Proof. We apply Lemma 1.2.9 to $f = \chi_A$. By the assumption (1), we obtain a sequence of dyadic cubes $\{Q^j\}$ such that

$$A \subset \bigcup_j Q^j \quad \text{except for a set of measure zero,}$$

$$\delta \leq \frac{|A \cap Q^j|}{|Q^j|} < 2^n\delta,$$

and, for any predecessor $\widetilde{Q^j}$ of Q^j,

$$\frac{|A \cap \widetilde{Q^j}|}{|\widetilde{Q^j}|} < \delta.$$

By the assumption (2), we have $\widetilde{Q^j} \subset B$ for each j. Hence,

$$A \subset \bigcup_j \widetilde{Q^j} \subset B \quad \text{except for a set of measure zero.}$$

We relabel $\{\widetilde{Q^j}\}$ so that they are nonoverlapping. Then, we get

$$|A| \leq \sum_j |A \cap \widetilde{Q^j}| \leq \delta \sum_j |\widetilde{Q^j}| \leq \delta|B|.$$

This yields the desired estimate. □

Now we are ready to prove the weak Harnack inequality for nonnegative supersolutions.

Theorem 1.2.11. *Let B_R be a ball in $\mathbb{R}^n$ and L be given by (1.2.1) in B_R, for some $a_{ij} \in C(B_R)$ satisfying (1.2.2). Suppose that $u \in L^\infty(B_R) \cap C^2(B_R)$ satisfies $u \geq 0$ and $Lu \leq f$ in B_R, for some $f \in L^n(B_R) \cap C(B_R)$. Then,*

$$R^{-\frac{n}{p}}\|u\|_{L^p(B_{2\tau R})} \leq C\left\{\inf_{B_{\tau R}} u + R\|f\|_{L^n(B_R)}\right\},$$

where $\tau = (8\sqrt{n})^{-1}$ and p and C are positive constants depending only on n, λ, and Λ.

The proof consists of several steps. In the first step, we prove that if a positive supersolution is small somewhere in Q_3, then it has an upper bound in a good portion of Q_1. This step is the key ingredient in the weak Harnack inequality. In the second step, we iterate to get a power decay of distribution functions for positive supersolutions.

Proof. We consider the case $R = 2\sqrt{n}$ and prove

$$\|u\|_{L^p(B_{1/2})} \leq C\left\{\inf_{B_{1/4}} u + \|f\|_{L^n(B_{2\sqrt{n}})}\right\}.$$

The proof consists of several steps.

Step 1. We prove that there exist constants $\varepsilon_0 > 0$, $\mu \in (0,1)$, and $M > 1$, depending only on n, λ, and Λ, such that if

$$\inf_{Q_3} u \leq 1, \quad \|f\|_{L^n(B_{2\sqrt{n}})} \leq \varepsilon_0, \tag{1}$$

then

$$|\{u \leq M\} \cap Q_1| > \mu.$$

The basic idea of the proof is to construct a function φ, which is very concave outside Q_1, such that if we correct u by φ, the lower contact set of $u + \varphi$ occurs in Q_1 and occupies a large portion of Q_1. In other words, we localize where the contact occurs by choosing suitable functions.

Note that $B_{1/4} \subset B_{1/2} \subset Q_1 \subset Q_3 \subset B_{2\sqrt{n}}$. For some large constant $\beta > 0$ to be determined and some $M > 0$, define

$$\varphi(x) = -M\left(1 - \frac{|x|^2}{4n}\right)^\beta \quad \text{in } B_{2\sqrt{n}}.$$

Then, $\varphi = 0$ on $\partial B_{2\sqrt{n}}$. We choose M, according to β, such that

$$\varphi \leq -2 \text{ in } Q_3. \tag{2}$$

Set

$$w = u + \varphi \quad \text{in } B_{2\sqrt{n}}.$$

We will prove, by choosing β large,

$$Lw \leq f + \eta \quad \text{in } B_{2\sqrt{n}}, \tag{3}$$

for some $\eta \in C_0^\infty(Q_1)$ and $0 \leq \eta \leq C$, for a positive constant C depending only on n, λ, and Λ. To do this, we calculate the Hessian matrix of φ. A straightforward calculation yields

$$\begin{aligned}\partial_{ij}\varphi(x) = \frac{M}{2n}\beta\delta_{ij}\left(1 - \frac{|x|^2}{4n}\right)^{\beta-1} \\ - \frac{M}{(2n)^2}\beta(\beta-1)x_i x_j\left(1 - \frac{|x|^2}{4n}\right)^{\beta-2},\end{aligned}$$

and hence

$$\begin{aligned}L\varphi &= \frac{M}{2n}\beta\left(1-\frac{|x|^2}{4n}\right)^{\beta-2}\left(\left(1-\frac{|x|^2}{4n}\right)a_{ij}\delta_{ij}-\frac{1}{2n}(\beta-1)a_{ij}x_ix_j\right)\\ &\le \frac{M}{2n}\beta\left(1-\frac{|x|^2}{4n}\right)^{\beta-2}\left(\left(1-\frac{|x|^2}{4n}\right)n\Lambda-\frac{1}{2n}(\beta-1)\lambda|x|^2\right).\end{aligned}$$

Therefore for $|x|\ge 1/4$, we have $L\varphi\le 0$ if we choose β large, depending only on n,λ, and Λ. Hence,

$$Lw\le f \quad \text{in } B_{2\sqrt{n}}\setminus Q_1.$$

This finishes the proof of (3).

Now we apply Theorem 1.2.5 to $-w$ in $B_{2\sqrt{n}}$. Note that $\inf_{Q_3} w\le -1$ by (1) and (2) and $w\ge 0$ on $\partial B_{2\sqrt{n}}$. In view of Remark 1.2.6(i), we obtain

$$\begin{aligned}1 &\le C\,\||f|+\eta\|_{L^n(B_{2\sqrt{n}}\cap\Gamma^-\cap\Omega_-)}\\ &\le C\|f\|_{L^n(B_{2\sqrt{n}})}+C|\Gamma^-\cap\Omega_-\cap Q_1|^{\frac{1}{n}},\end{aligned}$$

where Γ^- is the lower contact set of w and $\Omega_-=\{x\in B_{2\sqrt{n}}:\ w(x)<0\}$. Choosing ε_0 small enough, we get

$$\frac{1}{2}\le C|\Gamma^-\cap\Omega_-\cap Q_1|^{\frac{1}{n}}\le C\big|\{u\le M\}\cap Q_1\big|^{\frac{1}{n}}$$

since $w(x)\le 0$ implies $u(x)\le -\varphi(x)\le M$ in Ω_-.

Step 2. We prove that there exist positive constants ε_0, γ, and C, depending only on n,λ, and Λ, such that if (1) holds, then, for any $t>0$,

$$|\{u>t\}\cap Q_1|\le Ct^{-\gamma}.$$

In fact, under the assumption (1), we will prove, for any $k=1,2,\ldots$,

$$\Big|\{u>M^k\}\cap Q_1\Big|\le(1-\mu)^k, \tag{4}$$

where M and μ are as in Step 1.

For $k=1$, (4) is implied by Step 1. Now suppose (4) holds for $k-1$, for some $k\ge 2$. Set

$$A=\{u>M^k\}\cap Q_1,\quad B=\{u>M^{k-1}\}\cap Q_1.$$

Clearly, $A\subset B\subset Q_1$ and $|A|\le|\{u>M\}\cap Q_1|\le 1-\mu$ by Step 1. We claim that if $Q_r(x_0)$ is a cube in Q_1, with $r\in(0,1/2)$, such that

$$|A\cap Q_r(x_0)|>(1-\mu)|Q_r(x_0)|, \tag{5}$$

then $Q_{3r}(x_0)\cap Q_1\subset B$. Assuming the claim, we are in a position to apply Lemma 1.2.10 to get

$$|A|\le(1-\mu)|B|.$$

Then, (4) follows.

To prove the claim, we consider the transform

$$x = x_0 + ry \quad \text{for } y \in Q_1 \text{ and } x \in Q_r(x_0).$$

Set

$$\widetilde{a}_{ij}(y) = a_{ij}(x), \quad \widetilde{f}(y) = \frac{r^2}{M^{k-1}} f(x),$$

and

$$\widetilde{u}(y) = \frac{1}{M^{k-1}} u(x).$$

Then, $\widetilde{u} \geq 0$ and

$$\widetilde{a}_{ij}\partial_{y_i y_j}\widetilde{u} \leq \widetilde{f} \quad \text{in } B_{2\sqrt{n}}.$$

Moreover,

$$\|\widetilde{f}\|_{L^n(B_{2\sqrt{n}})} \leq \frac{r}{M^{k-1}}\|f\|_{L^n(B_{2\sqrt{n}})} \leq \|f\|_{L^n(B_{2\sqrt{n}})} \leq \varepsilon_0.$$

Writing (5) in terms of $\widetilde{u}$, we get

$$|\{\widetilde{u}(y) > M\} \cap Q_1| = r^{-n}\left|\{u(x) > M^k\} \cap Q_r(x_0)\right| > 1 - \mu,$$

and hence,

$$|\{\widetilde{u}(y) \leq M\} \cap Q_1| \leq \mu.$$

By applying what we proved in Step 1 to $\widetilde{u}$, we have

$$\inf_{Q_3} \widetilde{u} > 1.$$

Hence, $u > M^{k-1}$ in $Q_{3r}(x_0)$, and in particular $Q_{3r}(x_0) \cap Q_1 \subset B$. This finishes the proof of the claim.

Step 3. We prove that there exist positive constants γ and C, depending only on n, λ, and Λ, such that, for any $t > 0$,

$$|\{x \in Q_1 : u(x) > t\}| \leq Ct^{-\gamma}\left(\inf_{Q_3} u + \|f\|_{L^n(B_{2\sqrt{n}})}\right)^{\gamma}.$$

To prove this, we consider, for any $\delta > 0$,

$$u_\delta = \left(\inf_{Q_3} u + \delta + \frac{1}{\varepsilon_0}\|f\|_{L^n(B_{2\sqrt{n}})}\right)^{-1} u.$$

By applying Step 2 to u_δ and then letting $\delta \to 0$, we obtain the desired result. We point out that δ is introduced only for the case $f \equiv 0$ and $\inf_{Q_3} u = 0$.

Step 4. Now we prove

$$\|u\|_{L^p(Q_1)} \leq C\left\{\inf_{Q_3} u + \|f\|_{L^n(B_{2\sqrt{n}})}\right\},$$

where p and C are positive constants depending only on n, λ, and Λ.

Set, for any $t > 0$,

$$A(t) = \{x \in Q_1 : u(x) > t\}.$$

First, we have, for any $p > 0$ and $\tau > 0$,

$$\begin{aligned}\int_{Q_1} u^p\,dx &= p\int_0^\infty t^{p-1}|A(t)|\,dt\\ &= p\int_0^\tau t^{p-1}|A(t)|\,dt + p\int_\tau^\infty t^{p-1}|A(t)|\,dt = I + II.\end{aligned}$$

For I, we have easily

$$I \le p|Q_1|\int_0^\tau t^{p-1}dt = \tau^p.$$

For II, we get, by Step 3,

$$\begin{aligned}II &= p\int_\tau^\infty t^{p-1}|A(t)|\,dt\\ &\le Cp\int_\tau^\infty t^{p-\gamma-1}dt\left(\inf_{Q_3} u + \|f\|_{L^n(B_{2\sqrt{n}})}\right)^\gamma\\ &= C\tau^{p-\gamma}\left(\inf_{Q_3} u + \|f\|_{L^n(B_{2\sqrt{n}})}\right)^\gamma\end{aligned}$$

if we choose $p < \gamma$. Combining these two estimates, we obtain

$$\int_{Q_1} u^p\,dx \le C\left(\tau^p + \tau^{p-\gamma}\big(\inf_{Q_3} u + \|f\|_{L^n(B_{2\sqrt{n}})}\big)^\gamma\right).$$

Next, we choose

$$\tau = \inf_{Q_3} u + \|f\|_{L^n(B_{2\sqrt{n}})}.$$

This implies the desired result. □

We point out that p in Theorem 1.2.11 may be less than 1.

Now, we can prove the Harnack inequality due to Krylov and Safonov [**99**], [**100**].

Theorem 1.2.12. *Let B_R be a ball in $\mathbb{R}^n$ and L be given by (1.2.1) in B_R, for some $a_{ij} \in C(B_R)$ satisfying (1.2.2). Suppose that u is a nonnegative $L^\infty(B_R) \cap C^2(B_R)$-solution of*

$$Lu = f \quad \text{in } B_R,$$

for some $f \in L^n(B_R) \cap C(B_R)$. Then,

$$\sup_{B_{R/2}} u \le C\left\{\inf_{B_{R/2}} u + R\|f\|_{L^n(B_R)}\right\},$$

where C is a positive constant depending only on n, λ, and Λ.

Proof. By Theorem 1.2.7 and Theorem 1.2.11, we have

$$\sup_{B_{\tau R}} u \le C\left\{\inf_{B_{\tau R}} u + R\|f\|_{L^n(B_R)}\right\},$$

where $\tau = (8\sqrt{n})^{-1}$ and C is a positive constant depending only on n, λ, and Λ. Then, the desired result follows from a simple covering argument. □

The interior estimate of the Hölder semi-norms of solutions is a direct consequence of Theorem 1.2.12. We first prove a simple lemma.

Lemma 1.2.13. *Let ω and σ be nondecreasing functions in an interval $(0, R]$ and let γ and τ be constants in $(0, 1)$. Suppose, for any $r \in (0, R]$,*

$$\omega(\tau r) \le \gamma\omega(r) + \sigma(r).$$

Then, for any $\mu \in (0, 1)$ and $r \in (0, R]$,

$$\omega(r) \le C\left\{\left(\frac{r}{R}\right)^{\alpha}\omega(R) + \sigma(r^{\mu}R^{1-\mu})\right\},$$

where C is a positive constant depending only on γ, τ, and $\alpha = (1-\mu)\log\gamma/\log\tau$.

Proof. Fix an $r_1 \in (0, R]$. Then for any $r \in (0, r_1]$, we have

$$\omega(\tau r) \le \gamma\omega(r) + \sigma(r_1)$$

since σ is nondecreasing. We now iterate this inequality to get, for any positive integer k,

$$\omega(\tau^k r_1) \le \gamma^k\omega(r_1) + \sigma(r_1)\sum_{i=0}^{k-1}\gamma^i \le \gamma^k\omega(R) + \frac{\sigma(r_1)}{1-\gamma}.$$

For any $r \in (0, r_1]$, we choose a positive integer k such that

$$\tau^k r_1 < r \le \tau^{k-1} r_1.$$

Hence,

$$\omega(r) \le \omega(\tau^{k-1}r_1) \le \gamma^{k-1}\omega(R) + \frac{\sigma(r_1)}{1-\gamma}.$$

By writing $\gamma = \tau^{\log\gamma/\log\tau}$, we get

$$\omega(r) \le \frac{1}{\gamma}\left(\frac{r}{r_1}\right)^{\log\gamma/\log\tau}\omega(R) + \frac{\sigma(r_1)}{1-\gamma}.$$

By letting $r_1 = r^{\mu}R^{1-\mu}$, we obtain

$$\omega(r) \le \frac{1}{\gamma}\left(\frac{r}{R}\right)^{(1-\mu)(\log\gamma/\log\tau)}\omega(R) + \frac{\sigma(r^{\mu}R^{1-\mu})}{1-\gamma}.$$

This is the desired estimate. □

Now we are ready to prove the interior estimate of the Hölder semi-norms of solutions. For any set $\Omega \subset \mathbb{R}^n$ and any bounded function u in Ω, we define

$$\operatorname*{osc}_{\Omega} u = \sup_{\Omega} u - \inf_{\Omega} u.$$

Theorem 1.2.14. *Let B_R be a ball in $\mathbb{R}^n$ and L be given by (1.2.1) in B_R, for some $a_{ij} \in C(B_R)$ satisfying (1.2.2). Suppose that u is an $L^\infty(B_R) \cap C^2(B_R)$-solution of*

$$Lu = f \quad \text{in } B_R,$$

for some $f \in L^n(B_R) \cap C(B_R)$. Then, for any $r \in (0, R]$,

$$\operatorname*{osc}_{B_r} u \leq C \left(\frac{r}{R}\right)^\alpha \left\{ \operatorname*{osc}_{B_R} u + R\|f\|_{L^n(B_R)} \right\},$$

where $\alpha \in (0,1)$ and $C > 0$ are constants depending only on n, λ, and Λ. Moreover,

$$R^\alpha [u]_{C^\alpha(B_{R/2})} \leq C \left\{ |u|_{L^\infty(B_R)} + R\|f\|_{L^n(B_R)} \right\}.$$

Proof. Set, for any $r \in (0, R)$,

$$M(r) = \sup_{B_r} u, \quad m(r) = \inf_{B_r} u,$$

and

$$\omega(r) = M(r) - m(r).$$

Then, $M(r) < +\infty$ and $m(r) > -\infty$. We now prove, for any $r \in (0, R)$,

$$\omega(r) \leq C \left(\frac{r}{R}\right)^\alpha \left\{ \omega(R) + R\|f\|_{L^n(B_R)} \right\}.$$

In the following, we let $\tau = 1/2$. By applying Theorem 1.2.12 to $M(r) - u \geq 0$ in B_r, we get

$$\sup_{B_{\tau r}} (M(r) - u) \leq C \left\{ \inf_{B_{\tau r}} (M(r) - u) + r\|f\|_{L^n(B_r)} \right\};$$

i.e.,

$$M(r) - m(\tau r) \leq C \left\{ M(r) - M(\tau r) + r\|f\|_{L^n(B_r)} \right\}. \tag{1}$$

Similarly, by applying Theorem 1.2.12 to $u - m(r) \geq 0$ in B_r, we get

$$M(\tau r) - m(r) \leq C \left\{ m(\tau r) - m(r) + r\|f\|_{L^n(B_r)} \right\}. \tag{2}$$

Then by adding (1) and (2), we obtain

$$\omega(r) + \omega(\tau r) \leq C \left\{ \omega(r) - \omega(\tau r) + r\|f\|_{L^n(B_r)} \right\},$$

or

$$\omega(\tau r) \leq \gamma \omega(r) + r\|f\|_{L^n(B_r)},$$

for some constant $\gamma \in (0, 1)$.

Choosing μ satisfying

$$\alpha = (1-\mu)\frac{\log \gamma}{\log \tau} < \mu$$

and applying Lemma 1.2.13 with $\sigma(r) = r\|f\|_{L^n(B_R)}$, we obtain, for any $r \in (0, R)$,

$$\begin{aligned}\omega(r) &\leq C\left\{\left(\frac{r}{R}\right)^\alpha \omega(R) + r^\mu R^{1-\mu}\|f\|_{L^n(B_R)}\right\} \\ &\leq C\left(\frac{r}{R}\right)^\alpha \left\{\omega(R) + R\|f\|_{L^n(B_R)}\right\}.\end{aligned}$$

This is the desired first estimate. The second estimate follows readily. □

As an application of Theorem 1.2.14, we prove a Liouville type theorem.

Theorem 1.2.15. *Let L be given by* (1.2.1) *in $\mathbb{R}^n$, for some $a_{ij} \in C(\mathbb{R}^n)$ satisfying* (1.2.2). *Suppose that u is a $C^2(\mathbb{R}^n)$-solution of*

$$Lu = 0 \quad \text{in } \mathbb{R}^n.$$

If u is bounded, then u is constant.

Proof. By Theorem 1.2.14, we have, for any $0 < r < R$,

$$\operatorname*{osc}_{B_r} u \leq C\left(\frac{r}{R}\right)^\alpha \operatorname*{osc}_{B_R} u,$$

where $C > 0$ and $\alpha \in (0, 1)$ are constants depending only on n, λ, and Λ. For any fixed $r > 0$, let $R \to \infty$. We then conclude that u is constant in $\mathbb{R}^n$. □

We now point out an important fact. The constants C, p, and α in Theorem 1.2.11, Theorem 1.2.12, and Theorem 1.2.14 depend on the coefficients a_{ij} only through λ and Λ and in particular do not depend on the modulus of continuity of these functions. This fact plays an important role in the study of nonlinear elliptic differential equations.

1.2.3. The Hölder Continuity of Normal Derivatives.

An important application of the Harnack inequality is the following estimate of the Hölder semi-norms of normal derivatives of solutions on the boundary, due to Krylov [**97**]. In the following, we write $x = (x', x_n)$, with $x' \in \mathbb{R}^{n-1}$. We set, for any $r > 0$,

$$\begin{aligned}B_r^+ &= \{x \in B_r : x_n > 0\}, \\ \Sigma_r &= \{x \in B_r : x_n = 0\}.\end{aligned}$$

Theorem 1.2.16. *Let L be given by (1.2.1) in B_4^+, for some $a_{ij} \in C(B_4^+)$ satisfying (1.2.2). Suppose that u is a $C^1(B_4^+ \cup \Sigma_4) \cap C^2(B_4^+)$-solution of*

$$\begin{aligned} Lu &= f \quad \text{in } B_4^+, \\ u &= 0 \quad \text{on } \Sigma_4, \end{aligned}$$

for some $f \in L^\infty(B_4^+)$. Then,

$$[\partial_n u]_{C^\alpha(\Sigma_1)} \le C\left\{|\nabla u|_{L^\infty(B_4^+)} + |f|_{L^\infty(B_4^+)}\right\},$$

where $\alpha \in (0,1)$ and $C > 0$ are constants depending only on n, λ, and Λ.

Proof. Since $u = 0$ on Σ_4, we have

$$\partial_n u(x', 0) = \lim_{x_n \to 0} \frac{u(x)}{x_n}.$$

Set

$$v(x) = \frac{u(x)}{x_n}.$$

It is convenient to introduce some notations. For any $r \le 1$ and some $\delta > 0$ to be fixed (in fact, we can take $\delta = \lambda/(6n\Lambda) < 1/2$), set

$$\begin{aligned} Q(r) &= \{x \in \mathbb{R}^n : |x'| < r,\ 0 < x_n < \delta r\}, \\ Q^+(r) &= \{x \in \mathbb{R}^n : |x'| < r,\ \delta r/2 < x_n < \delta r\}. \end{aligned}$$

Two preliminary results are needed.

Lemma 1.2.17. *Assume $Lu \le f$ in $Q(r)$ with $u \ge 0$ and $u(x', 0) = 0$. Then,*

$$\inf_{Q^+(r)} v \le \frac{2}{\delta} \inf_{Q(r/2)} v + \frac{r}{\lambda} \sup_{Q(r)} |f|.$$

Proof. Set

$$m = \inf\{v(x) : x_n = \delta r,\ |x'| < r\}$$

and

$$w(x) = m x_n \left(\delta - 2\delta \frac{|x'|^2}{r^2} + \frac{x_n}{r}\right) - \frac{1}{2\lambda} x_n(\delta r - x_n) \sup_{Q(r)} |f|.$$

It is straightforward to verify that if δ is sufficiently small, then

(i) $w(x', 0) = 0$ for $|x'| < r$;

(ii) $w(x) \le 0$ on $\{|x'| = r, 0 \le x_n \le \delta r\}$, the sides of $Q(r)$;

(iii) $w(x) \le 2m\delta^2 r \le m\delta r$ on $\{|x'| \le r, x_n = \delta r\}$, the top of $Q(r)$;

(iv) $Lw \ge \sup |f| \ge f$ in $Q(r)$.

In particular, these all hold if $\delta = \lambda/(6n\Lambda)$. Note that such a δ is less than $1/2$. Parts (i)–(iii) hold obviously. For part (iv), we have, for $i, j = 1, \ldots, n-1$,

$$\begin{aligned}
\partial_{ij} w &= -\frac{4\delta m}{r^2} x_n \delta_{ij}, \\
\partial_{in} w &= -\frac{4\delta m}{r^2} x_i, \\
\partial_{nn} w &= \frac{2m}{r} + \frac{1}{\lambda} \sup |f|,
\end{aligned}$$

and hence,

$$Lw = \frac{a_{nn}}{\lambda} \sup |f| + \frac{2m}{r} a_{nn} - \frac{4\delta m}{r^2} \left(\sum_{i=1}^{n-1} a_{ii} x_n + 2 \sum_{i=1}^{n-1} a_{in} x_i \right).$$

So, (iv) follows easily.

Since $u \geq 0$ in $Q(r)$ and $u = x_n v \geq m\delta r$ on the top of $Q(r)$, then

$$\begin{aligned}
Lu &\leq Lw \quad \text{in } Q(r), \\
u &\geq w \quad \text{on } \partial Q(r).
\end{aligned}$$

Hence, by the maximum principle, $u \geq w$ in $Q(r)$, or, for any $x \in Q(r)$,

$$m \left(\delta - 2\delta \frac{|x'|^2}{r^2} + \frac{x_n}{r} \right) \leq \frac{u(x)}{x_n} + \frac{1}{2\lambda} (\delta r - x_n) \sup_{Q(r)} |f|.$$

By restricting $x \in Q(r/2)$, we obtain

$$\frac{1}{2} \delta m \leq \inf_{Q(r/2)} v + \frac{\delta r}{2\lambda} \sup_{Q(r)} |f|.$$

Since $m \geq \inf_{Q^+(r)} v$, this implies the desired result. □

The second lemma we need is a Harnack type inequality.

Lemma 1.2.18. *Suppose $Lu = f$ in B_4^+ and $u \geq 0$ in $Q(2r)$, with $r \leq 1$. Then,*

$$\sup_{Q^+(r)} v \leq C \left\{ \inf_{Q^+(r)} v + r \sup_{Q(2r)} |f| \right\},$$

where C is a positive constant depending only on n, λ, and Λ.

Proof. Note that $B_{\delta r/2}(x) \subset Q(2r)$, for any $x \in Q^+(r)$. By $u \geq 0$ in $Q(2r)$ and Theorem 1.2.12, we get

$$\sup_{B_{\delta r/8}(x)} u \leq C \left\{ \inf_{B_{\delta r/8}(x)} u + r^2 \sup_{Q(2r)} |f| \right\}.$$

Since a finite number (independent of r) of balls $B_{\delta r/8}(x)$, with $x \in Q^+(r)$, cover $Q^+(r)$, we conclude

$$\sup_{Q^+(r)} u \leq C\Big\{ \inf_{Q^+(r)} u + r^2 \sup_{Q(2r)} |f| \Big\}.$$

In $Q^+(r)$, we have $\delta r/2 \leq x_n \leq \delta r$. Since $u = x_n v$, this implies

$$\frac{1}{2}\delta r \sup_{Q^+(r)} v \leq \sup_{Q^+(r)} u \quad \text{and} \quad \inf_{Q^+(r)} u \leq \delta r \inf_{Q^+(r)} v.$$

We have the desired result. □

Now we continue the proof of Theorem 1.2.16. We assume $r \in (0,1]$ and set

$$m(r) = \inf_{Q(r)} v, \quad M(r) = \sup_{Q(r)} v,$$

and

$$\omega(r) = M(r) - m(r).$$

By Lemma 1.2.18, with u replaced by $u - m(2r)x_n \geq 0$ in $Q(2r)$, we obtain

$$\sup_{Q^+(r)} (v - m(2r)) \leq C\Big\{ \inf_{Q^+(r)} (v - m(2r)) + r \sup_{Q(2r)} |f| \Big\}.$$

By Lemma 1.2.17, we get

$$\begin{aligned} \sup_{Q^+(r)} (v - m(2r)) &\leq C\Big\{ \inf_{Q(r/2)} (v - m(2r)) + r \sup_{Q(2r)} |f| \Big\} \\ &= C\Big\{ m\left(\frac{r}{2}\right) - m(2r) + r \sup_{Q(2r)} |f| \Big\}. \end{aligned}$$

Repeating these inequalities with u replaced by $M(2r)x_n - u \geq 0$ in $Q(2r)$, we have

$$\sup_{Q^+(r)} \big(M(2r) - v\big) \leq C\Big\{ M(2r) - M\left(\frac{r}{2}\right) + r \sup_{Q(2r)} |f| \Big\}.$$

Adding these two inequalities, we obtain

$$M(2r) - m(2r) \leq C\Big\{ M(2r) - m(2r) - \Big(M\left(\frac{r}{2}\right) - m\left(\frac{r}{2}\right)\Big) + r \sup_{Q(2r)} |f| \Big\}.$$

Then, we have, for any $r \in (0,1]$,

$$\omega\left(\frac{r}{2}\right) \leq \gamma\omega(2r) + r \sup_{Q(2r)} |f|,$$

for some constant $\gamma \in (0, 1)$. By employing Lemma 1.2.13 as in the proof of Theorem 1.2.14, we obtain, for any $r \in (0, 1]$,

$$\omega(r) \le Cr^\alpha \Big\{ \omega(2) + \sup_{Q(2)} |f| \Big\},$$

where $\alpha \in (0, 1)$ and $C > 0$ are constants depending only on n, λ, and Λ. Hence, for any $x, \overline{x} \in Q(r)$,

$$\left| \frac{u(x)}{x_n} - \frac{u(\overline{x})}{\overline{x}_n} \right| \le Cr^\alpha \Big\{ \sup_{Q(2)} |\nabla u| + \sup_{Q(2)} |f| \Big\}.$$

Letting $x_n, \overline{x}_n \to 0$, we have, for any $r \in (0, 1]$,

$$\operatorname*{osc}_{\Sigma_r} \partial_n u \le Cr^\alpha \Big\{ \sup_{B_4^+} |\nabla u| + \sup_{B_4^+} |f| \Big\}.$$

This implies the desired estimate. □

1.3. The Schauder Theory

In this section, we review the Schauder theory for uniformly elliptic linear equations. Three main topics are a priori estimates in Hölder norms, the regularity of arbitrary solutions, and the solvability of the Dirichlet problem. Among these topics, a priori estimates are the most fundamental and form the basis for the existence and the regularity of solutions. We will review both the interior Schauder theory and the global Schauder theory and simply state results without proof. Refer to Chapter 6 of [**59**] for details.

We let Ω be a domain in $\mathbb{R}^n$, bounded most of the time, and let a_{ij}, b_i, and c be defined in Ω, with $a_{ij} = a_{ji}$. We consider the operator L given by

$$Lu = a_{ij}\partial_{ij}u + b_i\partial_i u + cu \quad \text{in } \Omega, \tag{1.3.1}$$

for any $u \in C^2(\Omega)$. The operator L is always assumed to be *strictly elliptic* in Ω; namely, for any $x \in \Omega$ and $\xi \in \mathbb{R}^n$,

$$a_{ij}\xi_i\xi_j \ge \lambda|\xi|^2, \tag{1.3.2}$$

for some positive constant λ.

1.3.1. The Interior Schauder Theory. In this subsection, we review the interior Schauder theory. Three main topics are interior a priori estimates in Hölder norms, the interior regularity of arbitrary solutions, and the solvability of the Dirichlet problem for continuous boundary values.

We first review Schauder interior estimates, which yield estimates of $C^{2,\alpha}$-norms of solutions in compact subsets in terms of the L^∞-norms of solutions and the C^α-norms of the nonhomogeneous terms, if coefficients are also C^α, for some $\alpha \in (0, 1)$.

Let B_R be a ball in $\mathbb{R}^n$, k be a nonnegative integer, and $\alpha \in (0,1)$ be a constant. We define

$$|u|^*_{C^{k,\alpha}(B_R)} = \sum_{i=0}^{k} R^i |\nabla^i u|_{L^\infty(B_R)} + R^{k+\alpha}[\nabla^k u]_{C^\alpha(B_R)}.$$

We note that factors of R result from scaling.

In the next result, we consider the operator L in the ball B_R.

Theorem 1.3.1. *Let $\alpha \in (0,1)$ be a constant, B_R be a ball in $\mathbb{R}^n$, and L be given by (1.3.1) in B_R, for some a_{ij}, b_i, $c \in C^\alpha(B_R)$ satisfying (1.3.2) and*

$$|a_{ij}|^*_{C^\alpha(B_R)} + R|b_i|^*_{C^\alpha(B_R)} + R^2|c|^*_{C^\alpha(B_R)} \le \Lambda,$$

for some positive constant Λ. Suppose that u is an $L^\infty(B_R) \cap C^{2,\alpha}(B_R)$-solution of

$$Lu = f \quad \text{in } B_R,$$

for some $f \in C^\alpha(B_R)$, with $|f|_{C^\alpha(B_R)} < \infty$. Then,

$$|u|^*_{C^{2,\alpha}(B_{R/2})} \le C\left\{|u|_{L^\infty(B_R)} + R^2|f|^*_{C^\alpha(B_R)}\right\},$$

where C is a positive constant depending only on n, α, λ, and Λ.

We now state a general interior estimate of the $C^{2,\alpha}$-norm of solutions.

Theorem 1.3.2. *Let $\alpha \in (0,1)$ be a constant, Ω be a bounded domain in $\mathbb{R}^n$, and L be given by (1.3.1) in Ω, for some a_{ij}, b_i, and $c \in C^\alpha(\Omega)$ satisfying (1.3.2) and*

$$|a_{ij}|_{C^\alpha(\Omega)} + |b_i|_{C^\alpha(\Omega)} + |c|_{C^\alpha(\Omega)} \le \Lambda,$$

for some positive constant Λ. Suppose that u is an $L^\infty(\Omega) \cap C^{2,\alpha}(\Omega)$-solution of

$$Lu = f \quad \text{in } \Omega,$$

for some $f \in C^\alpha(\Omega)$, with $|f|_{C^\alpha(\Omega)} < \infty$. Then, for any $\Omega' \Subset \Omega$,

$$|u|_{C^{2,\alpha}(\Omega')} \le C\left\{|u|_{L^\infty(\Omega)} + |f|_{C^\alpha(\Omega)}\right\},$$

where C is a positive constant depending only on n, α, λ, Λ, Ω', and $\operatorname{dist}(\Omega', \partial\Omega)$.

The assumption that $\nabla^2 u \in C^\alpha(\Omega)$ is not necessary and can be removed. In fact, it suffices to assume $u \in C^2(\Omega)$ and it follows as a consequence that $\nabla^2 u \in C^\alpha(\Omega)$.

An application of the interior estimates is the compactness of sequences of bounded solutions.

Corollary 1.3.3. *Let $\alpha \in (0,1)$ be a constant, Ω be a bounded domain in $\mathbb{R}^n$, and L be given by (1.3.1) in Ω, for some a_{ij}, b_i, $c \in C^\alpha(\Omega)$ with finite C^α-norms and satisfying (1.3.2). Suppose that $u_k \in L^\infty(\Omega) \cap C^{2,\alpha}(\Omega)$ satisfies, for $k = 1, 2, \ldots$,*

$$Lu_k = f \quad \text{in } \Omega,$$

for some $f \in C^\alpha(\Omega)$, with $|f|_{C^\alpha(\Omega)} < \infty$, and

$$\sup_{k \geq 1} |u_k|_{L^\infty(\Omega)} < \infty.$$

Then, there exist a subsequence $u_{k'}$ and a $u \in C^{2,\alpha}(\Omega)$ such that $u_{k'} \to u$ in C^2 in any compact subsets of Ω and $Lu = f$ in Ω.

With the interior estimates, we are able to solve the Dirichlet problem with continuous boundary values for uniformly elliptic differential equations.

Theorem 1.3.4. *Let $\alpha \in (0,1)$ be a constant, Ω be a bounded domain in $\mathbb{R}^n$ satisfying an exterior sphere condition at every boundary point, and L be given by (1.3.1) in Ω, for some a_{ij}, b_i, $c \in C^\alpha(\Omega)$ with finite $C^\alpha(\Omega)$-norms and satisfying (1.3.2) and $c \leq 0$. Then for any $f \in C^\alpha(\Omega)$, with $|f|_{C^\alpha(\Omega)} < \infty$, and any $\varphi \in C(\partial\Omega)$, there exists a (unique) solution $u \in C(\bar{\Omega}) \cap C^{2,\alpha}(\Omega)$ of the Dirichlet problem*

$$\begin{aligned} Lu &= f \quad \text{in } \Omega, \\ u &= \varphi \quad \text{on } \partial\Omega. \end{aligned}$$

Theorem 1.3.4 still holds if Ω satisfies an exterior cone condition at every boundary point instead of the exterior sphere condition.

Next, we state a general interior regularity of solutions of the uniformly elliptic equations.

Theorem 1.3.5. *Let k be a nonnegative integer, $\alpha \in (0,1)$ be a constant, Ω be a bounded domain in $\mathbb{R}^n$, and L be given by (1.3.1) in Ω, for some a_{ij}, b_i, $c \in C^{k,\alpha}(\Omega)$ with finite $C^{k,\alpha}(\Omega)$-norms and satisfying (1.3.2). Suppose that u is a $C^2(\Omega)$-solution of*

$$Lu = f \quad \text{in } \Omega,$$

for some $f \in C^{k,\alpha}(\Omega)$, with $|f|_{C^{k,\alpha}(\Omega)} < \infty$. Then, $u \in C^{k+2,\alpha}(\Omega)$ and, for any $\Omega' \Subset \Omega$,

$$|u|_{C^{k+2,\alpha}(\Omega')} \leq C \left\{ |u|_{L^\infty(\Omega)} + |f|_{C^{k,\alpha}(\Omega)} \right\},$$

where C is a positive constant depending only on n, k, α, λ, the $C^{k,\alpha}$-norms of a_{ij}, b_i, and c, Ω', and $\operatorname{dist}(\Omega', \partial\Omega)$. Moreover, if a_{ij}, b_i, c, and f are in $C^\infty(\Omega)$, then $u \in C^\infty(\Omega)$.

For $k = 0$, we have the basic result that any C^2-solutions are $C^{2,\alpha}$ as long as the coefficients and the nonhomogeneous terms are C^α.

1.3.2. The Global Schauder Theory. In this subsection, we review the global Schauder theory.

For any $r > 0$, set

$$B_r^+ = \{x = (x_1, \ldots, x_{n-1}, x_n) \in B_r : x_n > 0\},$$
$$\Sigma_r = \{x = (x_1, \ldots, x_{n-1}, x_n) \in B_r : x_n = 0\}.$$

We first state the boundary Schauder estimates. Let B_R be a ball centered at the origin in $\mathbb{R}^n$, k be a nonnegative integer, and $\alpha \in (0,1)$ be a constant. We define

$$|u|^*_{C^{k,\alpha}(B_R^+\cup\Sigma_R)} = \sum_{i=0}^{k} R^i |\nabla^i u|_{L^\infty(B_R^+\cup\Sigma_R)} + R^{k+\alpha}[\nabla^k u]_{C^\alpha(B_R^+\cup\Sigma_R)}.$$

We note that factors of R result from scaling.

Theorem 1.3.6. *Let $\alpha \in (0,1)$ be a constant, B_R^+ be an upper half-ball in $\mathbb{R}^n$, and L be given by (1.3.1) in B_R^+, for some a_{ij}, b_i, $c \in C^\alpha(B_R^+ \cup \Sigma_R)$ satisfying (1.3.2) and*

$$|a_{ij}|^*_{C^\alpha(B_R^+\cup\Sigma_R)} + R|b_i|^*_{C^\alpha(B_R^+\cup\Sigma_R)} + R^2|c|^*_{C^\alpha(B_R^+\cup\Sigma_R)} \le \Lambda,$$

for some positive constant Λ. Suppose that u is an $L^\infty(B_R^+)\cap C^{2,\alpha}(B_R^+\cup\Sigma_R)$-solution of

$$Lu = f \quad \text{in } B_R,$$
$$u = 0 \quad \text{on } \Sigma_R,$$

for some $f \in C^\alpha(B_R^+ \cup \Sigma_R)$, with $|f|_{C^\alpha(B_R^+\cup\Sigma_R)} < \infty$. Then,

$$|u|^*_{C^{2,\alpha}(B_{R/2}^+\cup\Sigma_{R/2})} \le C\left\{|u|_{L^\infty(B_R^+)} + R^2|f|^*_{C^\alpha(B_R^+\cup\Sigma_R)}\right\},$$

where C is a positive constant depending only on n, α, λ, and Λ.

We have a more general result.

Theorem 1.3.7. *Let $\alpha \in (0,1)$ be a constant, Ω be a bounded domain in $\mathbb{R}^n$ with a $C^{2,\alpha}$-boundary portion $\Sigma \subset \partial\Omega$, and L be given by (1.3.1) in Ω, for some a_{ij}, b_i, $c \in C^\alpha(\Omega \cup \Sigma)$ satisfying (1.3.2) and*

$$|a_{ij}|_{C^\alpha(\Omega\cup\Sigma)} + |b_i|_{C^\alpha(\Omega\cup\Sigma)} + |c|_{C^\alpha(\Omega\cup\Sigma)} \le \Lambda,$$

for some positive constant Λ. Suppose that u is an $L^\infty(\Omega) \cap C^{2,\alpha}(\Omega \cup \Sigma)$-solution of

$$Lu = f \quad \text{in } \Omega,$$
$$u = \varphi \quad \text{on } \Sigma,$$

for some $f \in C^\alpha(\Omega \cup \Sigma)$ *and* $\varphi \in C^{2,\alpha}(\Omega \cup \Sigma)$, *with* $|f|_{C^\alpha(\Omega\cup\Sigma)} < \infty$ *and* $|\varphi|_{C^{2,\alpha}(\Omega\cup\Sigma)} < \infty$. *Then, for any subset* $\Omega' \subset \Omega$ *with* $\operatorname{dist}(\Omega', \partial\Omega \setminus \Sigma) > 0$,

$$|u|_{C^{2,\alpha}(\Omega')} \le C\left\{|u|_{L^\infty(\Omega)} + |\varphi|_{C^{2,\alpha}(\Omega\cup\Sigma)} + |f|_{C^\alpha(\Omega\cup\Sigma)}\right\},$$

where C *is a positive constant depending only on* n, α, λ, Λ, Ω', *and* $\operatorname{dist}(\Omega', \partial\Omega \setminus \Sigma)$.

As a consequence, we have the global Schauder estimates.

Theorem 1.3.8. *Let* $\alpha \in (0,1)$ *be a constant,* Ω *be a bounded domain in* $\mathbb{R}^n$ *with a* $C^{2,\alpha}$*-boundary, and* L *be given by* (1.3.1) *in* Ω, *for some* a_{ij}, b_i, $c \in C^\alpha(\bar\Omega)$ *satisfying* (1.3.2) *and*

$$|a_{ij}|_{C^\alpha(\bar\Omega)} + |b_i|_{C^\alpha(\bar\Omega)} + |c|_{C^\alpha(\bar\Omega)} \le \Lambda,$$

for some positive constant Λ. *Suppose that* u *is a* $C^{2,\alpha}(\bar\Omega)$*-solution of*

$$\begin{aligned} Lu &= f \quad \text{in } \Omega, \\ u &= \varphi \quad \text{on } \partial\Omega, \end{aligned}$$

for some $f \in C^\alpha(\bar\Omega)$ *and* $\varphi \in C^{2,\alpha}(\bar\Omega)$. *Then,*

$$|u|_{C^{2,\alpha}(\bar\Omega)} \le C\left\{|u|_{L^\infty(\Omega)} + |\varphi|_{C^{2,\alpha}(\bar\Omega)} + |f|_{C^\alpha(\bar\Omega)}\right\},$$

where C *is a positive constant depending only on* n, α, λ, Λ, *and* Ω.

With the global estimates, we are able to solve the Dirichlet problem with $C^{2,\alpha}$-boundary values for uniformly elliptic differential equations.

Theorem 1.3.9. *Let* $\alpha \in (0,1)$ *be a constant,* Ω *be a bounded domain in* $\mathbb{R}^n$ *with a* $C^{2,\alpha}$*-boundary, and* L *be given by* (1.3.1) *in* Ω, *for some* a_{ij}, b_i, $c \in C^\alpha(\bar\Omega)$ *satisfying* (1.3.2) *and* $c \le 0$. *Then for any* $f \in C^\alpha(\bar\Omega)$ *and* $\varphi \in C^{2,\alpha}(\bar\Omega)$, *there exists a (unique)* $C^{2,\alpha}(\bar\Omega)$*-solution* u *of*

$$\begin{aligned} Lu &= f \quad \text{in } \Omega, \\ u &= \varphi \quad \text{on } \partial\Omega. \end{aligned}$$

We now discuss the regularity of solutions up to the boundary. First, we have the following result.

Theorem 1.3.10. *Let* k *be a nonnegative integer,* $\alpha \in (0,1)$ *be a constant,* Ω *be a bounded domain in* $\mathbb{R}^n$ *with a* $C^{k+2,\alpha}$*-boundary portion* $\Sigma \subset \partial\Omega$, *and* L *be given by* (1.3.1) *in* Ω, *for some* a_{ij}, b_i, $c \in C^{k,\alpha}(\Omega \cup \Sigma)$ *with finite* $C^{k,\alpha}(\Omega \cup \Sigma)$*-norms and satisfying* (1.3.2). *Suppose that* u *is an* $L^\infty(\Omega) \cap C(\Omega \cup \Sigma) \cap C^2(\Omega)$*-solution of*

$$\begin{aligned} Lu &= f \quad \text{in } \Omega, \\ u &= \varphi \quad \text{on } \Sigma, \end{aligned}$$

for some $f \in C^{k,\alpha}(\Omega \cup \Sigma)$ *and* $\varphi \in C^{k+2,\alpha}(\Omega \cup \Sigma)$, *with* $|f|_{C^{k,\alpha}(\Omega\cup\Sigma)} < \infty$ *and* $|\varphi|_{C^{k+2,\alpha}(\Omega\cup\Sigma)} < \infty$. *Then,* $u \in C^{k+2,\alpha}(\Omega\cup\Sigma)$ *and, for any* $\Omega' \subset \Omega$ *with* $\operatorname{dist}(\Omega', \partial\Omega \setminus \Sigma) > 0$,

$$|u|_{C^{k+2,\alpha}(\Omega'\cup\Sigma)} \leq C\left\{|u|_{L^\infty(\Omega)} + |\varphi|_{C^{k+2,\alpha}(\Omega\cup\Sigma)} + |f|_{C^{k,\alpha}(\Omega\cup\Sigma)}\right\},$$

where C *is a positive constant depending only on* n, α, λ, *the* $C^{k,\alpha}$*-norms of* a_{ij}, b_i, *and* c *in* $\Omega \cup \Sigma$, Ω', *and* $\operatorname{dist}(\Omega', \partial\Omega \setminus \Sigma)$.

Finally, we have the following global regularity.

Theorem 1.3.11. *Let* k *be a nonnegative integer,* $\alpha \in (0,1)$ *be a constant,* Ω *be a bounded domain in* $\mathbb{R}^n$ *with a* $C^{k+2,\alpha}$*-boundary, and* L *be given by* (1.3.1) *in* Ω, *for some* a_{ij}, b_i, $c \in C^{k,\alpha}(\bar{\Omega})$ *satisfying* (1.3.2). *Suppose that* u *is a* $C(\bar{\Omega}) \cap C^2(\Omega)$*-solution of*

$$\begin{aligned} Lu &= f \quad \text{in } \Omega, \\ u &= \varphi \quad \text{on } \partial\Omega, \end{aligned}$$

for some $f \in C^{k,\alpha}(\bar{\Omega})$ *and* $\varphi \in C^{k+2,\alpha}(\bar{\Omega})$. *Then,* $u \in C^{k+2,\alpha}(\bar{\Omega})$ *and*

$$|u|_{C^{k+2,\alpha}(\bar{\Omega})} \leq C\left\{|u|_{L^\infty(\Omega)} + |\varphi|_{C^{k+2,\alpha}(\bar{\Omega})} + |f|_{C^{k,\alpha}(\bar{\Omega})}\right\},$$

where C *is a positive constant depending only on* n, α, λ, Ω, *and the* $C^{k,\alpha}$*-norms of* a_{ij}, b_i, *and* c *in* $\bar{\Omega}$.

Part 1

Quasilinear Elliptic Equations

Chapter 2

Quasilinear Uniformly Elliptic Equations

In this chapter, we discuss quasilinear uniformly elliptic differential equations. We derive various a priori estimates for their solutions and solve the Dirichlet boundary-value problems.

In Section 2.1, we study basic properties of quasilinear uniformly elliptic differential equations. We prove a comparison principle and the higher regularity of solutions.

In Sections 2.2 and 2.3, we discuss gradient estimates of solutions of quasilinear uniformly elliptic differential equations. Here, the gradient estimates refer to estimates of the L^∞-norms of the first derivatives of solutions. In Section 2.2, we derive interior gradient estimates, while in Section 2.3, we derive boundary and global gradient estimates for solutions of the Dirichlet boundary-value problem. These two sections are based on the maximum principle.

In Sections 2.4 and 2.5, we derive estimates of the Hölder semi-norms of the gradients for quasilinear uniformly elliptic differential equations. Section 2.4 concerns interior estimates and Section 2.5 global estimates. Derivations are based on Krylov-Safonov's weak Harnack inequality.

In Section 2.6, we solve the Dirichlet boundary-value problems for quasilinear uniformly elliptic differential equations by the method of continuity. The estimates of the Hölder semi-norms of gradients play an essential role.

The three topics reviewed in Chapter 1 play different roles in this chapter. The maximum principle will be used to derive estimates of derivatives up to the first order, the Harnack inequality will be used to derive estimates of the Hölder semi-norms of derivatives of the first order, and the Schauder theory will be used to solve the linearized equations.

To illustrate ideas and techniques more clearly, we study a special class of quasilinear elliptic equations in this chapter. Refer to Chapters 13–15 of [**59**] for results on more general quasilinear elliptic equations.

2.1. Basic Properties

Let Ω be a domain in $\mathbb{R}^n$ and u be a C^2-function in Ω. The general quasilinear equation of the second order can be written as

$$a_{ij}(x,u,\nabla u)u_{ij}=f(x,u,\nabla u)\quad\text{in }\Omega.$$

This equation is linear with respect to the derivatives of the second order. When discussing quasilinear equations, we usually write $p=\nabla u$. The equation above is called *elliptic* at $(x,u,p)\in\Omega\times\mathbb{R}\times\mathbb{R}^n$ if, for any $\xi\in\mathbb{R}^n\setminus\{0\}$,

$$a_{ij}(x,u,p)\xi_i\xi_j>0;$$

it is *uniformly elliptic* at (x,u,p) if, for any $\xi\in\mathbb{R}^n$,

$$\lambda|\xi|^2\le a_{ij}(x,u,p)\xi_i\xi_j\le\Lambda|\xi|^2,$$

for some positive constants λ and Λ.

Throughout this chapter, we consider a special class of quasilinear elliptic equations given by

$$a_{ij}(x,\nabla u)u_{ij}=f(x)\quad\text{in }\Omega.$$

We always assume the *uniform ellipticity* in the following form: for any $x\in\Omega$ and any $p,\xi\in\mathbb{R}^n$,

$$\lambda|\xi|^2\le a_{ij}(x,p)\xi_i\xi_j\le\Lambda|\xi|^2,$$

for some positive constants λ and Λ. For simplicity, we write

$$Q(u)=a_{ij}(x,\nabla u)u_{ij}.$$

We first prove a comparison principle.

Theorem 2.1.1. *Let $a_{ij}\in C^1(\Omega\times\mathbb{R}^n)$ satisfy the uniform ellipticity. Suppose that $u,v\in C(\bar\Omega)\cap C^2(\Omega)$ satisfy $Q(u)\ge Q(v)$ in Ω and $u\le v$ on $\partial\Omega$. Then, $u\le v$ in Ω.*

Proof. We first write

$$\begin{aligned}Q(u)-Q(v)&=a_{ij}(x,\nabla u)u_{ij}-a_{ij}(x,\nabla v)v_{ij}\\&=a_{ij}(x,\nabla u)(u-v)_{ij}+\big(a_{ij}(x,\nabla u)-a_{ij}(x,\nabla v)\big)v_{ij}.\end{aligned}$$

Next,

$$\begin{aligned}
a_{ij}(x,\nabla u) - a_{ij}(x,\nabla v)
&= \int_0^1 \frac{d}{dt} a_{ij}\big(x, t\nabla u + (1-t)\nabla v\big)dt \\
&= \int_0^1 a_{ij,p_k}\big(x, t\nabla u + (1-t)\nabla v\big)dt \cdot \big(u-v\big)_k.
\end{aligned}$$

Hence,

$$Q(u) - Q(v) = \widetilde{a}_{ij}(u-v)_{ij} + \widetilde{b}_k(u-v)_k,$$

where

$$\begin{aligned}
\widetilde{a}_{ij} &= a_{ij}(x,\nabla u), \\
\widetilde{b}_k &= v_{ij}\int_0^1 a_{ij,p_k}\big(x, t\nabla u + (1-t)\nabla v\big)dt.
\end{aligned}$$

Then, we have

$$\begin{aligned}
\widetilde{a}_{ij}(u-v)_{ij} + \widetilde{b}_k(u-v)_k &\geq 0 \quad \text{in } \Omega, \\
u - v &\leq 0 \quad \text{on } \partial\Omega,
\end{aligned}$$

where $\widetilde{a}_{ij}$ satisfies, for any $x \in \Omega$ and any $\xi \in \mathbb{R}^n$,

$$\lambda|\xi|^2 \leq \widetilde{a}_{ij}(x)\xi_i\xi_j \leq \Lambda|\xi|^2.$$

By the maximum principle (for linear elliptic equations), we have $u \leq v$ in Ω. □

A simple consequence is the uniqueness of solutions of the Dirichlet problem.

Corollary 2.1.2. *Let $a_{ij} \in C^1(\Omega \times \mathbb{R}^n)$ satisfy the uniform ellipticity. Suppose that $u, v \in C(\bar{\Omega}) \cap C^2(\Omega)$ satisfy $Q(u) = Q(v)$ in Ω and $u = v$ on $\partial\Omega$. Then, $u = v$ in Ω.*

Next, we prove a Liouville type theorem. Compare this with Theorem 1.2.15.

Theorem 2.1.3. *Let $a_{ij} \in C(\mathbb{R}^n \times \mathbb{R}^n)$ satisfy the uniform ellipticity. Suppose that u is a $C^2(\mathbb{R}^n)$-solution of*

$$a_{ij}(x,\nabla u)u_{ij} = 0 \quad \text{in } \mathbb{R}^n.$$

If u is bounded, then u is constant.

Proof. We write the equation as

$$\widetilde{a}_{ij}u_{ij} = 0 \quad \text{in } \mathbb{R}^n,$$

where $\widetilde{a}_{ij}(x) = a_{ij}\big(x, \nabla u(x)\big)$ satisfies, for any $x \in \mathbb{R}^n$ and $\xi \in \mathbb{R}^n$,

$$\lambda|\xi|^2 \leq \widetilde{a}_{ij}(x)\xi_i\xi_j \leq \Lambda|\xi|^2.$$

By Theorem 1.2.15, we conclude that u is constant. □

To end this section, we discuss the higher regularity of solutions.

Proposition 2.1.4. *Let $\alpha \in (0,1)$ be a constant and $a_{ij} \in C(\Omega\times\mathbb{R}^n)$ satisfy the uniform ellipticity. Suppose that u is a $C^2(\Omega)$-solution of*

$$a_{ij}(x, \nabla u)u_{ij} = f \quad \text{in } \Omega,$$

for some $f \in C(\Omega)$. For any integer $m \geq 0$, if $a_{ij}(\cdot, v_1, \ldots, v_n) \in C^{m,\alpha}(\Omega)$ for any $v_1, \ldots, v_n \in C^{m,\alpha}(\Omega)$ and if $f \in C^{m,\alpha}(\Omega)$, then $u \in C^{m+2,\alpha}(\Omega)$. In particular, if $a_{ij} \in C^\infty(\Omega \times \mathbb{R}^n)$ and $f \in C^\infty(\Omega)$, then $u \in C^\infty(\Omega)$.

The proof is based on a bootstrap argument.

Proof. We first consider $m = 0$. Since u is in $C^{1,\alpha}(\Omega)$, then ∇u is in $C^\alpha(\Omega)$, and hence, the coefficient $a_{ij}(\cdot, \nabla u)$ is in $C^\alpha(\Omega)$. Now the interior Schauder regularity implies $u \in C^{2,\alpha}(\Omega)$. We can repeat this argument as long as the regularity of a_{ij} and f allows and hence obtain the desired result. □

A global version of the higher regularity also holds if we employ the global Schauder regularity.

Proposition 2.1.5. *Let $\alpha \in (0,1)$ be a constant and $a_{ij} \in C(\bar{\Omega}\times\mathbb{R}^n)$ satisfy the uniform ellipticity. Suppose that u is a $C^2(\bar{\Omega})$-solution of*

$$\begin{aligned} a_{ij}(x, \nabla u)u_{ij} &= f \quad \text{in } \Omega, \\ u &= \varphi \quad \text{on } \partial\Omega, \end{aligned}$$

for some $f \in C(\bar{\Omega})$ and $\varphi \in C(\bar{\Omega})$. For any integer $m \geq 0$, if $\partial\Omega \in C^{m+2,\alpha}$, $a_{ij}(\cdot, v_1, \ldots, v_n) \in C^{m,\alpha}(\bar{\Omega})$ for any $v_1, \ldots, v_n \in C^{m,\alpha}(\bar{\Omega})$, $f \in C^{m,\alpha}(\bar{\Omega})$, and $\varphi \in C^{m+2,\alpha}(\bar{\Omega})$, then $u \in C^{m+2,\alpha}(\bar{\Omega})$. In particular, if $\partial\Omega \in C^\infty$, $a_{ij} \in C^\infty(\bar{\Omega} \times \mathbb{R}^n)$, $f \in C^\infty(\bar{\Omega})$, and $\varphi \in C^\infty(\bar{\Omega})$, then $u \in C^\infty(\bar{\Omega})$.

In Propositions 2.1.4 and 2.1.5, we require u to be C^2 to start with. In terms of estimates, $C^{1,\alpha}$ is sufficient. The C^2-regularity only ensures the validity of the equation. In the next two sections, we will derive a priori estimates of the $C^{1,\alpha}$-norms of solutions.

2.2. Interior C^1-Estimates

In this section, we derive interior gradient estimates for quasilinear uniformly elliptic differential equations.

Let B_R be a ball in $\mathbb{R}^n$. Consider the quasilinear equation given by

$$a_{ij}(x, \nabla u)u_{ij} = f(x) \quad \text{in } B_R.$$

We always assume the *uniform ellipticity* in the following form: for any $x \in B_R$ and any $p, \xi \in \mathbb{R}^n$,

$$\lambda|\xi|^2 \leq a_{ij}(x,p)\xi_i\xi_j \leq \Lambda|\xi|^2,$$

for some positive constants λ and Λ.

Now, we derive interior estimates of first derivatives, also referred to as the gradient estimates. We achieve this by the classical *Bernstein method.* The basic idea is to derive a differential equation for $|\nabla u|^2$ and then apply the maximum principle. There are two classes of gradient estimates, global gradient estimates and interior gradient estimates. Global gradient estimates yield estimates of gradients ∇u in Ω in terms of ∇u on $\partial\Omega$, as well as u in Ω, while interior gradient estimates yield estimates of ∇u in compact subsets of Ω in terms of u in Ω. We discuss the interior gradient estimates in this section and the global gradient estimates in the next section.

Theorem 2.2.1. *Let $a_{ij} \in C^1(B_R \times \mathbb{R}^n)$ satisfy the uniform ellipticity and, for any $x \in B_R$ and $p \in \mathbb{R}^n$,*

$$R|\nabla_x a_{ij}(x,p)| \leq A_0, \quad |p||\nabla_p a_{ij}(x,p)| \leq A_1,$$

for some positive constants A_0 and A_1. Suppose that u is an $L^\infty(B_R) \cap C^3(B_R)$-solution of

$$a_{ij}(x, \nabla u)u_{ij} = f \quad \text{in } B_R,$$

for some $f \in C^1(B_R)$, with $|f|_{C^1(B_R)} < \infty$. Then,

$$R|\nabla u|_{L^\infty(B_{R/2})} \leq C\left\{|u|_{L^\infty(B_R)} + R^2|f|_{L^\infty(B_R)} + R^3|\nabla f|_{L^\infty(B_R)}\right\},$$

where C is a positive constant depending only on n, λ, Λ, A_0, and A_1.

Proof. Differentiating the equation with respect to x_k, we have

$$a_{ij}u_{ijk} + a_{ij,p_l}u_{ij}u_{kl} + a_{ij,x_k}u_{ij} = f_k.$$

By setting

$$L = a_{ij}\partial_{ij} + a_{ij,p_l}u_{ij}\partial_l,$$

we get

$$Lu_k = -a_{ij,x_k}u_{ij} + f_k.$$

We now consider $|\nabla u|^2$. Note that

$$(|\nabla u|^2)_i = 2u_k u_{ki},$$
$$(|\nabla u|^2)_{ij} = 2u_k u_{kij} + 2u_{ki}u_{kj}.$$

By multiplying Lu_k by $2u_k$ and summing over k, we obtain

$$L(|\nabla u|^2) = 2a_{ij}u_{ki}u_{kj} - 2a_{ij,x_k}u_k u_{ij} + 2u_k f_k.$$

Take a cutoff function $\eta \in C_0^2(B_R)$ such that $0 \leq \eta \leq 1$ in B_R, $\eta = 1$ in $B_{R/2}$, and

$$|\nabla\eta|^2 + |\nabla^2\eta| \leq \frac{c_0}{R^2},$$

for some positive constant c_0 depending only on n. We now consider $\eta^2|\nabla u|^2$. First,

$$\begin{aligned}L(\eta^2|\nabla u|^2) &= \eta^2 L(|\nabla u|^2) + 2a_{ij}(\eta^2)_i(|\nabla u|^2)_j + (L\eta^2)|\nabla u|^2\\ &= 2\eta^2 a_{ij}u_{ki}u_{kj} + 2\eta^2 u_k f_k + 8a_{ij}\eta\eta_i u_k u_{kj} - 2\eta^2 a_{ij,x_k}u_k u_{ij}\\ &\quad + (2\eta a_{ij}\eta_{ij} + 2a_{ij}\eta_i\eta_j)\,|\nabla u|^2 + 2\eta a_{ij,p_l}u_{ij}\eta_l|\nabla u|^2.\end{aligned}$$

By the uniform ellipticity, we have

$$2\eta^2 a_{ij}u_{ki}u_{kj} \geq 2\lambda\eta^2|\nabla^2 u|^2.$$

By the Cauchy inequality and the assumptions on $\nabla_x a_{ij}$ and $\nabla_p a_{ij}$, we get

$$|8a_{ij}\eta\eta_i u_k u_{kj}| \leq \frac{1}{2}\lambda\eta^2|\nabla^2 u|^2 + \frac{32\Lambda^2}{\lambda}|\nabla\eta|^2|\nabla u|^2,$$
$$|2\eta a_{ij,p_l}u_{ij}\eta_l|\nabla u|^2| \leq \frac{1}{2}\lambda\eta^2|\nabla^2 u|^2 + \frac{2A_1^2}{\lambda}|\nabla\eta|^2|\nabla u|^2,$$

and

$$|2\eta^2 a_{ij,x_k}u_k u_{ij}| \leq \frac{1}{2}\lambda\eta^2|\nabla^2 u|^2 + \frac{2A_0^2}{\lambda R^2}\eta^2|\nabla u|^2.$$

Hence,

$$\begin{aligned}L(\eta^2|\nabla u|^2) \geq{}& \frac{1}{2}\lambda\eta^2|\nabla^2 u|^2 + 2\eta^2 u_k f_k - \frac{1}{\lambda}\left(32\Lambda^2 + 2A_1^2\right)|\nabla\eta|^2|\nabla u|^2\\ &+ (2\eta a_{ij}\eta_{ij} + 2a_{ij}\eta_i\eta_j)\,|\nabla u|^2 - \frac{2A_0^2}{\lambda R^2}\eta^2|\nabla u|^2.\end{aligned}$$

By the Cauchy inequality again, we have

$$|2\eta^2 u_k f_k| \leq \eta^2\frac{|\nabla u|^2}{R^2} + \eta^2 R^2|\nabla f|^2,$$

and then

$$L(\eta^2|\nabla u|^2) \geq \frac{1}{2}\lambda\eta^2|\nabla^2 u|^2 - \frac{c_1}{R^2}|\nabla u|^2 - R^2|\nabla f|^2,$$

where c_1 is a positive constant depending only on n, λ, Λ, A_0, and A_1. To control ∇u in the right-hand side, we consider u^2. Note that

$$L(u^2) = 2a_{ij}u_iu_j + 2uf + 2uu_l a_{ij,p_l}u_{ij}.$$

By the uniform ellipticity and the assumption on $\nabla_p a_{ij}$, we have

$$L(u^2) \geq 2\lambda|\nabla u|^2 + 2uf - 2A_1|uu_{ij}|.$$

We now consider the function

$$v = \eta^2|\nabla u|^2 + \frac{c_1}{\lambda R^2}u^2.$$

Then,

$$Lv \geq \frac{1}{2}\lambda\eta^2|\nabla^2 u|^2 + \frac{c_1}{R^2}|\nabla u|^2 - R^2|\nabla f|^2 + \frac{2c_1}{\lambda R^2}uf - \frac{2c_1A_1}{\lambda R^2}|uu_{ij}|.$$

By the Cauchy inequality, we have

$$\frac{2c_1}{\lambda R^2}|uf| \leq \frac{c_1^2}{\lambda^2R^4}u^2 + f^2$$

and

$$\frac{2c_1A_1}{\lambda R^2}|uu_{ij}| \leq \frac{1}{2}\lambda\eta^2|\nabla^2 u|^2 + \frac{2c_1^2A_1^2}{\lambda^3R^4\eta^2}u^2.$$

Therefore,

$$\begin{aligned} Lv &\geq \frac{c_1}{R^2}|\nabla u|^2 - \frac{c_2}{R^4\eta^2}u^2 - f^2 - R^2|\nabla f|^2 \\ &\geq \frac{1}{R^2\eta^2}\left(c_1\eta^2|\nabla u|^2 - \frac{c_2}{R^2}u^2 - R^2f^2 - R^4|\nabla f|^2\right), \end{aligned}$$

where c_2 is a positive constant depending only on n, λ, Λ, A_0, and A_1. By the definition of v, we have

$$Lv \geq \frac{c_1}{R^2\eta^2}\left(v - C\left(\frac{u^2}{R^2} + R^2f^2 + R^4|\nabla f|^2\right)\right).$$

We assume that v attains its maximum at $x_0 \in \bar{B}_R$. If $\eta(x_0) \neq 0$, then $x_0 \in B_R$ and $Lv \leq 0$ at x_0. Therefore,

$$v \leq C\left(\frac{u^2}{R^2} + R^2f^2 + R^4|\nabla f|^2\right) \quad \text{at } x_0.$$

This estimate obviously holds if $\eta(x_0) = 0$. In conclusion, we obtain

$$\sup_{B_R}\left(\eta^2|\nabla u|^2 + \frac{c_1}{\lambda R^2}u^2\right) \leq C\sup_{B_R}\left(\frac{u^2}{R^2} + R^2f^2 + R^4|\nabla f|^2\right).$$

This implies the desired result. $\square$

2.3. Global C^1-Estimates

In this section, we derive boundary and global gradient estimates for solutions of the Dirichlet problem for quasilinear uniformly elliptic differential equations.

Let Ω be a bounded domain in $\mathbb{R}^n$ and u be a C^2-function in Ω. In the following, we consider the quasilinear equation given by

$$a_{ij}(x, \nabla u)u_{ij} = f(x) \quad \text{in } \Omega.$$

We always assume the *uniform ellipticity* in the following form: for any $x \in \Omega$ and any $p, \xi \in \mathbb{R}^n$,

$$\lambda|\xi|^2 \le a_{ij}(x, p)\xi_i\xi_j \le \Lambda|\xi|^2,$$

for some positive constants λ and Λ.

First, we derive an estimate of the sup-norm of u in Ω in terms of that on $\partial\Omega$.

Theorem 2.3.1. *Let Ω be a bounded domain in $\mathbb{R}^n$ and $a_{ij} \in C(\Omega \times \mathbb{R}^n)$ satisfy the uniform ellipticity. Suppose that u is a $C(\bar{\Omega}) \cap C^2(\Omega)$-solution of*

$$\begin{aligned} a_{ij}(x, \nabla u)u_{ij} &= f \quad \text{in } \Omega, \\ u &= \varphi \quad \text{on } \partial\Omega, \end{aligned}$$

for some $f \in C(\bar{\Omega})$ and $\varphi \in C(\partial\Omega)$. Then,

$$|u|_{L^\infty(\Omega)} \le |\varphi|_{L^\infty(\partial\Omega)} + C|f|_{L^\infty(\Omega)},$$

where C is a positive constant depending only on n, λ, Λ, and the diameter of Ω.

Proof. For the given u, we can regard $a_{ij}(x, \nabla u(x))$ as a function of x and then regard the equation of u as a linear equation. Hence, the desired result follows from Theorem 1.1.10. □

Next, we derive a boundary gradient estimate.

Theorem 2.3.2. *Let Ω be a bounded domain in $\mathbb{R}^n$ with a C^2-boundary and $a_{ij} \in C(\Omega \times \mathbb{R}^n)$ satisfy the uniform ellipticity. Suppose that u is a $C^1(\bar{\Omega}) \cap C^2(\Omega)$-solution of*

$$\begin{aligned} a_{ij}(x, \nabla u)u_{ij} &= f \quad \text{in } \Omega, \\ u &= \varphi \quad \text{on } \partial\Omega, \end{aligned}$$

for some $f \in C(\bar{\Omega})$ and $\varphi \in C^2(\bar{\Omega})$. Then,

$$\left|\frac{\partial u}{\partial \nu}\right|_{L^\infty(\partial\Omega)} \le C\left\{|u|_{L^\infty(\Omega)} + |\varphi|_{C^2(\bar{\Omega})} + |f|_{L^\infty(\Omega)}\right\},$$

where ν is the exterior unit normal vector to $\partial\Omega$ and C is a positive constant depending only on n, λ, Λ, the diameter of Ω, and the radius of exterior balls tangent to $\partial\Omega$.

Proof. The desired result follows from Theorem 1.1.14 if we regard the equation of u as a linear equation. □

Next, we derive a global gradient estimate.

Theorem 2.3.3. *Let Ω be a bounded domain in $\mathbb{R}^n$ with a C^1-boundary and $a_{ij} \in C^1(\bar{\Omega}\times\mathbb{R}^n)$ satisfy the uniform ellipticity and, for any $(x,p) \in \bar{\Omega}\times\mathbb{R}^n$,*

$$|\nabla_x a_{ij}(x,p)| \le A_0, \quad |p\cdot\nabla_p a_{ij}(x,p)| \le A_1,$$

for some positive constants A_0 and A_1. Suppose that u is a $C^1(\bar{\Omega})\cap C^3(\Omega)$-solution of

$$a_{ij}(x,\nabla u)u_{ij} = f \quad \text{in } \Omega,$$

for some $f \in C^1(\bar{\Omega})$. Then,

$$|\nabla u|_{L^\infty(\Omega)} \le |\nabla u|_{L^\infty(\partial\Omega)} + C\left\{|u|_{L^\infty(\Omega)} + |f|_{C^1(\bar{\Omega})}\right\},$$

where C is a positive constant depending only on n, λ, Λ, A_0, and A_1.

The proof is similar to that of Theorem 2.2.1. In fact, it is easier in the present case since no cutoff functions are involved.

Proof. We set

$$F_0 = |f|_{L^\infty(\Omega)}, \quad F_1 = |\nabla f|_{L^\infty(\Omega)},$$

and

$$M = |u|_{L^\infty(\Omega)}.$$

By setting

$$L = a_{ij}\partial_{ij} + a_{ij,p_l}u_{ij}\partial_l,$$

as in the proof of Theorem 2.2.1, we have

$$L(|\nabla u|^2) = 2a_{ij}u_{ki}u_{kj} - 2a_{ij,x_k}u_k u_{ij} + 2u_k f_k$$

and

$$L(u^2) = 2a_{ij}u_i u_j + 2uf + 2uu_l a_{ij,p_l}u_{ij}.$$

By a simple addition, we have

$$\begin{aligned} L\left(|\nabla u|^2 + Bu^2\right) = {} & 2a_{ij}u_{ki}u_{kj} + 2Ba_{ij}u_i u_j + 2Buu_l a_{ij,p_l}u_{ij} \\ & - 2a_{ij,x_k}u_k u_{ij} + 2u_k f_k + 2Buf, \end{aligned}$$

where B is a positive constant to be determined. By the uniform ellipticity, the assumptions on $\nabla_x a_{ij}$ and $\nabla_p a_{ij}$, and the bounds on f, $\nabla_x f$, and u, we have

$$\begin{aligned} L\left(|\nabla u|^2 + Bu^2\right) \geq\ & 2\lambda|\nabla^2 u|^2 + 2\lambda B|\nabla u|^2 - 2BA_1M|\nabla^2 u| \\ & - 2A_0|\nabla u||\nabla^2 u| - 2F_1|\nabla u| - 2BMF_0. \end{aligned}$$

By the Cauchy inequality, we have

$$\begin{aligned} 2BA_1M|\nabla^2 u| &\leq \lambda|\nabla^2 u|^2 + \frac{1}{\lambda}B^2A_1^2M^2, \\ 2A_0|\nabla u||\nabla^2 u| &\leq \lambda|\nabla^2 u|^2 + \frac{1}{\lambda}A_0^2|\nabla u|^2, \\ 2F_1|\nabla u| &\leq \lambda|\nabla u|^2 + \frac{1}{\lambda}F_1^2, \\ 2BMF_0 &\leq \lambda M^2 + \frac{B^2F_0^2}{\lambda}. \end{aligned}$$

Hence,

$$\begin{aligned} L\left(|\nabla u|^2 + Bu^2\right) \geq\ & \left(2\lambda B - \frac{A_0^2}{\lambda} - \lambda\right)|\nabla u|^2 \\ & - \frac{1}{\lambda}\left(B^2A_1^2M^2 + B^2F_0^2 + F_1^2\right) - \lambda M^2. \end{aligned}$$

In the following, we take B such that

$$2\lambda B - \frac{A_0^2}{\lambda} - \lambda = \frac{\lambda}{B};$$

i.e.,

$$2B^2 - \left(\frac{A_0^2}{\lambda^2} + 1\right)B - 1 = 0.$$

Then,

$$L\left(|\nabla u|^2 + Bu^2\right) \geq \frac{\lambda}{B}|\nabla u|^2 - \frac{1}{\lambda}\left(B^2A_1^2M^2 + B^2F_0^2 + F_1^2\right) - \lambda M^2.$$

Suppose that $|\nabla u|^2 + Bu^2$ attains its maximum in $\bar{\Omega}$ at some point $x_0 \in \bar{\Omega}$. If $x_0 \in \partial\Omega$, we have

$$(|\nabla u|^2 + Bu^2)(x_0) \leq \max_{\partial\Omega}|\nabla u|^2 + BM^2.$$

If $x_0 \in \Omega$, then $L\left(|\nabla u|^2 + Bu^2\right) \leq 0$ at x_0 and hence

$$\frac{\lambda}{B}|\nabla u(x_0)|^2 \leq \frac{1}{\lambda}\left(B^2A_1^2M^2 + B^2F_0^2 + F_1^2\right) + \lambda M^2.$$

Therefore,

$$(|\nabla u|^2 + Bu^2)(x_0) \leq \frac{B}{\lambda^2}\left(B^2A_1^2M^2 + B^2F_0^2 + F_1^2\right) + 2BM^2.$$

By combining these two cases, we obtain

$$\begin{aligned}(|\nabla u|^2 + Bu^2)(x_0) \\ \leq \max_{\partial\Omega} |\nabla u|^2 + \frac{B}{\lambda^2}\left(B^2A_1^2M^2 + B^2F_0^2 + F_1^2\right) + 2BM^2.\end{aligned}$$

Since $|\nabla u|^2 + Bu^2$ attains its maximum at x_0, we then have

$$\sup_{\Omega} |\nabla u|^2 \leq \max_{\partial\Omega} |\nabla u|^2 + \frac{B}{\lambda^2}\left(B^2A_1^2M^2 + B^2F_0^2 + F_1^2\right) + 2BM^2.$$

By taking the square root, we get

$$\sup_{\Omega} |\nabla u| \leq \max_{\partial\Omega} |\nabla u| + C\left\{M + F_0 + F_1\right\}.$$

By the definitions of M, F_0, and F_1, we obtain the desired result. □

2.4. Interior $C^{1,\alpha}$-Estimates

In this section, we derive interior estimates of the Hölder semi-norms of the gradients for quasilinear uniformly elliptic differential equations. Krylov-Safonov's weak Harnack inequality plays an essential role in our derivation.

Let B_R be a ball in $\mathbb{R}^n$. Consider the quasilinear equation given by

$$a_{ij}(x, \nabla u)u_{ij} = f(x) \quad \text{in } B_R.$$

We always assume the *uniform ellipticity* in the following form: for any $x \in B_R$ and any $p, \xi \in \mathbb{R}^n$,

$$\lambda|\xi|^2 \leq a_{ij}(x,p)\xi_i\xi_j \leq \Lambda|\xi|^2,$$

for some positive constants λ and Λ.

The following result is due to Ladyzhenskaya and Ural'tseva [**101**], [**102**].

Theorem 2.4.1. *Let $a_{ij} \in C^1(B_R \times \mathbb{R}^n)$ satisfy the uniform ellipticity and*

$$R|\nabla_x a_{ij}|_{L^\infty(B_R\times\mathbb{R}^n)} \leq A_0,$$

for some positive constant A_0. Suppose that u is a $C^3(B_R)$-solution of

$$a_{ij}(x, \nabla u)u_{ij} = f \quad \text{in } B_R,$$

for some $f \in C^1(B_R)$, with $|u|_{C^1(B_R)} < \infty$ and $|f|_{C^1(B_R)} < \infty$, and suppose that it satisfies

$$\sup_{B_R} |\nabla u| \cdot \sup_{B_R} |\nabla_p a_{ij}(\cdot, \nabla u)| \leq A_1^*,$$

for some positive constant A_1^*. Then, for some $\alpha \in (0,1)$,

$$R^\alpha [\nabla u]_{C^\alpha(B_{R/2})} \leq C\left\{|\nabla u|_{L^\infty(B_R)} + R^2|\nabla f|_{L^\infty(B_R)}\right\},$$

where α and C are positive constants depending only on n, λ, Λ, A_0, and A_1^*.

Proof. The proof consists of two steps.

Step 1. As shown in the proof of Theorem 2.2.1, we have

$$a_{ij}u_{ijl} + a_{ij,p_k}u_{ij}u_{kl} = -a_{ij,x_l}u_{ij} + f_l$$

and

$$\begin{aligned} a_{ij}(|\nabla u|^2)_{ij} &+ a_{ij,p_k}u_{ij}(|\nabla u|^2)_k \\ &= 2a_{ij}u_{ki}u_{kj} - 2a_{ij,x_k}u_{ij}u_k + 2u_kf_k. \end{aligned}$$

Set $K = |\nabla u|_{L^\infty(B_R)}$. Then,

$$|\nabla_p a_{ij}(\cdot, \nabla u)| \leq \frac{A_1^*}{K} \quad \text{in } B_R.$$

For some constant $\varepsilon \in (0,1)$ and $\ell = 1, \ldots, n$, we consider

$$v = \pm K u_\ell + \varepsilon|\nabla u|^2.$$

A simple addition yields

$$\begin{aligned} a_{ij}v_{ij} + a_{ij,p_k}u_{ij}v_k = 2\varepsilon a_{ij}u_{ki}u_{kj} &+ 2\varepsilon u_kf_k \pm Kf_\ell \\ &- 2\varepsilon a_{ij,x_k}u_{ij}u_k \mp K a_{ij,x_\ell}u_{ij}. \end{aligned}$$

By the uniform ellipticity, we have

$$2\varepsilon a_{ij}u_{ki}u_{kj} \geq 2\varepsilon\lambda|\nabla^2 u|^2.$$

By the Cauchy inequality and the assumption on $\nabla_x a_{ij}$, we get

$$|2\varepsilon a_{ij,x_k}u_{ij}u_k| \leq \frac{1}{2}\varepsilon\lambda|\nabla^2 u|^2 + \frac{2\varepsilon}{\lambda}\frac{A_0^2}{R^2}|\nabla u|^2$$

and

$$|K a_{ij,x_\ell}u_{ij}| \leq \frac{1}{2}\varepsilon\lambda|\nabla^2 u|^2 + \frac{1}{2\varepsilon\lambda}\frac{A_0^2}{R^2}K^2.$$

Hence,

$$a_{ij}v_{ij} + a_{ij,p_k}u_{ij}v_k \geq \varepsilon\lambda|\nabla^2 u|^2 - 3K|\nabla f| - \left(2\varepsilon + \frac{1}{2\varepsilon}\right)\frac{A_0^2}{\lambda R^2}K^2.$$

By the Cauchy inequality again, we get

$$|2a_{ij,p_k}u_{ij}v_k| \leq \varepsilon\lambda|\nabla^2 u|^2 + \frac{(A_1^*)^2}{\varepsilon\lambda K^2}|\nabla v|^2.$$

Hence,

$$a_{ij}v_{ij} + \frac{(A_1^*)^2}{\varepsilon\lambda K^2}|\nabla v|^2 \geq -3K|\nabla f| - \left(2\varepsilon + \frac{1}{2\varepsilon}\right)\frac{A_0^2}{\lambda R^2}K^2.$$

By the definitions of K and v, we have

$$-K^2 \le v \le 2K^2. \tag{1}$$

We set, for a positive constant μ to be determined,

$$\widetilde{v} = \frac{1}{\mu}\left(\exp\left\{\frac{\mu v}{K^2}\right\} - 1\right).$$

A straightforward calculation yields

$$\begin{aligned} K^2 a_{ij}\widetilde{v}_{ij} \exp\left\{-\frac{\mu v}{K^2}\right\} &+ \frac{(A_1^*)^2}{\varepsilon\lambda K^2}|\nabla v|^2 \\ &\ge \frac{\mu}{K^2} a_{ij} v_i v_j - 3K|\nabla f| - \left(2\varepsilon + \frac{1}{2\varepsilon}\right)\frac{A_0^2}{\lambda R^2}K^2 \\ &\ge \frac{\mu\lambda}{K^2}|\nabla v|^2 - 3K|\nabla f| - \left(2\varepsilon + \frac{1}{2\varepsilon}\right)\frac{A_0^2}{\lambda R^2}K^2. \end{aligned}$$

Now we take $\mu = (A_1^*)^2/(\varepsilon\lambda^2)$. Then,

$$a_{ij}\widetilde{v}_{ij} \ge -\frac{3}{K}e^{2\mu}|\nabla f| - e^{2\mu}\left(2\varepsilon + \frac{1}{2\varepsilon}\right)\frac{A_0^2}{\lambda R^2}.$$

For any $r > 0$, we have

$$a_{ij}\partial_{ij}\left(\sup_{B_r}\widetilde{v} - \widetilde{v}\right) \le \frac{3}{K}e^{2\mu}|\nabla f| + e^{2\mu}\left(2\varepsilon + \frac{1}{2\varepsilon}\right)\frac{A_0^2}{\lambda R^2} \quad \text{in } B_r.$$

In the following, we set $\tau = 1/8\sqrt{n}$. By the weak Harnack inequality provided by Theorem 1.2.11, we have

$$\begin{aligned} r^{-\frac{n}{p}}\left\|\sup_{B_r}\widetilde{v} - \widetilde{v}\right\|_{L^p(B_{\tau r})} &\le C\left(\sup_{B_r}\widetilde{v} - \sup_{B_{\tau r}}\widetilde{v}\right) \\ &\quad + C_1\left\{\frac{r^2}{K}|\nabla f|_{L^\infty(B_r)} + \frac{r^2}{R^2}\right\}, \end{aligned}$$

where p and C are positive constants depending only on n, λ, and Λ and C_1 is a positive constant depending only on n, λ, A_0, A_1^*, and ε. By (1), v/K^2 is bounded. Since $\widetilde{v}$ is increasing with respect to v/K^2, we have, for any $x_1, x_2 \in B_R$,

$$\frac{c_1}{K^2}\big(v(x_1) - v(x_2)\big) \le \widetilde{v}(x_1) - \widetilde{v}(x_2) \le \frac{c_2}{K^2}\big(v(x_1) - v(x_2)\big),$$

where c_1 and c_2 are two positive constants depending only on μ. Hence,

$$\begin{aligned} r^{-\frac{n}{p}}\left\|\sup_{B_r} v - v\right\|_{L^p(B_{\tau r})} &\le C\Big\{\sup_{B_r} v - \sup_{B_{\tau r}} v \\ &\quad + Kr^2|\nabla f|_{L^\infty(B_r)} + \frac{r^2}{R^2}K^2\Big\}, \end{aligned}$$

where C is a positive constant depending only on n, λ, Λ, ε, A_0, and A_1^*. Later on, we choose $\varepsilon = (10n)^{-1}$. Note that

$$\sup_{B_r} v - \sup_{B_{\tau r}} v \leq \operatorname{osc}_{B_r} v - \operatorname{osc}_{B_{\tau r}} v.$$

Next, we assume

$$r^{-\frac{n}{p}} \| \sup_{B_r} v - v \|_{L^p(B_{\tau r})} \geq c_* \operatorname{osc}_{B_{\tau r}} v, \tag{2}$$

for some positive constant c_*. Then,

$$\operatorname{osc}_{B_{\tau r}} v \leq C \left\{ \operatorname{osc}_{B_r} v - \operatorname{osc}_{B_{\tau r}} v + K r^2 |\nabla f|_{L^\infty(B_R)} + \frac{r^2}{R^2} K^2 \right\}, \tag{3}$$

where C is a positive constant depending only on n, λ, Λ, c_*, A_0, and A_1^*.

Step 2. For $i = 1, \ldots, n$, set

$$v_{\pm i} = \pm K u_i + \varepsilon |\nabla u|^2.$$

We note that there are $2n$ functions here. Then, we set

$$\omega(r) = \sum_{i=1}^n \left(\operatorname{osc}_{B_r} v_{+i} + \operatorname{osc}_{B_r} v_{-i} \right).$$

It is easy to check that, for any $r \in (0, R)$,

$$(2 - 4n\varepsilon) K \sum_{i=1}^n \operatorname{osc}_{B_r} u_i \leq \omega(r) \leq (2 + 4n\varepsilon) K \sum_{i=1}^n \operatorname{osc}_{B_r} u_i. \tag{4}$$

For an arbitrarily fixed $r \in (0, R)$, we take an $\ell \in \{1, \ldots, n\}$ such that, for any $i = 1, \ldots, n$,

$$\operatorname{osc}_{B_{\tau r}} u_\ell \geq \operatorname{osc}_{B_{\tau r}} u_i. \tag{5}$$

Then,

$$\operatorname{osc}_{B_{\tau r}} |\nabla u|^2 \leq 2nK \operatorname{osc}_{B_{\tau r}} u_\ell, \tag{6}$$

and hence

$$(1 - 2n\varepsilon) K \operatorname{osc}_{B_{\tau r}} u_\ell \leq \operatorname{osc}_{B_{\tau r}} v_{\pm \ell} \leq (1 + 2n\varepsilon) K \operatorname{osc}_{B_{\tau r}} u_\ell. \tag{7}$$

Next, for any $x_+, x_- \in B_{\tau r}$, we get

$$K u_\ell(x_+) + \varepsilon |\nabla u(x_+)|^2 = v_{+\ell}(x_+) \leq \sup_{B_{\tau r}} v_{+\ell},$$

$$-K u_\ell(x_-) + \varepsilon |\nabla u(x_-)|^2 = v_{-\ell}(x_-) \leq \sup_{B_{\tau r}} v_{-\ell}.$$

Hence, for some $x_+, x_- \in B_{\tau r}$, we have

$$K \sup_{B_{\tau r}} u_\ell + \varepsilon|\nabla u(x_+)|^2 \le \sup_{B_{\tau r}} v_{+\ell},$$
$$-K \inf_{B_{\tau r}} u_\ell + \varepsilon|\nabla u(x_-)|^2 \le \sup_{B_{\tau r}} v_{-\ell}.$$

Then, for any $x \in B_{\tau r}$,

$$\begin{aligned}
&\sup_{B_{\tau r}} v_{+\ell} - v_{+\ell}(x) + \sup_{B_{\tau r}} v_{-\ell} - v_{-\ell}(x) \\
&\quad \ge K\left(\sup_{B_{\tau r}} u_\ell - \inf_{B_{\tau r}} u_\ell\right) + \varepsilon(|\nabla u(x_+)|^2 + |\nabla u(x_-)|^2 - 2|\nabla u(x)|^2) \\
&\quad \ge K \operatorname*{osc}_{B_{\tau r}} u_\ell - 2\varepsilon \operatorname*{osc}_{B_{\tau r}} |\nabla u|^2 \ge (1-4n\varepsilon)K \operatorname*{osc}_{B_{\tau r}} u_\ell,
\end{aligned}$$

where we used (6) in the last step. For the positive constant p as in (2), we have

$$\begin{aligned}
&\left\|\sup_{B_{\tau r}} v_{+\ell} - v_{+\ell}\right\|_{L^p(B_{\tau r})} + \left\|\sup_{B_{\tau r}} v_{-\ell} - v_{-\ell}\right\|_{L^p(B_{\tau r})} \\
&\qquad \ge c_p \left\|\sup_{B_{\tau r}} v_{+\ell} - v_{+\ell} + \sup_{B_{\tau r}} v_{-\ell} - v_{-\ell}\right\|_{L^p(B_{\tau r})},
\end{aligned}$$

where c_p is a positive constant depending only on p and hence only on n, λ, and Λ. We point out that p may be less than one. Then,

$$\begin{aligned}
&r^{-\frac{n}{p}}\left\|\sup_{B_{\tau r}} v_{+\ell} - v_{+\ell}\right\|_{L^p(B_{\tau r})} + r^{-\frac{n}{p}}\left\|\sup_{B_{\tau r}} v_{-\ell} - v_{-\ell}\right\|_{L^p(B_{\tau r})} \\
&\quad \ge (1-4n\varepsilon)c_p K \operatorname*{osc}_{B_{\tau r}} u_\ell.
\end{aligned}$$

Without loss of generality, we assume

$$r^{-\frac{n}{p}}\left\|\sup_{B_{\tau r}} v_{+\ell} - v_{+\ell}\right\|_{L^p(B_{\tau r})} \ge \frac{1}{2}(1-4n\varepsilon)c_p K \operatorname*{osc}_{B_{\tau r}} u_\ell.$$

Then, by (7), we have

$$r^{-\frac{n}{p}}\left\|\sup_{B_{\tau r}} v_{+\ell} - v_{+\ell}\right\|_{L^p(B_{\tau r})} \ge \frac{1-4n\varepsilon}{2(1+2n\varepsilon)} c_p \operatorname*{osc}_{B_{\tau r}} v_{+\ell} = \frac{1}{4} c_p \operatorname*{osc}_{B_{\tau r}} v_{+\ell}$$

if we take $\varepsilon = (10n)^{-1}$. Hence, (2) is satisfied for $c_* = c_p/4$, which is a positive constant depending only on n, λ, and Λ. By (3), we have

$$\operatorname*{osc}_{B_{\tau r}} v_{+\ell} \le C\left\{\operatorname*{osc}_{B_r} v_{+\ell} - \operatorname*{osc}_{B_{\tau r}} v_{+\ell} + Kr^2|\nabla f|_{L^\infty(B_R)} + \frac{r^2}{R^2}K^2\right\},$$

and hence, by the definition of ω,

$$\operatorname*{osc}_{B_{\tau r}} v_{+\ell} \le C\left\{\omega(r) - \omega(\tau r) + Kr^2|\nabla f|_{L^\infty(B_R)} + \frac{r^2}{R^2}K^2\right\}.$$

Next, by (5), (6), and (7), we have, for any $i = 1, \ldots, n$,

$$\begin{aligned}\operatorname*{osc}_{B_{\tau r}} v_{\pm i} &\leq K \operatorname*{osc}_{B_{\tau r}} u_i + 2n\varepsilon K \operatorname*{osc}_{B_{\tau r}} u_\ell \\ &\leq (1+2n\varepsilon) K \operatorname*{osc}_{B_{\tau r}} u_\ell \leq \frac{1+2n\varepsilon}{1-2n\varepsilon} \operatorname*{osc}_{B_{\tau r}} v_{+\ell} = \frac{3}{2} \operatorname*{osc}_{B_{\tau r}} v_{+\ell}.\end{aligned}$$

Then, we obtain, for any $i = 1, \ldots, n$,

$$\operatorname*{osc}_{B_{\tau r}} v_{\pm i} \leq C \left\{ \omega(r) - \omega(\tau r) + K r^2 |\nabla f|_{L^\infty(B_R)} + \frac{r^2}{R^2} K^2 \right\}.$$

Now a simple addition yields

$$\omega(\tau r) \leq C \left\{ \omega(r) - \omega(\tau r) + K r^2 |\nabla f|_{L^\infty(B_R)} + \frac{r^2}{R^2} K^2 \right\}.$$

Hence,

$$\omega(\tau r) \leq \gamma \omega(r) + K r^2 |\nabla f|_{L^\infty(B_R)} + K^2 \frac{r^2}{R^2},$$

for a positive contant $\gamma = C(1+C)^{-1} < 1$. Lemma 1.2.13 implies, for any $r < R$,

$$\omega(r) \leq C \left(\frac{r}{R} \right)^\alpha \left\{ \omega(R) + K R^2 |\nabla f|_{L^\infty(B_R)} + K^2 \right\},$$

where $\alpha \in (0,1)$ is a constant depending only on n, λ, Λ, A_0, and A_1. By (4), we have, for any $r < R$,

$$\sum_{i=1}^n \operatorname*{osc}_{B_r} u_i \leq C \left(\frac{r}{R} \right)^\alpha \left\{ \sum_{i=1}^n \operatorname*{osc}_{B_R} u_i + R^2 |\nabla f|_{L^\infty(B_R)} + K \right\}.$$

In conclusion, we obtain, for any $i = 1, \ldots, n$ and any $r < R$,

$$\operatorname*{osc}_{B_r} u_i \leq C \left(\frac{r}{R} \right)^\alpha \left\{ |\nabla u|_{L^\infty(B_R)} + R^2 |\nabla f|_{L^\infty(B_R)} \right\}.$$

This yields the desired estimate. □

The original proof by Ladyzhenskaya and Ural'tseva is to reduce the differential inequality for $|\nabla u|^2$ to a divergence form and then employ de Giorgi's estimate of the Hölder semi-norms of the weak solutions. The present proof, based on [**146**], is to retain the nondivergence form and use the weak Harnack inequality provided by Theorem 1.2.11 instead. Later on, we will employ a similar method to derive estimates of the interior Hölder semi-norms of the second derivatives of solutions to fully nonlinear uniformly elliptic differential equations. (See Theorem 5.4.1.)

We now prove a Liouville type theorem.

Theorem 2.4.2. *Let $a_{ij} \in C^1(\mathbb{R}^n)$ satisfy the uniform ellipticity and*

$$\sup_{p\in\mathbb{R}^n} \big(|p||\nabla_p a_{ij}(p)|\big) < \infty.$$

Suppose that u is a $C^3(\mathbb{R}^n)$-solution of

$$a_{ij}(\nabla u)u_{ij} = 0 \quad \text{in } \mathbb{R}^n.$$

If u has at most a linear growth in $\mathbb{R}^n$, then u is an affine function.

Here, u has at most a linear growth in $\mathbb{R}^n$ if, for any $x \in \mathbb{R}^n$,

$$|u(x)| \leq c_0 + c_1|x|,$$

for some nonnegative constants c_0 and c_1.

Proof. Since $a_{ij} = a_{ij}(p)$, we take $A_0 = 0$ in Theorem 2.2.1 and Theorem 2.4.1. Set

$$A_1 = \sup_{p\in\mathbb{R}^n} \big(|p||\nabla_p a_{ij}(p)|\big).$$

By Theorem 2.2.1, we obtain, for any $R > 0$,

$$R|\nabla u|_{L^\infty(B_{R/2})} \leq C|u|_{L^\infty(B_R)},$$

where C is a positive constant depending only on n, λ, Λ, and A_1. By the linear growth of u, ∇u is bounded in $\mathbb{R}^n$. Set $K = |\nabla u|_{L^\infty(\mathbb{R}^n)}$ and

$$A_1^* = K|\nabla_p a_{ij}|_{L^\infty(B_K)}.$$

Next, by Theorem 2.4.1, we have, for any $R > 0$,

$$[\nabla u]_{C^\alpha(B_{R/2})} \leq \frac{C}{R^\alpha}|\nabla u|_{L^\infty(B_R)} = \frac{CK}{R^\alpha},$$

where $\alpha \in (0,1)$ and $C > 0$ are constants depending only on n, λ, Λ, and A_1^*. By letting $R \to \infty$, we conclude that ∇u is constant in $\mathbb{R}^n$. □

With the interior $C^{1,\alpha}$-estimate, we can derive an interior $C^{2,\alpha}$-estimate. For simplicity, we consider the unit ball.

Theorem 2.4.3. *Let $\alpha \in (0,1)$ be a constant and $a_{ij} \in C^1(B_1 \times \mathbb{R}^n)$ satisfy the uniform ellipticity and, for any $x \in B_1$ and $p \in \mathbb{R}^n$,*

$$|\nabla_x a_{ij}(x,p)| \leq A_0,$$

for some positive constant A_0. Suppose that u is a $C^{2,\alpha}(B_1)$-solution of

$$a_{ij}(x, \nabla u)u_{ij} = f \quad \text{in } B_1,$$

for some $f \in C^\alpha(B_1)$, with $|u|_{C^{1,\alpha}(B_1)} < \infty$ and $|f|_{C^\alpha(B_1)} < \infty$. Then,

$$|\nabla^2 u|_{C^\alpha(B_{1/2})} \leq C_*,$$

where C_* is a positive constant depending only on n, λ, Λ, A_0, $[\nabla u]_{C^\alpha(B_1)}$, $|f|_{C^\alpha(B_1)}$, and $|\nabla_p a_{ij}|_{L^\infty(B_1\times B_K)}$, with $K = |\nabla u|_{L^\infty(B_1)}$.

Proof. With $K = |\nabla u|_{L^\infty(B_1)}$, we first note that

$$[a_{ij}(\cdot,\nabla u)]_{C^\alpha(B_1)} \le \sup_{p\in B_K} [a_{ij}(\cdot,p)]_{C^\alpha(B_1)} + |\nabla_p a_{ij}|_{L^\infty(B_1\times B_K)}[\nabla u]_{C^\alpha(B_1)}.$$

The mean value theorem implies, for any $p\in\mathbb{R}^n$,

$$[a_{ij}(\cdot,p)]_{C^\alpha(B_1)} \le 2|\nabla_x a_{ij}(\cdot,p)|_{L^\infty(B_1)} \le 2A_0.$$

Hence,

$$[a_{ij}(\cdot,\nabla u)]_{C^\alpha(B_1)} \le \Lambda_0,$$

where Λ_0 is a positive constant depending only on n, λ, Λ, α, A_0, $[\nabla u]_{C^\alpha(B_1)}$, and $|\nabla_p a_{ij}|_{L^\infty(B_1\times B_K)}$. Then by the interior Schauder estimate, we conclude

$$|\nabla^2 u|_{C^\alpha(B_{1/2})} \le C\left\{|u|_{L^\infty(B_1)} + |f|_{C^\alpha(B_1)}\right\},$$

where C is a positive constant depending only on n, λ, Λ, α, and Λ_0. This implies the desired result. □

2.5. Global $C^{1,\alpha}$-Estimates

In this section, we derive global estimates of the Hölder semi-norms of the gradients for quasilinear uniformly elliptic differential equations.

Let Ω be a bounded domain in $\mathbb{R}^n$ and u be a C^2-function in Ω. In the following, we consider the quasilinear equation given by

$$a_{ij}(x,\nabla u)u_{ij} = f(x) \quad \text{in } \Omega.$$

We always assume the *uniform ellipticity* in the following form: for any $x\in\Omega$ and any $p,\xi\in\mathbb{R}^n$,

$$\lambda|\xi|^2 \le a_{ij}(x,p)\xi_i\xi_j \le \Lambda|\xi|^2,$$

for some positive constants λ and Λ.

Theorem 2.5.1. *Let Ω be a bounded domain in $\mathbb{R}^n$ with a C^2-boundary $\partial\Omega$, and let $a_{ij}\in C^1(\bar\Omega\times\mathbb{R}^n)$ be uniformly elliptic and satisfy*

$$|\nabla_x a_{ij}|_{L^\infty(\Omega\times\mathbb{R}^n)} \le A_0,$$

for some positive constant A_0. Suppose that u is a $C^2(\bar\Omega)\cap C^3(\Omega)$-solution of

$$\begin{aligned} a_{ij}(x,\nabla u)u_{ij} &= f \quad \text{in } \Omega,\\ u &= \varphi \quad \text{on } \partial\Omega, \end{aligned} \tag{1}$$

for some $f\in C^1(\bar\Omega)$ and $\varphi\in C^2(\bar\Omega)$. Assume

$$\sup_\Omega |\nabla u|\cdot\sup_\Omega |\nabla_p a_{ij}(\cdot,\nabla u)| \le A_1^*,$$

for some positive constant A_1^*. Then, for some $\alpha \in (0,1)$,

$$[\nabla u]_{C^\alpha(\bar\Omega)} \le C\left\{|\nabla u|_{L^\infty(\Omega)} + |f|_{C^1(\bar\Omega)} + |\varphi|_{C^2(\bar\Omega)}\right\},$$

where α and C are positive constants depending only on n, λ, Λ, A_0, A_1^*, and Ω.

Proof. The proof consists of several steps. In the following, a universal constant is a positive constant depending only on n, λ, Λ, A_0, A_1^*, and Ω. We set

$$N = |\nabla u|_{L^\infty(\Omega)} + |f|_{C^1(\bar\Omega)} + |\varphi|_{C^2(\bar\Omega)}.$$

Step 1. We estimate the C^β-norm of $\nabla u|_{\partial\Omega}$, for a constant $\beta \in (0,1)$ depending only on n, λ, and Λ, and prove, for any $x_0, x_1 \in \partial\Omega$,

$$|\nabla u(x_1) - \nabla u(x_0)| \le C_1 N |x_1 - x_0|^\beta. \tag{2}$$

For a fixed point $x_0 \in \partial\Omega$, we flatten the boundary $\partial\Omega$ near x_0. In other words, we consider a neighborhood U of x_0 which contains a ball centered at x_0, with a universal radius, and C^2-diffeomorphisms

$$x = \chi(y), \quad y = \chi^{-1}(x) = \eta(x) \quad \text{for any } x \in U \text{ and } y \in A = \eta(U),$$

so that $\eta(x_0) = 0$, and

$$\begin{aligned} \eta(U \cap \Omega) &= B_4^+ = \{y \in \mathbb{R}^n : |y| < 4, y_n > 0\}, \\ \eta(U \cap \partial\Omega) &= \Sigma_4 = \{y \in \mathbb{R}^n : |y| < 4, y_n = 0\}. \end{aligned}$$

We also require that $|\eta|_{C^2(\bar U)}$ and $|\chi|_{C^2(\bar A)}$ are bounded by a universal constant. We rewrite the equation as

$$a_{ij}(x, \nabla u)(u - \varphi)_{ij} = f - a_{ij}\varphi_{ij}$$

and consider the function

$$\widetilde{u}(y) = u(x) - \varphi(x) \quad \text{for any } y \in B_4^+ \cup \Sigma_4.$$

Then, $\widetilde{u}$ vanishes on Σ_4 and satisfies the equation

$$a_{ij}\eta_i^k \eta_j^l \widetilde{u}_{kl} = f - a_{ij}\varphi_{ij} - a_{ij}\eta_{ij}^k \widetilde{u}_k, \tag{3}$$

where $\eta = (\eta^1, \ldots, \eta^n)$ and η_i^k, η_{ij}^k, $a_{ij}(\cdot, \nabla u)$, φ_{ij}, and f are evaluated at $\chi(y)$. We prove (2) by showing, for any $y \in \Sigma_1$ and any $i = 1, \ldots, n$,

$$|\widetilde{u}_i(y) - \widetilde{u}_i(0)| \le CN|y|^\beta, \tag{4}$$

where

$$\Sigma_1 = \{y : |y| < 1, y_n = 0\}.$$

Since $\widetilde{u}|_{\Sigma_1} = 0$, then $\widetilde{u}_i|_{\Sigma_1} = 0$ for $i = 1, \ldots, n-1$. Hence (4) holds trivially for $i = 1, \ldots, n-1$. To prove (4) for $i = n$, we note that (3) and the boundary condition of $\widetilde{u}$ imply

$$\widetilde{L}\widetilde{u} \equiv \widetilde{a}_{kl}\widetilde{u}_{kl} = \widetilde{f} \quad \text{in } B_4^+,$$
$$\widetilde{u} = 0 \quad \text{on } \Sigma_4^+,$$

where

$$\widetilde{a}_{kl} = a_{ij}\eta_i^k\eta_j^l$$

and

$$\widetilde{f} = f - a_{ij}\varphi_{ij} - a_{ij}\eta_{ij}^k\widetilde{u}_k.$$

Hence, $\widetilde{L}$ is uniformly elliptic with ellipticity constants $C^{-1}\lambda$ and $C\Lambda$, for a universal constant $C \geq 1$. Since $|\chi|_{C^2(\bar{A})}$ and $|\eta|_{C^2(\bar{U})}$ are bounded, we then have

$$|\widetilde{f}|_{L^\infty(B_4^+)} \leq CN,$$

where C is a universal constant. By Theorem 1.2.16, there exists a constant $\beta \in (0,1)$, depending only on n, λ, and Λ, such that

$$|\partial_n\widetilde{u}|_{C^\beta(\Sigma_1)} \leq C\left\{|\nabla\widetilde{u}|_{L^\infty(B_4^+)} + |\widetilde{f}|_{L^\infty(B_4^+)}\right\},$$

or, for any $y_0, y_1 \in \Sigma_1$,

$$|\partial_n\widetilde{u}(y_0) - \partial_n\widetilde{u}(y_1)| \leq C|y_0 - y_1|^\beta\left\{|\nabla\widetilde{u}|_{L^\infty(B_4^+)} + |\widetilde{f}|_{L^\infty(B_4^+)}\right\},$$

for some universal constant C. This implies (4) for $i = n$.

Step 2. We prove, for any $x \in \bar{\Omega}$ and $x_0 \in \partial\Omega$,

$$|\nabla u(x) - \nabla u(x_0)| \leq C_2N|x - x_0|^{\frac{\beta}{1+\beta}}. \tag{5}$$

We set $K = |\nabla u|_{L^\infty(\Omega)}$. Consider, for $i = 1, \ldots, n$ and for $\varepsilon > 0$ to be determined,

$$v = \pm Ku_i + \varepsilon|\nabla u|^2.$$

Then, $-K^2 \leq v \leq 2K^2$. Set

$$\widetilde{v} = \frac{1}{\mu}\left(\exp\left\{\frac{\mu v}{K^2}\right\} - 1\right),$$

with $\mu = (A_1^*)^2/(\varepsilon\lambda^2)$. Then, for any $x_1, x_2 \in \bar{\Omega}$,

$$\frac{c_1}{K^2}(v(x_1) - v(x_2)) \leq \widetilde{v}(x_1) - \widetilde{v}(x_2) \leq \frac{c_2}{K^2}(v(x_1) - v(x_2)), \tag{6}$$

where c_1 and c_2 are two positive constants depending only on μ. Moreover,

$$|v(x_1) - v(x_2)| \leq 3K|\nabla u(x_1) - \nabla u(x_2)|, \tag{7}$$

and

$$(8)\qquad \begin{aligned} v(x_1) - v(x_2) \geq \pm K\big(u_i(x_1) - u_i(x_2)\big) \\ - 2nK\varepsilon \max_{1\leq k\leq n} |u_k(x_1) - u_k(x_2)|. \end{aligned}$$

As shown in the proof of Theorem 2.4.1, we have

$$a_{ij}\widetilde{v}_{ij} \geq -\frac{3}{K}e^{2\mu}|\nabla f| - e^{2\mu}\left(2\varepsilon + \frac{1}{2\varepsilon}\right)\frac{A_0^2}{\lambda} \geq -\frac{C_0}{K}N.$$

In the following, we set

$$L = a_{ij}\big(x, \nabla u(x)\big)\partial_{ij}.$$

Then,

$$L\widetilde{v} \geq -\frac{C_0}{K}N.$$

Fix a point $x_0 \in \partial\Omega$. By (2), (6), and (7), we have, for any $x \in \partial\Omega$,

$$\widetilde{v}(x) - \widetilde{v}(x_0) \leq \frac{C}{K}N|x - x_0|^\beta$$

and, for any $x \in \Omega$,

$$\widetilde{v}(x) - \widetilde{v}(x_0) \leq \frac{C}{K}N.$$

By Theorem 1.1.15, we obtain, for any $x \in \Omega$,

$$(9)\qquad \widetilde{v}(x) - \widetilde{v}(x_0) \leq \frac{C}{K}N|x - x_0|^{\frac{\beta}{1+\beta}}.$$

We point out that Theorem 1.1.15 is formulated for solutions of uniformly elliptic linear equations. However, its proof clearly yields a one-sided estimate for subsolutions. Hence, (6) and (9) imply, for any $x \in \Omega$,

$$v(x) - v(x_0) \leq CKN|x - x_0|^{\frac{\beta}{1+\beta}}.$$

Next, by (8), we have, for each fixed $x \in \Omega$,

$$\pm\big(u_i(x) - u_i(x_0)\big) - 2n\varepsilon \max_{1\leq k\leq n} |u_k(x) - u_k(x_0)| \leq CN|x - x_0|^{\frac{\beta}{1+\beta}}.$$

By taking the maximum over $i = 1, \ldots, n$ and then taking $\varepsilon = 1/(4n)$, we obtain (5).

Step 3. Take α to be the minimum of α in Theorem 2.4.1 and $\beta/(1+\beta)$ in Step 2. We prove, for any $x, y \in \bar{\Omega}$,

$$(10)\qquad |\nabla u(x) - \nabla u(y)| \leq CN|x - y|^\alpha.$$

We first recall Theorem 2.4.1, the interior estimates. For any $B_R(x_0) \subset \Omega$ and any $x, y \in B_{R/2}(x_0)$, we have

$$(11)\qquad R^\alpha \frac{|\nabla u(x) - \nabla u(y)|}{|x - y|^\alpha} \leq C\left\{|\nabla u|_{L^\infty(B_R)} + R^2|\nabla f|_{L^\infty(\Omega)}\right\}.$$

For any $x, y \in \Omega$, set $d_x = \text{dist}(x, \Omega)$ and $d_y = \text{dist}(y, \Omega)$ and take $x_0, y_0 \in \partial\Omega$ such that $|x - x_0| = d_x$ and $|y - y_0| = d_y$. Suppose $d_y \leq d_x$. First, we assume $|x - y| < d_x/2$. Then, $y \in B_{d_x/2}(x) \subset B_{d_x}(x) \subset \Omega$. Set

$$w = u - u(x_0) - \nabla u(x_0)(x - x_0).$$

Then, w satisfies

$$a_{ij}\big(x, \nabla w + \nabla u(x_0)\big)w_{ij} = f \quad \text{in } \Omega.$$

We apply (11) to w in $B_{d_x}(x)$ and get

$$d_x^\alpha \frac{|\nabla u(x) - \nabla u(y)|}{|x - y|^\alpha} \leq C\left\{|\nabla u - \nabla u(x_0)|_{L^\infty(B_{d_x}(x))} + d_x^2 |\nabla f|_{L^\infty(\Omega)}\right\}.$$

By (5), we obtain

$$|\nabla u - \nabla u(x_0)|_{L^\infty(B_{d_x}(x))} \leq C_2 N d_x^\alpha,$$

and hence

$$|\nabla u(x) - \nabla u(y)| \leq C|x - y|^\alpha \left\{C_2 N + |\nabla f|_{L^\infty(\Omega)}\right\} \leq CN|x - y|^\alpha.$$

Next, we assume $d_y \leq d_x \leq 2|x - y|$. Then,

$$|x_0 - y_0| \leq d_x + |x - y| + d_y \leq 5|x - y|.$$

By (5) again, we have

$$\begin{aligned}
&|\nabla u(x) - \nabla u(y)| \\
&\quad \leq |\nabla u(x) - \nabla u(x_0)| + |\nabla u(x_0) - \nabla u(y_0)| + |\nabla u(y_0) - \nabla u(y)| \\
&\quad \leq CN\left\{d_x^\alpha + |x_0 - y_0|^\alpha + d_y^\alpha\right\} \\
&\quad \leq CN|x - y|^\alpha.
\end{aligned}$$

This finishes the proof of (10). □

We now state a global $C^{2,\alpha}$-estimate of solutions.

Theorem 2.5.2. *Let $\alpha \in (0, 1)$ be a constant, Ω be a bounded domain in $\mathbb{R}^n$ with a $C^{2,\alpha}$-boundary $\partial\Omega$, $a_{ij} \in C^1(\bar{\Omega} \times \mathbb{R}^n)$ be uniformly elliptic and satisfy, for any $x \in \Omega$ and $p \in \mathbb{R}^n$,*

$$|\nabla_x a_{ij}(x, p)| \leq A_0,$$

for some positive constant A_0. Suppose that u is a $C^3(\bar{\Omega})$-solution of

$$\begin{aligned}
a_{ij}(x, \nabla u)u_{ij} &= f \quad \text{in } \Omega, \\
u &= \varphi \quad \text{on } \partial\Omega,
\end{aligned} \tag{1}$$

for some $f \in C^\alpha(\bar\Omega)$ *and* $\varphi \in C^{2,\alpha}(\bar\Omega)$. *Then,*

$$|u|_{C^{2,\alpha}(\bar\Omega)} \le C_*,$$

where C_* *is a positive constant depending only on* n, λ, Λ, A_0, Ω, $[\nabla u]_{C^\alpha(\bar\Omega)}$, $|f|_{C^\alpha(\bar\Omega)}$, $|\varphi|_{C^{2,\alpha}(\bar\Omega)}$, *and* $|\nabla_p a_{ij}|_{L^\infty(\Omega\times B_K)}$, *with* $K = |\nabla u|_{L^\infty(\Omega)}$.

The proof is similar to that of Theorem 2.4.3 and is omitted.

2.6. Dirichlet Problems

In this section, we employ the method of continuity to solve the Dirichlet problem for quasilinear uniformly elliptic differential equations given by

$$\begin{aligned} a_{ij}(x,\nabla u)u_{ij} &= f \quad \text{in } \Omega,\\ u &= \varphi \quad \text{on } \partial\Omega. \end{aligned}$$

The global $C^{1,\alpha}$-estimates we derived in the previous section play an essential role.

The method of continuity consists of three steps.

Step 1. Prove that the given Dirichlet problem, labeled as $(*)$, can be embedded in a family of Dirichlet problems $\{(*)_t\}$ indexed by a parameter t in a bounded closed interval, say $[0,1]$, with $(*)_1$ corresponding to the given Dirichlet problem and $(*)_0$ corresponding to a Dirichlet problem which can be solved.

Step 2. Prove that the set of $t \in [0,1]$ for which the Dirichlet problem $(*)_t$ can be solved is open; that is, if $(*)_{t_0}$ can be solved, then there exists a small neighborhood of t_0, say $|t-t_0| < \varepsilon(t_0)$, such that $(*)_t$ can be solved for all t in this neighborhood.

Step 3. Prove that the set of $t \in [0,1]$ for which $(*)_t$ can be solved is closed.

Steps 1–3 imply that the set of t for which $(*)_t$ can be solved is the whole segment $0 \le t \le 1$ and hence contains $t=1$. Therefore, the given Dirichlet problem, corresponding to $(*)_1$, is solved.

Step 1 is usually fulfilled by connecting the given Dirichlet problem with a Dirichlet problem which can be solved easily. Step 2, referred to as the "openness", is based on the implicit function theorem or an iteration process. It reduces the solvability of nonlinear equations to that of linear equations. Step 3, referred to as the "closedness", is based on a priori estimates for solutions independent of the parameter t.

Now we solve the Dirichlet problem for quasilinear uniformly elliptic differential equations by the method of continuity.

Theorem 2.6.1. *Let $\alpha \in (0,1)$ be a constant, $\Omega \subset \mathbb{R}^n$ be a bounded domain in $\mathbb{R}^n$ with a $C^{3,\alpha}$-boundary, and $a_{ij} \in C^2(\bar{\Omega} \times \mathbb{R}^n)$ be uniformly elliptic and satisfy, for any $x \in \Omega$ and $p \in \mathbb{R}^n$,*

$$|\nabla_x a_{ij}(x,p)| \leq A_0, \quad |p||\nabla_p a_{ij}(x,p)| \leq A_1,$$

for some positive constants A_0 and A_1. Then for any $f \in C^{1,\alpha}(\bar{\Omega})$ and $\varphi \in C^{3,\alpha}(\bar{\Omega})$, there exists a unique solution $u \in C^{3,\alpha}(\bar{\Omega})$ of

$$\begin{aligned} a_{ij}(x, \nabla u) u_{ij} &= f \quad \text{in } \Omega, \\ u &= \varphi \quad \text{on } \partial\Omega. \end{aligned} \tag{1}$$

Proof. We need only prove the existence of a $C^{2,\alpha}(\bar{\Omega})$-solution of (1). Then, the uniqueness follows from Corollary 2.1.2 and the higher regularity $u \in C^{3,\alpha}(\bar{\Omega})$ from Proposition 2.1.5. With $u \in C^3(\bar{\Omega})$, we can apply Theorems 2.3.1, 2.3.2, 2.3.3, 2.5.1, and 2.5.2 to conclude

$$|u|_{C^{2,\alpha}(\bar{\Omega})} \leq C_*, \tag{2}$$

where C_* is a positive constant depending only on n, λ, Λ, A_0, A_1, $\|f\|_{C^1(\bar{\Omega})}$, $\|\varphi\|_{C^{2,\alpha}(\bar{\Omega})}$, and $|\nabla_p a_{ij}|_{L^\infty(\Omega \times B_K)}$, with K given by

$$C\left\{|\varphi|_{C^2(\bar{\Omega})} + |f|_{C^1(\bar{\Omega})}\right\},$$

for some positive constant C depending only on n, λ, Λ, A_0, and A_1. We point out that Theorem 2.5.1 yields an estimate of $[\nabla u]_{C^{\alpha'}(\bar{\Omega})}$ for some $\alpha' \in (0,1)$, possibly smaller than the given α. Then, Theorem 2.5.2 holds for such α' and hence implies an estimate of $[\nabla u]_{C^{\alpha}(\bar{\Omega})}$ by a simple interpolation. Another application of Theorem 2.5.2 yields (2).

For each $t \in [0,1]$, consider a family of problems

$$\begin{aligned} a_{ij}(x, \nabla u) u_{ij} &= tf \quad \text{in } \Omega, \\ u &= t\varphi \quad \text{on } \partial\Omega. \end{aligned} \tag{3}$$

For $t = 0$, (3) corresponds to a Dirichlet problem whose unique solution is given by $u = 0$. The equation in (3) is uniformly elliptic by the assumption. It follows that any $C^{2,\alpha}$-solutions of (3) are $C^3(\bar{\Omega})$ and satisfy the estimate (2).

Letting $v = u - t\varphi$, (3) is equivalent to

$$\begin{aligned} a_{ij}(x, \nabla v + t\nabla\varphi) v_{ij} + t a_{ij}(x, \nabla v + t\nabla\varphi)\varphi_{ij} &= tf \quad \text{in } \Omega, \\ v &= 0 \quad \text{on } \partial\Omega. \end{aligned} \tag{4}$$

Moreover, any solutions of (4) satisfy the estimate (2).

Next, we set

$$\mathcal{X} = \{v \in C^{2,\alpha}(\bar{\Omega}) : \ v = 0 \text{ on } \partial\Omega\}$$

and

$$Q(v,t) = a_{ij}(\cdot, \nabla v + t\nabla\varphi)v_{ij} + ta_{ij}(\cdot, \nabla v + t\nabla\varphi)\varphi_{ij} - tf.$$

Solving (4) is equivalent to finding a function $v \in \mathcal{X}$ such that $Q(v,t) = 0$ in Ω.

Set

$$I = \{t \in [0,1] : \text{there exists a } v \in \mathcal{X} \text{ such that } Q(v,t) = 0\}.$$

We note that $0 \in I$. To prove $1 \in I$, we will prove that I is both open and closed in $[0,1]$. For the openness, we note that $Q : \mathcal{X} \times [0,1] \to C^\alpha(\bar{\Omega})$ is of class C^1 and its Frèchet derivative with respect to $v \in \mathcal{X}$ is given by

$$Q_v(v,t)w = a_{ij}(\cdot, \nabla v + t\nabla\varphi)w_{ij} + (v_{ij} + t\varphi_{ij})a_{ij,p_k}w_k.$$

This is a uniformly elliptic linear operator with C^α-coefficients. By the Schauder theory, $Q_v(v,t)$ is an invertible operator from $\mathcal{X}$ to $C^\alpha(\bar{\Omega})$. Suppose $t_0 \in I$; i.e., $Q(v^{t_0}, t_0) = 0$ for some $v^{t_0} \in \mathcal{X}$. By the implicit function theorem, for any t close to t_0, there is a unique $v^t \in \mathcal{X}$, close to v^{t_0} in the $C^{2,\alpha}$-norm, satisfying $Q(v^t, t) = 0$. Hence $t \in I$ for all such t, and therefore I is open. For the closedness, we note by (2) that any solution v in $\mathcal{X}$ of $Q(v,t) = 0$ in $\bar{\Omega}$ satisfies a uniform $C^{2,\alpha}(\bar{\Omega})$-estimate, independent of t; i.e.,

$$|v^t|_{C^{2,\alpha}(\bar{\Omega})} \le C_*, \text{ independent of } t.$$

Hence, the closedness of I follows from the compactness in $C^2(\bar{\Omega})$ of bounded sets in $C^{2,\alpha}(\bar{\Omega})$, a consequence of the Arzela-Ascoli theorem. Therefore, I is the whole unit interval. The function v^1 is then our desired solution of (4) corresponding to $t = 1$. □

In Theorem 2.6.1, we proved the existence of $C^{3,\alpha}$-solutions in $\bar{\Omega}$ if boundary values are $C^{3,\alpha}$ on $\partial\Omega$. By an approximation, we can conclude the existence of solutions under weaker conditions. In particular, when boundary values are only continuous, we have solutions which are continuous up to the boundary.

Theorem 2.6.2. *Let $\Omega \subset \mathbb{R}^n$ be a bounded domain in $\mathbb{R}^n$ with a C^3-boundary and $a_{ij} \in C^2(\bar{\Omega} \times \mathbb{R}^n)$ be uniformly elliptic and satisfy, for any $x \in \bar{\Omega}$ and $p \in \mathbb{R}^n$,*

$$|\nabla_x a_{ij}(x,p)| \le A_0, \quad |p||\nabla_p a_{ij}(x,p)| \le A_1,$$

for some positive constants A_0 and A_1. Then for any $f \in C(\bar\Omega) \cap C^1(\Omega)$ and $\varphi \in C(\bar\Omega)$, there exists a unique solution $u \in C(\bar\Omega) \cap C^2(\Omega)$ of

$$\begin{aligned} a_{ij}(x, \nabla u) &= f \quad \text{in } \Omega, \\ u &= \varphi \quad \text{on } \partial\Omega. \end{aligned} \tag{1}$$

Theorem 2.6.2 can be proved as a corollary of Theorem 2.6.1 by an approximation. Next, we employ a different approximation which does not require Theorem 2.6.1. In the following proof, we do not need boundary estimates of ∇u in Theorem 2.3.2 or global Hölder estimates of ∇u in Theorem 2.5.1. We can also relax the requirement on $\partial\Omega$ and assume instead that Ω satisfies a uniform exterior sphere condition.

Proof. Let $\{\eta_m\}$ be a sequence of $C_0^2(\Omega)$-functions satisfying $0 \le \eta_m \le 1$ and $\eta_m(x) = 1$ for any $x \in \Omega$ with $\text{dist}(x, \partial\Omega) \ge 1/m$. For each $m \ge 1$ and any $u \in C^2(\Omega)$, consider

$$Q_m(u) = \eta_m a_{ij}(\cdot, \nabla u) u_{ij} + (1 - \eta_m)\Delta u \quad \text{in } \Omega.$$

Then, Q_m is a uniformly elliptic operator. In fact, the uniform ellipticity holds since, for any $x \in \Omega$ and any $p, \xi \in \mathbb{R}^n$,

$$\min\{1, \lambda\}|\xi|^2 \le \big(\eta_m(x) a_{ij}(x, p) + (1 - \eta_m(x))\delta_{ij}\big)\xi_i\xi_j \le \max\{1, \Lambda\}|\xi|^2.$$

Fix an $\alpha \in (0,1)$ and take $f_m \in C^{1,\alpha}(\bar\Omega)$ and $\varphi_m \in C^{2,\alpha}(\bar\Omega)$ such that

$$\begin{aligned} f_m &\to f \quad \text{uniformly in } \bar\Omega \text{ as } m \to \infty, \\ \varphi_m &\to \varphi \quad \text{uniformly in } \bar\Omega \text{ as } m \to \infty, \end{aligned}$$

and, for any $\Omega' \Subset \Omega$,

$$f_m \to f \quad \text{in } C^1(\bar\Omega') \text{ as } m \to \infty.$$

Now, we consider the Dirichlet problem

$$\begin{aligned} \eta_m a_{ij}(x, \nabla u) u_{ij} + (1 - \eta_m)\Delta u &= f_m \quad \text{in } \Omega, \\ u &= \varphi_m \quad \text{on } \partial\Omega. \end{aligned} \tag{2}$$

We first prove the existence of a $C^{2,\alpha}$-solution of (2) by the method of continuity.

We fix an m and proceed to derive a $C^{2,\alpha}$-estimate for any $C^{2,\alpha}$-solution u_m of (2). By Theorem 2.3.1, we have

$$|u_m|_{L^\infty(\Omega)} \le |\varphi_m|_{L^\infty(\partial\Omega)} + C|f_m|_{L^\infty(\Omega)}, \tag{3}$$

where C is a positive constant depending only on n, λ, Λ, and Ω. It is obvious that the right-hand side of (3) can be made independent of m by the uniform convergence of f_m and φ_m. Moreover, by Theorem 1.1.12, we have, for any $x \in \Omega$ and $x_0 \in \partial\Omega$,

$$|u_m(x) - u_m(x_0)| \le \omega(|x - x_0|), \tag{4}$$

where ω is a nondecreasing continuous function on $(0, D)$, with $D = \operatorname{diam}(\Omega)$ and $\lim_{r\to 0}\omega(r) = 0$, depending only on n, λ, Λ, Ω, $|f_m|_{L^\infty(\Omega)}$, $|\varphi_m|_{L^\infty(\partial\Omega)}$, and the modulus of continuity of φ_m on $\partial\Omega$. Again, ω can be made independent of m. We point out that (3) and (4) are the only global estimates related to quasilinear equations which are needed in this proof.

Next, we consider $\Omega_1 \Subset \Omega_2 \Subset \Omega_3 \Subset \Omega$ such that $\operatorname{supp}\eta_m \subset \Omega_1$. Then in $\Omega \setminus \bar\Omega_1$, the equation in (2) reduces to

$$\Delta u_m = f_m \quad \text{in } \Omega \setminus \bar\Omega_1.$$

Hence, by the boundary Schauder estimates, we have

$$\begin{aligned} |u_m|_{C^{2,\alpha}(\bar\Omega\setminus\Omega_2)} &\le C\left\{|u_m|_{L^\infty(\Omega\setminus\Omega_1)} + |\varphi_m|_{C^{2,\alpha}(\bar\Omega\setminus\Omega_1)} + |f_m|_{C^\alpha(\bar\Omega\setminus\Omega_1)}\right\} \\ &\le C\left\{|u_m|_{L^\infty(\Omega)} + |\varphi_m|_{C^{2,\alpha}(\bar\Omega)} + |f_m|_{C^\alpha(\bar\Omega)}\right\}, \end{aligned} \tag{5}$$

where C is a positive constant depending only on n, α, λ, Λ, $\Omega \setminus \bar\Omega_1$, and $\operatorname{dist}(\Omega_1, \partial\Omega_2)$. We note that this constant C depends on m. By using Theorem 2.2.1, Theorem 2.4.1, and Theorem 2.4.3 successively, we obtain

$$|u_m|_{C^{2,\alpha}(\Omega_3)} \le C_{*m}, \tag{6}$$

where C_{*m} is a positive constant depending only on n, λ, Λ, A_0, A_1, $|u_m|_{L^\infty(\Omega)}$, $|f_m|_{C^1(\bar\Omega)}$, and $|\nabla_p a_{ij}|_{L^\infty(\Omega\times B_{K_m})}$, with K_m given by

$$C_m\left\{|u_m|_{L^\infty(\Omega)} + |f_m|_{C^1(\bar\Omega)}\right\},$$

for some positive constant C_m depending only on n, m, λ, Λ, A_0, A_1, Ω_3, and Ω. We point out that Theorem 2.4.1 yields an estimate of $[\nabla u]_{C^{\alpha'}(\Omega_3)}$ for some $\alpha' \in (0,1)$, possibly smaller than the given α. Then, Theorem 2.4.3 holds for such α' and hence implies an estimate of $[\nabla u]_{C^\alpha(\Omega_3)}$ by a simple interpolation. Another application of Theorem 2.4.3 yields (6). By combining (3), (5), and (6), we obtain

$$|u_m|_{C^{2,\alpha}(\bar\Omega)} \le C'_{*m},$$

where C'_{*m} is a positive constant depending only on $|\varphi_m|_{C^{2,\alpha}(\bar\Omega)}$ and C_{*m} in (6). In general, the constant C'_{*m} depends on m. For each m, we employ the method of continuity as in the proof of Theorem 2.6.1 to conclude the existence of a $C^{2,\alpha}(\bar\Omega)$-solution u_m of (2). We note that (4) is not used in this step.

Next, we discuss the limit of u_m as $m \to \infty$. For any fixed subdomain $\Omega' \Subset \Omega$, we take Ω'' such that $\Omega' \Subset \Omega'' \Subset \Omega$ and $\eta_m = 1$ in Ω'' for sufficiently large m. Then, for such m, we have

$$a_{ij}(x, \nabla u_m)\partial_{ij}u_m = f_m \quad \text{in } \Omega''.$$

By using Theorem 2.2.1, Theorem 2.4.1, and Theorem 2.4.3 successively, we obtain

$$|u_m|_{C^{2,\alpha}(\Omega')} \le C_*, \tag{7}$$

where C_* is a positive constant depending only on n, λ, Λ, A_0, A_1, Ω', Ω'', $|u_m|_{L^\infty(\Omega)}$, $|f_m|_{C^1(\bar\Omega'')}$, and $|\nabla_p a_{ij}|_{L^\infty(\Omega\times B_{K_m})}$, with K given by

$$C\left\{|u_m|_{L^\infty(\Omega)} + |f_m|_{C^1(\bar\Omega'')}\right\},$$

for some positive constant C depending only on n, λ, Λ, A_0, A_1, Ω', and Ω''. Note that C_* can be made independent of m. By (3), (4), (7), and the Arzela-Ascoli theorem, there exists a subsequence of $\{u_m\}$ convergent in C^2 in any subset $\Omega' \Subset \Omega$ to a function $u \in C^{2,\alpha}(\Omega)$, which also satisfies

$$|u(x) - \varphi(x_0)| \le \omega(|x - x_0|),$$

where ω is the function as in (4). Hence, u can be extended to a $C(\bar\Omega)$-function and satisfies (1), by passing the limit. □

Chapter 3

Mean Curvature Equations

In this chapter, we discuss the equation of the prescribed mean curvature, or the mean curvature equation. We derive various a priori estimates for its solutions and solve the Dirichlet boundary-value problems.

In Section 3.1, we introduce principal curvatures of hypersurfaces in Euclidean spaces and discuss distance functions to hypersurfaces.

In Section 3.2, we derive global estimates up to first derivatives of solutions of the mean curvature equation. We prove the global estimates of L^∞-norms by Alexandrov's maximum principle and the boundary gradient estimates by constructing appropriate barrier functions, respectively. The global gradient estimates are based on Bernstein's method. We point out that the mean curvature equation is not uniformly elliptic. The structure of the equation plays an important role in the derivation.

In Section 3.3, we derive interior gradient estimates of solutions of the mean curvature equation. Again, we make use of the structure of the equation essentially. As an application of the interior gradient estimates, we prove a Liouville type theorem for the minimal surface equation.

In Section 3.4, we discuss the Dirichlet boundary-value problems for the mean curvature equation. First, we solve the Dirichlet problem by the method of continuity under a suitable condition on the boundary mean curvature. Then, we present a nonsolvability result if such a condition is not satisfied.

3.1. Principal Curvatures

In this section, we introduce principal curvatures of hypersurfaces in Euclidean spaces and discuss how they are related to distance functions to hypersurfaces.

We first introduce hypersurfaces in $\mathbb{R}^{n+1}$. A subset Σ in $\mathbb{R}^{n+1}$ is a C^l-*hypersurface* if, for any $p_0 \in \Sigma$, there exist a domain $\Omega \subset \mathbb{R}^n$, a domain $U \subset \mathbb{R}^{n+1}$ containing p_0, and a C^l-immersion $\mathbf{r} : \Omega \to U$ such that $\Sigma \cap U = \mathbf{r}(\Omega)$. Recall that a mapping $\mathbf{r}$ is an *immersion* if the matrix $(\mathbf{r}_1, \ldots, \mathbf{r}_n)$ has a full rank. Here and hereafter, $\mathbf{r}_i = \mathbf{r}_{x_i}$ and $\mathbf{r}_{ij} = \mathbf{r}_{x_i x_j}$. The pair $(\Omega, \mathbf{r})$ is called a *local representation* of Σ. The *tangent space* $T_{p_0}\Sigma$ of Σ at p_0 is the subspace in $\mathbb{R}^{n+1}$ spanned by $\mathbf{r}_1, \ldots, \mathbf{r}_n$, which is n-dimensional due to the immersion of $\mathbf{r}$. A vector normal to $T_{p_0}\Sigma$ is called a *normal vector* of Σ at p_0.

Let Σ be a C^2-hypersurface in $\mathbb{R}^{n+1}$ and $(\Omega, \mathbf{r})$ be a local representation of Σ. We define, for $i, j = 1, \ldots, n$,

$$g_{ij} = \mathbf{r}_i \cdot \mathbf{r}_j.$$

The matrix (g_{ij}) is positive definite under the assumption that $\mathbf{r}$ is an immersion. Its associated bilinear form is called the *first fundamental form* of Σ. Next, let ν be a unit normal vector to Σ. Define

$$h_{ij} = \mathbf{r}_{ij} \cdot \nu.$$

Its associated bilinear form is called the *second fundamental form* of Σ. We usually write the first and second fundamental forms as

$$I = g_{ij} dx_i dx_j, \quad II = h_{ij} dx_i dx_j.$$

We call g_{ij} and h_{ij} the coefficients of the first and second fundamental forms associated with the local representation $(\Omega, \mathbf{r})$, respectively.

Definition 3.1.1. Let Σ be a C^2-hypersurface in $\mathbb{R}^{n+1}$, $(\Omega, \mathbf{r})$ be a local representation of Σ, and $p \in \Sigma$ be a point with $p = \mathbf{r}(x)$. A number κ is a *principal curvature* of Σ at p with respect to a unit normal vector ν if κ is an eigenvalue of $(h_{ij}(x))$ with respect to $(g_{ij}(x))$; namely,

$$\det \big(h_{ij}(x) - \kappa g_{ij}(x)\big) = 0,$$

where g_{ij} and h_{ij} are the coefficients of the first and second fundamental forms associated with the local representation $(\Omega, \mathbf{r})$, respectively.

We point out that principal curvatures depend on the choice of normal vectors.

We now prove a simple result.

Lemma 3.1.2. *Principal curvatures are well-defined real numbers.*

Proof. Let Σ be a C^2-hypersurface in $\mathbb{R}^{n+1}$ and $p \in \Sigma$ be a fixed point. Assume that $(\Omega, \mathbf{r})$ and $(\widetilde{\Omega}, \widetilde{\mathbf{r}})$ are two local representations of Σ with $p = \mathbf{r}(x_0) = \widetilde{\mathbf{r}}(y_0)$, for some $x_0 \in \Omega$ and $y_0 \in \widetilde{\Omega}$. By shrinking Ω and $\widetilde{\Omega}$ appropriately, there is a C^2-diffeomorphism $x : \widetilde{\Omega} \to \Omega$ with $x(y_0) = x_0$ and $\widetilde{\mathbf{r}}(y) = \mathbf{r}(x(y))$. Then,

$$\begin{aligned}\widetilde{\mathbf{r}}_{y_i} &= \mathbf{r}_{x_k} \partial_{y_i} x_k,\\ \widetilde{\mathbf{r}}_{y_i y_j} &= \mathbf{r}_{x_k x_l} \partial_{y_i} x_k \partial_{y_j} x_l + \mathbf{r}_{x_k} \partial_{y_i y_j} x_k.\end{aligned}$$

Let ν be a unit normal vector and g_{ij}, h_{ij} and $\widetilde{g}_{ij}$, $\widetilde{h}_{ij}$ be the coefficients of the first and second fundamental forms associated with $(\Omega, \mathbf{r})$ and $(\widetilde{\Omega}, \widetilde{\mathbf{r}})$, respectively. Then,

$$\widetilde{g}_{ij} = \widetilde{\mathbf{r}}_{y_i} \cdot \widetilde{\mathbf{r}}_{y_j} = \mathbf{r}_{x_k} \cdot \mathbf{r}_{x_l} \partial_{y_i} x_k \partial_{y_j} x_l = g_{kl} \partial_{y_i} x_k \partial_{y_j} x_l,$$

and

$$\begin{aligned}\widetilde{h}_{ij} &= \widetilde{\mathbf{r}}_{y_i y_j} \cdot \nu = \left(\mathbf{r}_{x_k x_l} \partial_{y_i} x_k \partial_{y_j} x_l + \mathbf{r}_{x_k} \partial_{y_i y_j} x_k\right) \cdot \nu\\ &= \mathbf{r}_{x_k x_l} \cdot \nu \partial_{y_i} x_k \partial_{y_j} x_l = h_{kl} \partial_{y_i} x_k \partial_{y_j} x_l,\end{aligned}$$

where we used $\mathbf{r}_{x_k} \cdot \nu = 0$. Hence, it is easy to see that

$$\det\left((h_{ij} - \kappa g_{ij})\right) = 0$$

if and only if

$$\det\left((\widetilde{h}_{ij} - \kappa \widetilde{g}_{ij})\right) = 0.$$

Therefore, κ is independent of representations of Σ. Since both (g_{ij}) and (h_{ij}) are symmetric matrices, any such κ has to be real. □

Next, we introduce an important combination of principal curvatures.

Definition 3.1.3. Let Σ be a C^2-hypersurface in $\mathbb{R}^{n+1}$, with principal curvatures $\kappa_1, \ldots, \kappa_n$ corresponding to a unit normal vector ν. The *mean curvature* H of Σ corresponding to ν is the sum of the principal curvatures corresponding to ν; i.e.,

$$H = \sum_{i=1}^{n} \kappa_i.$$

There are other versions of curvatures. For example, the *Gauss curvature* K is the product of principal curvatures; i.e.,

$$K = \prod_{i=1}^{n} \kappa_i.$$

In this chapter, we discuss mean curvatures only.

We now express the mean curvature in terms of the first and the second fundamental forms.

Proposition 3.1.4. *Let Σ be a C^2-hypersurface in $\mathbb{R}^{n+1}$ and ν be its unit normal vector. Suppose that g_{ij} and h_{ij} are coefficients of the first fundamental form of Σ and the second fundamental form of Σ corresponding to ν, respectively. Then, the mean curvature H of Σ corresponding to ν is given by*

$$H = g^{ij} h_{ij},$$

where (g^{ij}) is the inverse matrix of (g_{ij}).

Proof. Set $G = (g_{ij})$ and $A = (h_{ij})$. Then, G is invertible. Let κ be a principal curvature of Σ. Then, $\det(A - \kappa G) = 0$, or equivalently

$$\det(AG^{-1} - \kappa I) = 0.$$

In other words, the principal curvature is an eigenvalue of AG^{-1}. The mean curvature H, being the sum of all principal curvatures, is given by

$$H = \operatorname{tr}(AG^{-1}).$$

This implies the desired result. □

Next, we calculate the mean curvature of graphs in $\mathbb{R}^{n+1}$.

Lemma 3.1.5. *Let Ω be a domain in $\mathbb{R}^n$ and Σ be a graph in $\mathbb{R}^{n+1}$ given by $x_{n+1} = u(x)$ for some $u \in C^2(\Omega)$. Then, the mean curvature $H(x)$ of Σ at $(x, u(x))$ corresponding to the upward unit normal vector of Σ is given by*

$$H(x) = \frac{1}{\sqrt{1 + |\nabla u(x)|^2}} \left(\Delta u(x) - \frac{u_i(x) u_j(x)}{1 + |\nabla u(x)|^2} u_{ij}(x) \right).$$

Proof. We set

$$\mathbf{r}(x) = (x, u(x)).$$

Then, for $i, j = 1, \ldots, n$,

$$\mathbf{r}_i = (\mathbf{e}_i, u_i), \quad \mathbf{r}_{ij} = (0, u_{ij}),$$

where $\mathbf{e}_i$ is the unit vector along the x_i-axis. Now, we take

$$\nu = \frac{1}{\sqrt{1 + |\nabla u|^2}} (-\nabla u, 1).$$

Then, ν is the upward unit normal vector of Σ, and

$$g_{ij} = \mathbf{r}_i \cdot \mathbf{r}_j = \delta_{ij} + u_i u_j,$$
$$h_{ij} = \mathbf{r}_{ij} \cdot \nu = \frac{u_{ij}}{\sqrt{1 + |\nabla u|^2}}.$$

A straightforward calculation yields that the component g^{ij} of the inverse matrix $(g_{ij})^{-1}$ is given by

$$g^{ij} = \delta_{ij} - \frac{u_i u_j}{1 + |\nabla u|^2}.$$

By Proposition 3.1.4, we have

$$H = g^{ij}h_{ij} = \frac{1}{\sqrt{1+|\nabla u|^2}}\left(\Delta u - \frac{u_i u_j}{1+|\nabla u|^2}u_{ij}\right).$$

This is the desired identity. □

Remark 3.1.6. The mean curvature can also be expressed by

$$H(x) = \operatorname{div}\left(\frac{\nabla u(x)}{\sqrt{1+|\nabla u(x)|^2}}\right).$$

Remark 3.1.7. We now examine a special case in the proof of Lemma 3.1.5. Let x_0 be a point in Ω. If $\nabla u(x_0) = 0$, then at $(x_0, u(x_0))$, $\nu = (0, 1)$ and

$$g_{ij} = \delta_{ij}, \quad h_{ij} = u_{ij}.$$

As a consequence, the principal curvatures of Σ at $(x_0, u(x_0))$ is simply the eigenvalues of the Hessian matrix $\nabla^2 u(x_0)$.

Let Ω be a domain in $\mathbb{R}^n$ and $u \in C^2(\Omega)$. Suppose that the graph of u in $\mathbb{R}^{n+1}$ has a mean curvature $H(x)$ at the point $\big(x, u(x)\big)$, for $x \in \Omega$. Here, the mean curvature is calculated corresponding to the upward unit normal vector. Then, u satisfies

$$\Delta u - \frac{u_i u_j}{1+|\nabla u|^2}u_{ij} = H(x)\sqrt{1+|\nabla u|^2} \quad \text{in } \Omega.$$

This is the *equation of prescribed mean curvature*, or, simply, the *mean curvature equation*. If $H \equiv 0$, the mean curvature equation has the form

$$\Delta u - \frac{u_i u_j}{1+|\nabla u|^2}u_{ij} = 0 \quad \text{in } \Omega.$$

This is referred to as the *minimal surface equation*. Its solutions are the critical points of the area functional

$$\mathcal{A}(u) = \int_\Omega \sqrt{1+|\nabla u|^2}\, dx.$$

Their graphs are called *minimal surfaces*.

We now write the mean curvature equation as

$$a_{ij}(\nabla u)u_{ij} = H(x)\sqrt{1+|\nabla u|^2} \quad \text{in } \Omega,$$

where, for any $p \in \mathbb{R}^n$,

$$a_{ij}(p) = \delta_{ij} - \frac{p_i p_j}{1+|p|^2}.$$

It is easy to check that, for any $p \in \mathbb{R}^n$ and any $\xi \in \mathbb{R}^n$,

$$\frac{1}{1+|p|^2}|\xi|^2 \le a_{ij}(p)\xi_i\xi_j \le |\xi|^2.$$

For any $p \in \mathbb{R}^n \setminus \{0\}$, the left equality is attained if $\xi = p$ and the right equality is attained if $\xi \perp p$. The difficulty in discussions of the mean curvature equation arises from the lack of the uniform ellipticity. Although elliptic, the mean curvature equation has its ellipticity constants determined by the gradients of solutions. These ellipticity constants are controlled only after the C^1-norms of solutions are derived.

In the rest of this section, we discuss relations between the distance to a hypersurface and the mean curvature of this surface.

Let Ω be a bounded domain in $\mathbb{R}^n$. The *distance function* d to $\partial\Omega$ is defined by

$$d(x) = \text{dist}(x, \partial\Omega) \quad \text{for any } x \in \mathbb{R}^n.$$

It is easy to check that d is a Lipschitz function. In fact, for any $x, y \in \mathbb{R}^n$, we take $z \in \partial\Omega$ such that $d(y) = |y - z|$. Then,

$$d(x) \leq |x - z| \leq d(y) + |x - y|.$$

By interchanging x and y, we obtain

$$|d(x) - d(y)| \leq |x - y|.$$

In general, distance functions are not smooth globally. For example, the distance function to a sphere is not C^1 at the center. In the next result, we will relate the regularity of the distance function to that of the boundary $\partial\Omega$, at least in a region close to the boundary.

In the following, we set, for some positive constant $\mu > 0$,

$$\Omega_\mu = \{x \in \Omega : d(x) < \mu\},$$

where d is the distance function to $\partial\Omega$.

Lemma 3.1.8. *Let Ω be a bounded domain with a C^k-boundary, for some $k \geq 2$. Then, there exists a positive constant μ, depending on Ω, such that $d \in C^k(\Omega_\mu)$.*

Proof. The assumption of the C^2-regularity of the boundary implies that Ω satisfies a *uniform interior sphere* condition; namely, at each $y_0 \in \partial\Omega$, there exists a ball B such that $\bar{B} \cap (\mathbb{R}^n \setminus \Omega) = \{y_0\}$ and the radius of the ball B is bounded from below by a positive constant independent of y_0, which we take to be μ. For each $x \in \Omega_\mu$, there exists a unique $y = y(x) \in \partial\Omega$ such that $|x - y| = d(x)$. In fact, we have

$$x = y + \nu(y)d, \tag{1}$$

where $\nu(y)$ is the inner unit normal vector at y.

The relation (1) determines y and d as functions of x. To discuss the regularity of these functions, we fix an $x_0 \in \Omega_\mu$ and assume $y_0 = y(x_0)$ is the origin such that $\mathbf{e_n}$ is the inner unit normal vector of $\partial\Omega$ at y_0. Then, $\mathbb{R}^{n-1} \times \{0\}$ is the tangent plane of $\partial\Omega$ at y_0. Locally, we express $\partial\Omega$ as a function $y_n = \rho(y')$ for y' in a ball $B'_r \subset \mathbb{R}^{n-1}$. Then, $\nabla\rho(0) = 0$. By an appropriate rotation in $\mathbb{R}^{n-1}$, we assume $\nabla^2\rho(0)$ is diagonal. This implies

$$\nabla^2\rho(0) = \operatorname{diag}(\kappa_1, \ldots, \kappa_{n-1}),$$

where $\kappa_1, \ldots, \kappa_{n-1}$ are principal curvatures of $\partial\Omega$ at y_0. Since μ is the radius of an interior tangent ball, it is straightforward to check that, for $i = 1, \ldots, n-1$,

$$\kappa_i \leq \frac{1}{\mu}.$$

Next, for each $y' \in B'_r$, we view the inner unit normal vector $\nu(y)$ at $y = \big(y', \rho(y')\big)$ as $\bar\nu(y')$, a function of y'. Then, for $i = 1, \ldots, n-1$,

$$\bar\nu_i(y') = -\frac{\rho_i(y')}{\sqrt{1 + |\nabla\rho(y')|^2}},$$

and

$$\bar\nu_n(y') = \frac{1}{\sqrt{1 + |\nabla\rho(y')|^2}}.$$

Hence, for any $i, j = 1, \ldots, n-1$,

$$\partial_j\bar\nu_i(0) = -\kappa_i\delta_{ij}. \tag{2}$$

We now view the relation (1) as a map $x = x(y', d) : B'_r \times (0, \mu) \to \mathbb{R}^n$ for $y = \big(y', \rho(y')\big)$. Then, $x \in C^{k-1}\big(B'_r \times (0, \mu)\big)$ and its Jacobian matrix is given by

$$\frac{\partial x}{\partial(y', d)} = \begin{pmatrix} \delta_{ij} + \partial_{y_i}\nu_j d & \partial_{y_i}\rho + \partial_{y_i}\nu_n d \\ \nu_j & \nu_n \end{pmatrix}. \tag{3}$$

In particular,

$$\left.\frac{\partial x}{\partial(y', d)}\right|_{(0,d)} = \operatorname{diag}(1 - \kappa_1 d, \ldots, 1 - \kappa_{n-1}d, 1). \tag{4}$$

Hence, the Jacobian of x at $\big(0, d(x_0)\big)$ is given by

$$\det\left(\left.\frac{\partial x}{\partial(y', d)}\right|_{(0,d(x_0))}\right) = \big(1 - \kappa_1 d(x_0)\big)\cdots\big(1 - \kappa_{n-1}d(x_0)\big) > 0$$

since $d(x_0) < \mu$. By the inverse mapping theorem, the mapping y' is C^{k-1} in $B_s(x_0)$ for some $s > 0$. Note that $|\nu|^2 = 1$ and, for $i = 1, \ldots, n-1$,

$$\sum_{j=1}^{n-1}(\delta_{ij} + \partial_{y_i}\nu_j d)\nu_j + (\partial_{y_i}\rho + \partial_{y_i}\nu_n d)\nu_n = 0.$$

Hence, (3) implies

$$\nabla d(x) = \nu\big(y(x)\big) = \bar\nu\big(y'(x)\big). \tag{5}$$

Hence, ∇d is C^{k-1} in $B_s(x_0)$. Then, $d \in C^k(B_s(x_0))$, and thus $d \in C^k(\Omega_\mu)$. □

We now calculate derivatives of the distance functions and relate the distance function to the principal curvatures of the boundary.

Consider a bounded domain Ω in $\mathbb{R}^n$ with a C^2-boundary, which can be viewed as a C^2-hypersurface in $\mathbb{R}^n$. Then, we have the well-defined principal curvatures, and in particular the *mean curvature* $H_{\partial\Omega}$, of $\partial\Omega$ with respect to the inner unit normal vector.

Lemma 3.1.9. *Let Ω be a bounded domain with a C^2-boundary and μ be the positive constant as in Lemma 3.1.8. Then, for any $x \in \Omega_\mu$,*

$$\nabla d(x) = \nu(y),$$

and the eigenvalues of $\nabla^2 u(x)$ are given by

$$-\frac{\kappa_1(y)}{1-\kappa_1(y)d(x)}, \ldots, -\frac{\kappa_{n-1}(y)}{1-\kappa_{n-1}(y)d(x)}, 0,$$

where $y \in \partial\Omega$ is the unique point such that $d(x) = |x-y|$, ν is the inner unit normal vector to $\partial\Omega$, and $\kappa_1, \ldots, \kappa_{n-1}$ are the principal curvatures of $\partial\Omega$ with respect to ν. In particular,

$$\Delta d(x) \le -H_{\partial\Omega}(y),$$

where $H_{\partial\Omega}$ is the mean curvature of $\partial\Omega$ with respect to ν.

Proof. First, (5) in the proof of Lemma 3.1.8 is the identity concerning the gradient of d. Next, we adopt the setting in the proof of Lemma 3.1.8 to calculate the Hessian matrix $\nabla^2 d$.

Assume $\partial\Omega$ in a neighborhood of $0 \in \partial\Omega$ is given by a C^2-function $x_n = \rho(x')$ such that $\rho(0) = 0$, $\nabla\rho(0) = 0$, and $\nabla^2\rho(0)$ is diagonal. Then,

$$\nabla^2\rho(0) = (\kappa_1, \ldots, \kappa_{n-1}),$$

where $\kappa_1, \ldots, \kappa_{n-1}$ are the principal curvatures of $\partial\Omega$ at 0. Then, for any x along the x_n-axis with $d(x) < \mu$,

$$\nabla d(x) = (0, \ldots, 0, 1).$$

In the following, we fix such an x. Hence, $d_{in}(x) = 0$ for $i = 1, \ldots, n$. Next, for any $i, j = 1, \ldots, n-1$, we have

$$d_{ij}(x) = \partial_{x_j}\nu_i(0) = \partial_{y_k}\bar{\nu}_i(0)\partial_{x_j}y_k(x) = \frac{-\kappa_i}{1-\kappa_i d(x)}\delta_{ij},$$

by (2) and (4) in the proof of Lemma 3.1.8. Therefore, the eigenvalues of $\nabla^2 u(x)$ are given by

$$-\frac{\kappa_1}{1-\kappa_1 d(x)}, \ldots, -\frac{\kappa_{n-1}}{1-\kappa_{n-1}d(x)}, 0.$$

In particular,

$$\Delta d(x) = -\sum_{i=1}^{n-1} \frac{\kappa_i}{1-\kappa_i d(x)} \le -\sum_{i=1}^{n-1} \kappa_i.$$

The expression in the right-hand side is the negative mean curvature $-H_{\partial\Omega}$ of $\partial\Omega$ at 0. □

If $H_{\partial\Omega} \ge 0$ on $\partial\Omega$, then $\Delta d \le 0$ in Ω_μ. Set

$$M(u) = (1+|\nabla u|^2)\Delta u - u_i u_j u_{ij}.$$

This is called the *minimal surface operator*. Since $|\nabla d| = 1$, then $d_i d_{ij} = 0$, and hence

$$M(d) = 2\Delta d \le 0.$$

In other words, d is a supersolution of the minimal surface operator. This fact plays an important role in the derivation of the boundary gradient estimates for the mean curvature equation in the next section.

3.2. Global Estimates

Let Ω be a domain in $\mathbb{R}^n$. The mean curvature equation has the form

$$\Delta u - \frac{u_i u_j}{1+|\nabla u|^2} u_{ij} = H\sqrt{1+|\nabla u|^2} \quad \text{in } \Omega,$$

where H is a given function in Ω. This means that the hypersurface given by $x_{n+1} = u(x)$ in $\mathbb{R}^{n+1}$ has its mean curvature given by H. In this section, we derive estimates of solutions and their first-order derivatives.

We first derive a sup-norm estimate for solutions, due to Bakelman [**6**].

Theorem 3.2.1. *Let Ω be a bounded domain in $\mathbb{R}^n$ and $H \in C(\Omega)$ satisfy*

$$\|H\|_{L^n(\Omega)} < n\left(\int_{\mathbb{R}^n} (1+|p|^2)^{-\frac{n+2}{2}} dp\right)^{\frac{1}{n}}.$$

Suppose that u is a $C(\bar{\Omega}) \cap C^2(\Omega)$-solution of

$$\Delta u - \frac{u_i u_j}{1+|\nabla u|^2} u_{ij} = H\sqrt{1+|\nabla u|^2} \quad \text{in } \Omega.$$

Then,

$$\sup_{\Omega} |u| \le \max_{\partial\Omega} |u| + C\,\mathrm{diam}(\Omega),$$

where C is a positive constant depending only on n and $\|H\|_{L^n(\Omega)}$.

Proof. We set, for any $x \in \Omega$ and any $p \in \mathbb{R}^n$,

$$a_{ij}(p) = (1+|p|^2)\delta_{ij} - p_i p_j$$

and

$$f(x,p) = H(x)(1+|p|^2)^{\frac{3}{2}}.$$

Then, we write the mean curvature equation as

$$a_{ij}(\nabla u)u_{ij} = f(x, \nabla u) \quad \text{in } \Omega.$$

We first consider the special case $H \equiv 0$, in which the equation has the form

$$a_{ij}(\nabla u)u_{ij} = 0.$$

The maximum principle implies

$$\sup_{\Omega} |u| \leq \max_{\partial\Omega} |u|.$$

This is the desired result.

Next, we discuss the general case. A simple calculation yields

$$D \equiv \det\big(a_{ij}(p)\big) = (1 + |p|^2)^{n-1}$$

and

$$D^* \equiv \sqrt[n]{D} = (1 + |p|^2)^{\frac{n-1}{n}}.$$

We then have

$$\frac{|f(x,p)|}{nD^*} = \frac{|H(x)|(1 + |p|^2)^{\frac{3}{2}}}{n(1 + |p|^2)^{\frac{n-1}{n}}} = \frac{1}{n}|H(x)|(1 + |p|^2)^{\frac{n+2}{2n}}.$$

Now we set

$$h(p) = (1 + |p|^2)^{-\frac{n+2}{2n}}.$$

Then,

$$\int_{\mathbb{R}^n} h^n(p)dp = \int_{\mathbb{R}^n} (1 + |p|^2)^{-\frac{n+2}{2}} dp < \infty.$$

Suppose Γ^+ is the upper contact set of u, and set

$$M = \sup_{\Omega} u - \max_{\partial\Omega} u^+$$

and $d = \operatorname{diam}(\Omega)$. Note that $\nabla^2 u$ is nonpositive in Γ^+. Then, $-a_{ij}u_{ij} \geq 0$ in Γ^+, and hence $-f(x, \nabla u) \geq 0$ in Γ^+. Therefore,

$$\frac{f^-(x, \nabla u)}{nD^*} = \frac{|H(x)|}{nh(\nabla u)} \quad \text{in } \Gamma^+ \cap \Omega^+.$$

By applying Lemma 1.2.4 to h^n, we obtain

$$\begin{aligned}\int_{B_{M/d}} h^n dp &\leq \int_{\Gamma^+\cap\Omega^+} h^n(\nabla u)\left(\frac{f^-}{nD^*}\right)^n dx \\ &= \int_{\Gamma^+\cap\Omega^+} \frac{|H(x)|^n}{n^n} dx \leq \int_{\Omega} \frac{|H(x)|^n}{n^n} dx < \int_{\mathbb{R}^n} h^n dp.\end{aligned}$$

Therefore, $M/d \leq C$ for some positive constant C, depending only on n and $\|H\|_{L^n(\Omega)}$. This implies

$$\sup_{\Omega} u \leq \max_{\partial\Omega} u^+ + Cd.$$

A similar argument yields the lower bound of u. We then have the desired result. $\square$

We now derive a boundary gradient estimate, due to Serrin [**134**].

Theorem 3.2.2. *Let Ω be a bounded domain in $\mathbb{R}^n$ with a C^2-boundary and $H \in C^1(\bar{\Omega})$ satisfy, for any $y \in \partial\Omega$,*

$$|H(y)| \leq H_{\partial\Omega}(y),$$

where $H_{\partial\Omega}$ is the mean curvature of $\partial\Omega$ corresponding to the inner unit normal vector to $\partial\Omega$. Suppose that u is a $C^1(\bar{\Omega}) \cap C^2(\Omega)$-solution of

$$\Delta u - \frac{u_i u_j}{1+|\nabla u|^2} u_{ij} = H\sqrt{1+|\nabla u|^2} \quad \text{in } \Omega,$$
$$u = \varphi \quad \text{on } \partial\Omega,$$

for some $\varphi \in C^2(\bar{\Omega})$. Then,

$$\max_{\partial\Omega} |\nabla u| \leq \max_{\partial\Omega} |\nabla\varphi| + \exp\left\{C\big(1+|\nabla\varphi|^3_{C^1(\bar{\Omega})}\big)\big(1+|H|_{C^1(\bar{\Omega})}\big)\sup_{\Omega}|u|\right\},$$

where C is a positive constant depending only on n and Ω.

Proof. We set, for any $v \in C^2(\Omega)$,

$$Q(v) = (1+|\nabla v|^2)\Delta v - v_i v_j v_{ij} - H(1+|\nabla v|^2)^{\frac{3}{2}}.$$

Hence, $Q(u) = 0$ in Ω.

Let d be the distance function to $\partial\Omega$. Then d is C^2 in $\Omega_\mu = \{x \in \Omega : d(x) < \mu\}$, for some positive constant μ. Let ψ be a strictly increasing function defined in $[0, \mu)$, with

$$\psi(0) = 0. \tag{1}$$

Set

$$w = \psi(d) \quad \text{in } \Omega_\mu.$$

We now calculate $Q(\varphi + w)$. First,

$$w_i = \psi' d_i, \quad w_{ij} = \psi'' d_i d_j + \psi' d_{ij}.$$

Then, $|\nabla w| = \psi'$ and

$$\begin{aligned} Q(\varphi + w) = {} & (1+|\nabla\varphi + \psi'\nabla d|^2)(\Delta\varphi + \psi'' + \psi'\Delta d) \\ & - (\varphi_i + \psi' d_i)(\varphi_j + \psi' d_j)(\varphi_{ij} + \psi'' d_i d_j + \psi' d_{ij}) \\ & - H(1+|\nabla\varphi + \psi'\nabla d|^2)^{\frac{3}{2}}. \end{aligned}$$

We write

$$Q(\varphi + w) = I + II + III,$$

where

$$\begin{aligned} I &= \psi''(1 + |\nabla\varphi + \psi'\nabla d|^2) - \psi''(\varphi_i + \psi' d_i)(\varphi_j + \psi' d_j) d_i d_j, \\ II &= (1 + |\nabla\varphi + \psi'\nabla d|^2)(\Delta\varphi + \psi'\Delta d) \\ &\quad - (\varphi_i + \psi' d_i)(\varphi_j + \psi' d_j)(\varphi_{ij} + \psi' d_{ij}), \\ III &= -H(1 + |\nabla\varphi + \psi'\nabla d|^2)^{\frac{3}{2}}. \end{aligned}$$

We now estimate I, II, and III. First, we write I as

$$I = \psi''\left(1 + |\nabla\varphi + \psi'\nabla d|^2 - \left[(\nabla\varphi + \psi'\nabla d)\cdot\nabla d\right]^2\right).$$

The Cauchy inequality implies

$$\left[(\nabla\varphi + \psi'\nabla d)\cdot\nabla d\right]^2 \leq |\nabla\varphi + \psi'\nabla d|^2$$

since $|\nabla d| = 1$. By requiring

$$\psi'' < 0, \tag{2}$$

we obtain

$$I \leq \psi''.$$

For II, we have

$$\begin{aligned} II &= (1 + |\nabla\varphi|^2 + 2\psi'\nabla d\cdot\nabla\varphi + \psi'^2)(\Delta\varphi + \psi'\Delta d) \\ &\quad - (\varphi_i\varphi_j + 2\psi'\varphi_i d_j + \psi'^2 d_i d_j)(\varphi_{ij} + \psi' d_{ij}). \end{aligned}$$

Since $d_i d_{ij} = 0$, then

$$\begin{aligned} II &= (1 + \psi'^2)\psi'\Delta d + (\Delta\varphi + 2\nabla d\cdot\nabla\varphi\Delta d - d_i d_j\varphi_{ij})\psi'^2 \\ &\quad + (|\nabla\varphi|^2\Delta d - \varphi_i\varphi_j d_{ij} + 2\nabla d\cdot\nabla\varphi\Delta\varphi - 2d_i\varphi_j\varphi_{ij})\psi' \\ &\quad + (\Delta\varphi + |\nabla\varphi|^2\Delta\varphi - \varphi_i\varphi_j\varphi_{ij}). \end{aligned}$$

In the following, we rewrite

$$II + III = \widetilde{II} + \widetilde{III},$$

where

$$\begin{aligned} \widetilde{II} &= (\Delta\varphi + 2\nabla d\cdot\nabla\varphi\Delta d - d_i d_j\varphi_{ij})\psi'^2 \\ &\quad + (|\nabla\varphi|^2\Delta d - \varphi_i\varphi_j d_{ij} + 2\nabla d\cdot\nabla\varphi\Delta\varphi - 2d_i\varphi_j\varphi_{ij})\psi' \\ &\quad + (\Delta\varphi + |\nabla\varphi|^2\Delta\varphi - \varphi_i\varphi_j\varphi_{ij}) \end{aligned}$$

and

$$\widetilde{III} = (1 + \psi'^2)\psi'\Delta d - H(1 + |\nabla\varphi + \psi'\nabla d|^2)^{\frac{3}{2}}.$$

It is easy to check that $\widetilde{II}$ satisfies

$$\widetilde{II} \leq c_0(|\nabla\varphi| + |\nabla^2\varphi|)\psi'^2 + c_1|\nabla\varphi|(|\nabla\varphi| + |\nabla^2\varphi|)\psi' + c_2(1 + |\nabla\varphi|^2)|\nabla^2\varphi|,$$

where c_0, c_1, and c_2 are positive constants depending only on the C^2-norm of d in Ω_μ. Next, for any $x \in \Omega_\mu$, let $y \in \partial\Omega$ be the unique point such that $d(x) = |x - y|$. Lemma 3.1.9 implies

$$\Delta d(x) \leq -H_{\partial\Omega}(y).$$

By the assumption on H, we have

$$\Delta d(x) \leq H(y). \tag{3}$$

Then,

$$\Delta d(x) \leq (H(y) - H(x)) + H(x) \leq d|\nabla H|_{L^\infty} + H(x),$$

and hence

$$\widetilde{III} \leq (1 + \psi'^2)d\psi'|\nabla H|_{L^\infty} + H\psi'(1 + \psi'^2) - H(1 + |\nabla\varphi + \psi'\nabla d|^2)^{\frac{3}{2}}.$$

For the last two terms, we note that

$$\begin{aligned}
&\psi'(1 + \psi'^2) - (1 + |\nabla\varphi + \psi'\nabla d|^2)^{\frac{3}{2}} \\
&\quad = (1 + \psi'^2)\left(\psi' - \sqrt{1 + |\nabla\varphi + \psi'\nabla d|^2}\right) \\
&\qquad + \left((1 + \psi'^2) - (1 + |\nabla\varphi + \psi'\nabla d|^2)\right)\sqrt{1 + |\nabla\varphi + \psi'\nabla d|^2} \\
&\quad = -\frac{1 + |\nabla\varphi|^2 + 2\psi'\nabla d \cdot \nabla\varphi}{\psi' + \sqrt{1 + |\nabla\varphi + \psi'\nabla d|^2}} \cdot (1 + \psi'^2) \\
&\qquad - (|\nabla\varphi|^2 + 2\psi'\nabla d \cdot \nabla\varphi)\sqrt{1 + |\nabla\varphi + \psi'\nabla d|^2}.
\end{aligned}$$

It is then straightforward to check that

$$|\psi'(1 + \psi'^2) - (1 + |\nabla\varphi + \psi'\nabla d|^2)^{\frac{3}{2}}| \leq (1 + |\nabla\varphi|^3)(c_0\psi'^2 + c_1\psi' + c_2),$$

by adjusting c_0, c_1, and c_2. Therefore,

$$\widetilde{III} \leq (1 + \psi'^2)d\psi'|\nabla H|_{L^\infty} + |H|_{L^\infty}(1 + |\nabla\varphi|^3)(c_0\psi'^2 + c_1\psi' + c_2).$$

Next, we require

$$\psi' \geq 1 \quad \text{and} \quad d\psi' \leq 1. \tag{4}$$

By combining estimates of I, $\widetilde{II}$, and $\widetilde{III}$, we obtain

$$Q(\varphi + w) \leq \psi'' + L\psi'^2,$$

where

$$L = C\left\{1 + |\nabla H|_{L^\infty} + (1 + |\nabla\varphi|^3_{L^\infty})|H|_{L^\infty} + (1 + |\nabla\varphi|^2_{L^\infty})|\nabla\varphi|_{C^1}\right\}$$

and C is a positive constant depending only on the C^2-norm of d. All norms in L are evaluated in Ω_μ. We point out that an extra 1 (the first term in the parenthesis) is inserted in the expression of L for a later purpose. In addition, we require

$$\psi'' + L\psi'^2 \le 0. \tag{5}$$

By examining the arguments above, we also have

$$Q(\varphi - w) \ge -(\psi'' + L\psi'^2).$$

In getting this, we employ instead of (3)

$$\Delta d(x) \le -H(y).$$

Now, we collect requirements (1), (2), (4), and (5) on ψ. In summary, we need to find a constant $d_0 \in (0, \mu]$ and a function ψ such that

$$\begin{aligned}
&\psi'' + L\psi'^2 \le 0 \quad \text{on } (0, d_0),\\
&\psi'' < 0,\ \psi' > 0,\ d\psi' \le 1 \quad \text{on } (0, d_0),\\
&\psi(0) = 0,\ \psi(d_0) \ge M,\ \psi'(d_0) \ge 1,
\end{aligned}$$

where $M = \sup_\Omega |u - \varphi|$. To this end, we first solve

$$\psi'' + L\psi'^2 = 0.$$

With $\psi'(0) = A/L$ for some positive constant A, we have

$$\psi'(d) = \frac{A}{L(1 + Ad)}.$$

Next, with $\psi(0) = 0$, we obtain

$$\psi(d) = \frac{1}{L}\log(1 + Ad).$$

Obviously, $\psi' > 0$, $\psi'' < 0$, and $d\psi' \le 1$ if $L \ge 1$. Note that $\psi(d_0) = M$ is equivalent to $\log(1 + Ad_0) = LM$, or

$$1 + Ad_0 = \exp\{LM\}.$$

Then, $\psi'(d_0) = 1$ is reduced to $A = L(1 + Ad_0)$, or

$$A = L\exp\{LM\}.$$

We take A as above and then

$$d_0 = \frac{\exp\{LM\} - 1}{L\exp\{LM\}}.$$

Hence,

$$d_0 \le \frac{1}{L} \le \frac{1}{C} \le \mu,$$

by taking C sufficiently large. Such a ψ satisfies all the requirements we imposed.

Now we set

$$\Omega_0 = \{x \in \Omega : d(x) < d_0\}.$$

We have constructed a function w in $\bar{\Omega}_0$ such that

$$\begin{aligned} Q(\varphi - w) \geq 0 \geq Q(\varphi + w) \quad &\text{in } \Omega_0, \\ w = 0 \quad &\text{on } \partial\Omega_0 \cap \partial\Omega, \\ w \geq \sup_{\Omega} |u - \varphi| \quad &\text{on } \partial\Omega_0 \cap \Omega. \end{aligned}$$

Therefore,

$$\begin{aligned} Q(\varphi - w) \geq Q(u) \geq Q(\varphi + w) \quad &\text{in } \Omega_0, \\ \varphi - w \leq u \leq \varphi + w \quad &\text{on } \partial\Omega_0. \end{aligned}$$

By the maximum principle, we obtain

$$\varphi - w \leq u \leq \varphi + w \quad \text{in } \Omega_0.$$

Since $\varphi - w = u = \varphi + w$ on $\partial\Omega$, we can take normal derivatives on $\partial\Omega$ and get

$$\frac{\partial \varphi}{\partial \nu} - \frac{\partial w}{\partial \nu} \leq \frac{\partial u}{\partial \nu} \leq \frac{\partial \varphi}{\partial \nu} + \frac{\partial w}{\partial \nu} \quad \text{on } \partial\Omega,$$

where ν is the inner unit normal vector to $\partial\Omega$. Hence,

$$\left|\frac{\partial u}{\partial \nu}\right| \leq \left|\frac{\partial \varphi}{\partial \nu}\right| + \frac{\partial w}{\partial \nu} \quad \text{on } \partial\Omega.$$

Note that

$$\left.\frac{\partial w}{\partial \nu}\right|_{\partial\Omega} = \psi'(0) = \frac{A}{L} = \exp\{LM\}.$$

We have the desired estimate by the definition of L. □

If boundary values are only continuous, we can estimate the modulus of continuity of solutions near the boundary by a similar method.

Theorem 3.2.3. *Let Ω be a bounded domain in $\mathbb{R}^n$ with a C^2-boundary and $H \in C^1(\bar{\Omega})$ satisfy, for any $y \in \partial\Omega$,*

$$|H(y)| \leq H_{\partial\Omega}(y),$$

where $H_{\partial\Omega}$ is the mean curvature of $\partial\Omega$ with respect to the inner unit normal vector to $\partial\Omega$. Suppose that u is a $C(\bar{\Omega}) \cap C^2(\Omega)$-solution of

$$\begin{aligned} \Delta u - \frac{u_i u_j}{1 + |\nabla u|^2} u_{ij} = H\sqrt{1 + |\nabla u|^2} \quad &\text{in } \Omega, \\ u = \varphi \quad &\text{on } \partial\Omega, \end{aligned}$$

for some $\varphi \in C(\partial\Omega)$. Then, for any $x \in \Omega$ and any $x_0 \in \partial\Omega$,

$$|u(x) - u(x_0)| \leq \omega(|x - x_0|),$$

where ω is a nondecreasing function on $(0, \mathrm{diam}(\Omega))$, with $\lim_{r\to 0}\omega(r) = 0$, depending only on Ω, $\sup_\Omega |u|$, $\max_{\partial\Omega}|\varphi|$, $|H|_{C^1(\bar\Omega)}$, and the modulus of continuity of φ on $\partial\Omega$.

Proof. We fix an $x_0 \in \partial\Omega$. For any constant $\varepsilon > 0$, we take a constant $\delta > 0$ such that, for any $x \in \partial\Omega \cap B_\delta(x_0)$,

$$|\varphi(x) - \varphi(x_0)| < \varepsilon.$$

We note that δ can be chosen independent of x_0 by the uniform continuity of φ on $\partial\Omega$. We define

$$\varphi^\pm(x) = \varphi(x_0) \pm \left(\varepsilon + \frac{2}{\delta^2}\max_{\partial\Omega}|\varphi||x - x_0|^2\right).$$

Then, $\varphi^\pm \in C^2(\bar\Omega)$ and

$$\varphi^- \le \varphi \le \varphi^+ \quad \text{on } \partial\Omega.$$

Set

$$Q(v) = (1 + |\nabla v|^2)\Delta v - v_i v_j v_{ij} - H\sqrt{1 + |\nabla v|^2}.$$

As in the proof of Theorem 3.2.2, we can construct a function $w \in C(\bar\Omega_0) \cap C^2(\Omega_0)$ such that

$$\begin{aligned} Q(\varphi^- - w) \ge Q(u) \ge Q(\varphi^+ + w) \quad &\text{in } \Omega_0,\\ \varphi^- - w \le u \le \varphi^+ + w \quad &\text{on } \partial\Omega_0, \end{aligned}$$

where $\Omega_0 = \{x \in \Omega : d(x) < d_0\}$ for some constant $d_0 > 0$. By the maximum principle, we obtain

$$\varphi^- - w \le u \le \varphi^+ + w \quad \text{in } \Omega_0.$$

Hence, for any $x \in \Omega_0$,

$$|u(x) - \varphi(x_0)| \le \varepsilon + w(x) + \frac{2}{\delta^2}\max_{\partial\Omega}|\varphi||x - x_0|^2.$$

Therefore, there exists a positive constant $\delta' \le \min\{d_0, \delta\}$ such that, for any $x \in \Omega \cap B_{\delta'}(x_0)$,

$$|u(x) - \varphi(x_0)| \le 2\varepsilon.$$

We note that δ' can be chosen independent of x_0. This implies the desired result. □

We now derive a global gradient estimate.

Theorem 3.2.4. *Let Ω be a bounded domain in $\mathbb{R}^n$ with a C^1-boundary and $H \in C^1(\bar\Omega)$. Suppose that u is a $C^1(\bar\Omega) \cap C^3(\Omega)$-solution of*

$$\Delta u - \frac{u_i u_j}{1 + |\nabla u|^2}u_{ij} = H\sqrt{1 + |\nabla u|^2} \quad \text{in } \Omega.$$

Then,

$$\sup_\Omega |\nabla u| \le \Big(\max_{\partial\Omega} |\nabla u| + \sup_\Omega |H| + 2\Big) \cdot \exp\Big\{c_0\Big(\sup_\Omega |\nabla H|^{\frac{1}{2}} + 1\Big)\operatorname{osc}_\Omega u\Big\},$$

where c_0 *is a universal positive constant.*

Proof. We set, for any $x \in \Omega$ and any $p \in \mathbb{R}^n$,

$$a_{ij}(p) = \delta_{ij} - \frac{p_i p_j}{1+|p|^2}$$

and

$$f(x,p) = H(x)\sqrt{1+|p|^2}.$$

Then, we write the mean curvature equation as

$$a_{ij}(\nabla u)u_{ij} = f(x, \nabla u) \quad \text{in } \Omega. \tag{1}$$

Without loss of generality, we assume $u \ge 0$; otherwise, we consider $u - \inf_\Omega u$ instead.

Let γ and ψ be two positive functions defined in $[0,\infty)$, with positive derivatives. These functions will be determined. Set

$$v = \gamma(u)\psi(|\nabla u|^2) \quad \text{in } \Omega.$$

We assume v attains its maximum at $x_0 \in \bar{\Omega}$. Then,

$$\gamma(u)\psi(|\nabla u|^2) \le \gamma\big(u(x_0)\big)\psi\big(|\nabla u(x_0)|^2\big) \quad \text{in } \Omega. \tag{2}$$

We will estimate $\nabla u(x_0)$.

If $x_0 \in \partial\Omega$, then

$$|\nabla u(x_0)| \le \max_{\partial\Omega} |\nabla u|. \tag{3}$$

Next, we consider the case $x_0 \in \Omega$. Then,

$$(\log v)_i(x_0) = 0, \quad \big((\log v)_{ij}(x_0)\big) \le 0.$$

Now, we evaluate $a_{ij}(\log v)_{ij}$ at x_0.

A simple differentiation yields

$$(\log v)_i = \frac{2\psi'}{\psi}u_k u_{ki} + \frac{\gamma'}{\gamma}u_i$$

and

$$(\log v)_{ij} = \frac{2\psi'}{\psi}(u_k u_{kij} + u_{ki}u_{kj}) + 4\Big(\frac{\psi''}{\psi} - \frac{\psi'^2}{\psi^2}\Big)u_k u_l u_{ki} u_{lj} + \frac{\gamma'}{\gamma}u_{ij} + \Big(\frac{\gamma''}{\gamma} - \frac{\gamma'^2}{\gamma^2}\Big)u_i u_j.$$

Hence,

$$\begin{aligned} a_{ij}(\log v)_{ij} &= \frac{2\psi'}{\psi}(u_k a_{ij} u_{kij} + a_{ij} u_{ki} u_{kj}) \\ &\quad + 4\left(\frac{\psi''}{\psi} - \frac{\psi'^2}{\psi^2}\right) a_{ij} u_k u_l u_{ki} u_{lj} \\ &\quad + \frac{\gamma'}{\gamma} a_{ij} u_{ij} + \left(\frac{\gamma''}{\gamma} - \frac{\gamma'^2}{\gamma^2}\right) a_{ij} u_i u_j. \end{aligned}$$

To eliminate the third derivatives of u in the expression above, we differentiate (1) and get

$$a_{ij} u_{kij} + a_{ij,p_l} u_{kl} u_{ij} = \partial_k f,$$

where

$$a_{ij,p_l} = -\frac{\delta_{li} p_j + \delta_{lj} p_i}{1 + |p|^2} + \frac{2p_i p_j p_l}{(1 + |p|^2)^2}.$$

A simple substitution yields

$$a_{ij}(\log v)_{ij} = I + II + \left(\frac{\gamma''}{\gamma} - \frac{\gamma'^2}{\gamma^2}\right) a_{ij} u_i u_j, \tag{4}$$

where

$$I = \frac{2\psi'}{\psi}(-u_k a_{ij,p_l} u_{kl} u_{ij} + a_{ij} u_{ki} u_{kj}) + 4\left(\frac{\psi''}{\psi} - \frac{\psi'^2}{\psi^2}\right) a_{ij} u_k u_l u_{ki} u_{lj}$$

and

$$II = \frac{2\psi'}{\psi} u_k \partial_k f + \frac{\gamma'}{\gamma} f.$$

In the following, all calculations are made at x_0. Our goal is to eliminate second derivatives of u in I and II.

To simplify our calculation, we first choose an appropriate coordinate such that

$$\nabla u(x_0) = (u_1(x_0), 0, \ldots, 0),$$

with $u_1(x_0) > 0$. In particular, for $i = 2, \ldots, n$,

$$u_i(x_0) = 0.$$

Hence,

$$I = \frac{2\psi'}{\psi}(-u_1 a_{ij,p_l} u_{1l} u_{ij} + a_{ij} u_{ki} u_{kj}) + 4\left(\frac{\psi''}{\psi} - \frac{\psi'^2}{\psi^2}\right) a_{ij} u_1^2 u_{1i} u_{1j}.$$

Moreover,

$$\begin{aligned} a_{11} &= \frac{1}{1+u_1^2}, \quad a_{ii} = 1 \ (i \neq 1), \quad a_{ij} = 0 \ (i \neq j), \\ a_{11,p_1} &= -\frac{2u_1}{(1+u_1^2)^2}, \quad a_{11,p_l} = 0 \ (l \neq 1), \\ a_{1j,p_j} &= -\frac{u_1}{1+u_1^2} \ (j \neq 1), \quad a_{1j,p_l} = 0 \ (j \neq 1, l \neq j), \\ a_{ij,p_l} &= 0 \ (i \neq 1, j \neq 1). \end{aligned}$$

In fact, we can write

$$a_{11,p_1} = -\frac{2u_1}{1+u_1^2} a_{11}$$

and, for $i = 2, \dots, n$,

$$a_{1i,p_i} = -\frac{u_1}{1+u_1^2} a_{ii}.$$

Then,

$$\begin{aligned} -u_1 a_{ij,p_l} u_{1l} u_{ij} &= -u_1 a_{11,p_1} u_{11}^2 - 2\sum_{j=2}^n u_1 a_{1j,p_j} u_{1j}^2 \\ &= \frac{2u_1^2}{1+u_1^2} a_{11} u_{11}^2 + \sum_{j=2}^n \frac{2u_1^2}{1+u_1^2} a_{1j} u_{1j}^2 = \frac{2u_1^2}{1+u_1^2} \sum_{i=1}^n a_{ii} u_{1i}^2. \end{aligned}$$

Hence,

$$I = \frac{2\psi'}{\psi}\left(\frac{2u_1^2}{1+u_1^2}\sum_{i=1}^n a_{ii}u_{1i}^2 + \sum_{i,k=1}^n a_{ii}u_{ki}^2\right) + 4\left(\frac{\psi''}{\psi} - \frac{\psi'^2}{\psi^2}\right)\sum_{i=1}^n a_{ii}u_1^2 u_{1i}^2.$$

By expanding the summation in terms of i and k and using the expressions of a_{ii}, we obtain

$$(5) \qquad \begin{aligned} I = &\left\{\frac{2\psi'}{\psi}\frac{1+3u_1^2}{1+u_1^2} + 4\left(\frac{\psi''}{\psi} - \frac{\psi'^2}{\psi^2}\right)u_1^2\right\}\frac{u_{11}^2}{1+u_1^2} \\ &+ \left\{\frac{2\psi'}{\psi}\frac{2+3u_1^2}{1+u_1^2} + 4\left(\frac{\psi''}{\psi} - \frac{\psi'^2}{\psi^2}\right)u_1^2\right\}\sum_{i=2}^n u_{1i}^2 + \frac{2\psi'}{\psi}\sum_{i,k=2}^n u_{ik}^2. \end{aligned}$$

Concerning II, we have

$$II = \frac{2\psi'}{\psi}u_1\partial_1 f + \frac{\gamma'}{\gamma}f.$$

The expression of f implies

$$\begin{aligned} \partial_1 f &= H_1\sqrt{1+u_1^2} + \frac{H}{\sqrt{1+u_1^2}}u_l u_{1l} \\ &= H_1\sqrt{1+u_1^2} + \frac{H}{\sqrt{1+u_1^2}}u_1 u_{11}. \end{aligned}$$

Then,

$$II = \frac{H}{\sqrt{1+u_1^2}}\left(\frac{\gamma'}{\gamma}(1+u_1^2) + \frac{2\psi'}{\psi}u_1^2 u_{11}\right) + \frac{2\psi'}{\psi}u_1 H_1\sqrt{1+u_1^2}. \tag{6}$$

Next, we eliminate second derivatives of u from I and II.

The expression of $(\log v)_i$ and the condition $v_i(x_0) = 0$ imply

$$\frac{2\psi'}{\psi}u_1 u_{1i} = -\frac{\gamma'}{\gamma}u_i.$$

Hence,

$$\begin{aligned} u_{11} &= -\frac{\gamma'\psi}{2\gamma\psi'}, \\ u_{1i} &= 0 \quad \text{for } i = 2, \ldots, n. \end{aligned} \tag{7}$$

By substituting (7) in (5) and (6), we have

$$I = \left\{\frac{2\psi'}{\psi}\frac{1+3u_1^2}{(1+u_1^2)^2} + \left(\frac{\psi''}{\psi} - \frac{\psi'^2}{\psi^2}\right)\frac{4u_1^2}{1+u_1^2}\right\}\frac{\psi^2}{\psi'^2}\frac{\gamma'^2}{4\gamma^2} + \frac{2\psi'}{\psi}\sum_{i,k=2}^n u_{ik}^2$$

and

$$II = \frac{\gamma'}{\gamma}\frac{H}{\sqrt{1+u_1^2}} + \frac{2\psi'}{\psi}u_1 H_1\sqrt{1+u_1^2}.$$

By substituting the expressions above in (4), we obtain, at x_0,

$$\begin{aligned} a_{ij}(\log v)_{ij} \geq & \left\{\frac{2\psi'}{\psi}\frac{1+3u_1^2}{(1+u_1^2)^2} + \left(\frac{\psi''}{\psi} - \frac{\psi'^2}{\psi^2}\right)\frac{4u_1^2}{1+u_1^2}\right\}\frac{\psi^2}{\psi'^2}\frac{\gamma'^2}{4\gamma^2} \\ & + \frac{\gamma'}{\gamma}\frac{H}{\sqrt{1+u_1^2}} + \frac{2\psi'}{\psi}u_1 H_1\sqrt{1+u_1^2} \\ & + \left(\frac{\gamma''}{\gamma} - \frac{\gamma'^2}{\gamma^2}\right)\frac{u_1^2}{1+u_1^2}. \end{aligned} \tag{8}$$

The right-hand side depends only on u, u_1, H, and H_1 at x_0. We now choose γ and ψ. Set, for some α to be determined,

$$\gamma(t) = e^{\alpha t}, \quad \psi(t) = t. \tag{9}$$

Then,

$$\gamma'(t) = \alpha e^{\alpha t}, \quad \gamma''(t) = \alpha^2 e^{\alpha t},$$

and

$$\psi'(t) = 1, \quad \psi''(t) = 0.$$

Hence,

$$\frac{\gamma'}{\gamma}(u) = \alpha, \quad \frac{\gamma''}{\gamma}(u) = \alpha^2,$$

and

$$\frac{\psi'}{\psi}(|\nabla u|^2) = \frac{1}{u_1^2}, \quad \frac{\psi''}{\psi}(|\nabla u|^2) = 0.$$

By substitutions in (8) and straightforward calculations, we have

$$a_{ij}(\log v)_{ij} \geq \frac{\alpha^2 u_1^2(u_1^2-1)}{2(u_1^2+1)^2} + \frac{2H_1}{u_1}\sqrt{1+u_1^2} + \frac{\alpha H}{\sqrt{1+u_1^2}}.$$

Since $a_{ij}(\log v)_{ij}(x_0) \leq 0$, we obtain

$$\frac{\alpha^2 u_1^2(u_1^2-1)}{2(u_1^2+1)^2} + \frac{2H_1}{u_1}\sqrt{1+u_1^2} + \frac{\alpha H}{\sqrt{1+u_1^2}} \leq 0,$$

or

$$\sqrt{1+u_1^2}\left(\frac{\alpha^2 u_1^2(u_1^2-1)}{2(u_1^2+1)^2} + \frac{2H_1}{u_1}\sqrt{1+u_1^2}\right) \leq -\alpha H.$$

If $u_1^2 \geq 3$, then

$$\frac{u_1^2(u_1^2-1)}{2(u_1^2+1)^2} = \frac{1}{2}\left(1-\frac{1}{1+u_1^2}\right)\left(1-\frac{2}{1+u_1^2}\right) \geq \frac{3}{16},$$

and

$$\frac{1+u_1^2}{u_1^2} = 1+\frac{1}{u_1^2} \leq \frac{4}{3}.$$

This implies

$$u_1\left(\frac{3}{16}\alpha^2 - \frac{4}{\sqrt{3}}|\nabla H|\right) \leq \alpha|H|.$$

By taking

$$\alpha = c_0 + c_1 \sup_\Omega |\nabla H|^{\frac{1}{2}},$$

for some large universal constants c_0 and c_1, we have

$$\frac{3}{16}\alpha^2 - \frac{4}{\sqrt{3}}|\nabla H| \geq \alpha \geq 1.$$

In fact, it suffices to take $c_0 = 6$ and $c_1 = 4$. Hence, if $u_1(x_0) \geq \sqrt{3}$, then

$$u_1(x_0) \leq \sup_\Omega |H|.$$

Therefore,

$$|\nabla u(x_0)| \leq \sup_\Omega |H| + 2. \tag{10}$$

By combining (3) and (10), we obtain

$$|\nabla u(x_0)| \leq \max_{\partial\Omega} |\nabla u| + \sup_\Omega |H| + 2.$$

With the choice of γ and ψ in (9), (2) becomes

$$|\nabla u| \leq |\nabla u(x_0)| \exp\left\{\frac{1}{2}\alpha \sup_\Omega u\right\} \quad \text{in } \Omega.$$

We then have the desired result. □

We now make a comment on the proof. Besides γ and the known quantities such that H and H_1, the expression in the right-hand side of (8) is a function of u_1 since $\psi = \psi(u_1^2)$. Our goal is to choose γ and ψ appropriately so that (8) yields an upper bound of u_1. For this goal, the choice of γ and ψ in (9) is by no means unique. In the next section, we will modify the arguments above and derive interior gradient estimates by choosing different γ and ψ.

3.3. Interior Gradient Estimates

In this section, we discuss the interior gradient estimates of solutions of the mean curvature equation. As we pointed out, the mean curvature equation is uniformly elliptic only after the gradient estimate is established. In general, gradient estimates are difficult to prove for nonuniformly elliptic quasilinear equations. The structure of equations plays an important role.

Theorem 3.3.1. *Suppose that u is an $L^\infty(B_R) \cap C^3(B_R)$-solution of*

$$\Delta u - \frac{u_i u_j}{1+|\nabla u|^2} u_{ij} = H\sqrt{1+|\nabla u|^2} \quad \text{in } B_R,$$

for some $H \in C^1(B_R)$ with $|H|_{C^1(B_R)} < \infty$. Then,

$$|\nabla u(0)| \le \exp\left\{C\left(1+\frac{\omega^2}{R^2}\right)+C\left(\omega+\frac{\omega^2}{R}\right)|H|_{L^\infty(B_R)}+C\omega^2|\nabla H|_{L^\infty(B_R)}\right\},$$

where $\omega = \sup_{B_R} u - \inf_{B_R} u$ and C is a positive constant depending only on n.

Proof. We set, for any $x \in B_R$ and any $p \in \mathbb{R}^n$,

$$a_{ij}(p) = \delta_{ij} - \frac{p_i p_j}{1+|p|^2}$$

and

$$f(x,p) = H(x)\sqrt{1+|p|^2}.$$

Then, we write the mean curvature equation as

$$a_{ij}(\nabla u)u_{ij} = f(x, \nabla u) \quad \text{in } B_R.$$

Without loss of generality, we assume $u \ge 0$; otherwise, we consider $u - \inf_{B_R} u$ instead.

Let γ and ψ be two positive functions defined in $[0,\infty)$, with positive derivatives. These functions will be determined. Let η be a nonnegative function in B_R with $\eta = 0$ on ∂B_R. Set

$$v = \eta\gamma(u)\psi(|\nabla u|^2) \quad \text{in } B_R.$$

We assume v attains its maximum at $x_0 \in B_R$. Then,

$$\eta\gamma(u)\psi(|\nabla u|^2) \leq \eta(x_0)\gamma\big(u(x_0)\big)\psi\big(|\nabla u(x_0)|^2\big) \quad \text{in } B_R.$$

In the following, we assume $\eta(x_0) \neq 0$. Then,

$$(\log v)_i(x_0) = 0, \quad \big((\log v)_{ij}(x_0)\big) \leq 0.$$

Now we evaluate $a_{ij}(\log v)_{ij}$ at x_0.

A simple differentiation yields

$$(\log v)_i = \frac{2\psi'}{\psi}u_k u_{ki} + \frac{\gamma'}{\gamma}u_i + \frac{\eta_i}{\eta}$$

and

$$\begin{aligned}(\log v)_{ij} = {} & \frac{2\psi'}{\psi}(u_k u_{kij} + u_{ki}u_{kj}) + 4\Big(\frac{\psi''}{\psi} - \frac{\psi'^2}{\psi^2}\Big)u_k u_l u_{ki} u_{lj} \\ & + \frac{\gamma'}{\gamma}u_{ij} + \Big(\frac{\gamma''}{\gamma} - \frac{\gamma'^2}{\gamma^2}\Big)u_i u_j + \Big(\frac{\eta_{ij}}{\eta} - \frac{\eta_i\eta_j}{\eta^2}\Big).\end{aligned}$$

Comparing with similar expressions in the proof of Theorem 3.2.4, we note that the only extra terms here are those involving η. We now proceed as in the proof of Theorem 3.2.4 and calculate at x_0.

We first choose an appropriate coordinate such that

$$\nabla u(x_0) = (u_1(x_0), 0, \ldots, 0),$$

with $u_1(x_0) > 0$. In particular, for $i = 2, \ldots, n$,

$$u_i(x_0) = 0.$$

As in the proof of Theorem 3.2.4, we obtain

$$a_{ij}(\log v)_{ij} = I + II + \Big(\frac{\gamma''}{\gamma} - \frac{\gamma'^2}{\gamma^2}\Big)a_{ij}u_i u_j + a_{ij}\Big(\frac{\eta_{ij}}{\eta} - \frac{\eta_i\eta_j}{\eta^2}\Big), \tag{1}$$

where

$$\begin{aligned} I = {} & \Big\{\frac{2\psi'}{\psi}\frac{1+3u_1^2}{1+u_1^2} + 4\Big(\frac{\psi''}{\psi} - \frac{\psi'^2}{\psi^2}\Big)u_1^2\Big\}\frac{u_{11}^2}{1+u_1^2} \\ & + \Big\{\frac{2\psi'}{\psi}\frac{2+3u_1^2}{1+u_1^2} + 4\Big(\frac{\psi''}{\psi} - \frac{\psi'^2}{\psi^2}\Big)u_1^2\Big\}\sum_{i=2}^n u_{1i}^2 + \frac{2\psi'}{\psi}\sum_{i,k=2}^n u_{ik}^2 \end{aligned} \tag{2}$$

and

$$II = \frac{H}{\sqrt{1+u_1^2}}\Big(\frac{\gamma'}{\gamma}(1+u_1^2) + \frac{2\psi'}{\psi}u_1^2 u_{11}\Big) + \frac{2\psi'}{\psi}u_1 H_1\sqrt{1+u_1^2}. \tag{3}$$

Next, we eliminate second derivatives from I and II.

The expression of $(\log v)_i$ and the condition $v_i(x_0) = 0$ imply

$$\frac{2\psi'}{\psi}u_1 u_{1i} = -\Big(\frac{\gamma'}{\gamma}u_i + \frac{\eta_i}{\eta}\Big).$$

Hence,

$$\tag{4} \begin{aligned} u_{11} &= -\frac{\psi}{2\psi'}\left(\frac{\gamma'}{\gamma} + \frac{\eta_1}{u_1\eta}\right), \\ u_{1i} &= -\frac{\psi}{2\psi'}\frac{\eta_i}{u_1\eta} \quad \text{for } i = 2, \ldots, n. \end{aligned}$$

By substituting (4) in (2) and (3), we have

$$\tag{5} \begin{aligned} I = &\left\{\frac{2\psi'}{\psi}\frac{1+3u_1^2}{1+u_1^2} + 4\left(\frac{\psi''}{\psi} - \frac{\psi'^2}{\psi^2}\right)u_1^2\right\}\frac{\psi^2}{4(1+u_1^2)\psi'^2}\left(\frac{\gamma'}{\gamma} + \frac{\eta_1}{u_1\eta}\right)^2 \\ &+ \left\{\frac{2\psi'}{\psi}\frac{2+3u_1^2}{1+u_1^2} + 4\left(\frac{\psi''}{\psi} - \frac{\psi'^2}{\psi^2}\right)u_1^2\right\}\frac{\psi^2}{4u_1^2\psi'^2}\sum_{i=2}^{n}\frac{\eta_i^2}{\eta^2} + \frac{2\psi'}{\psi}\sum_{i,k=2}^{n} u_{ik}^2 \end{aligned}$$

and

$$\tag{6} II = \frac{H}{\sqrt{1+u_1^2}}\left(\frac{\gamma'}{\gamma} - \frac{\eta_1}{\eta}u_1\right) + \frac{2\psi'}{\psi}u_1 H_1\sqrt{1+u_1^2}.$$

We now choose γ and ψ. Set $M = \sup_{B_R} u$ and

$$\gamma(t) = 1 + \frac{t}{M}, \quad \psi(t) = \log t.$$

Then,

$$\gamma'(t) = \frac{1}{M}, \quad \gamma''(t) = 0,$$

and

$$\psi'(t) = \frac{1}{t}, \quad \psi''(t) = -\frac{1}{t^2}.$$

Hence, $1 \leq \gamma(u) \leq 2$ since $0 \leq u \leq M$, and

$$\frac{\psi'}{\psi}(|\nabla u|^2) = \frac{1}{2u_1^2 \log u_1}, \quad \frac{\psi''}{\psi}(|\nabla u|^2) = -\frac{1}{2u_1^4 \log u_1}.$$

A straightforward calculation yields

$$\begin{aligned} &\frac{2\psi'}{\psi}\frac{1+3u_1^2}{1+u_1^2} + 4\left(\frac{\psi''}{\psi} - \frac{\psi'^2}{\psi^2}\right)u_1^2 \\ &\quad = \frac{1}{u_1^2 \log u_1}\left(1 - \frac{2}{1+u_1^2} - \frac{1}{\log u_1}\right). \end{aligned}$$

Hence, if $u_1 \geq e^3$, then

$$\frac{2\psi'}{\psi}\frac{1+3u_1^2}{1+u_1^2} + 4\left(\frac{\psi''}{\psi} - \frac{\psi'^2}{\psi^2}\right)u_1^2 \geq \frac{1}{2u_1^2 \log u_1} > 0.$$

In particular, the summation corresponding to i from 2 to n in (5) has a positive coefficient. By keeping only the first term in the right-hand side of (5) and using $1+u_1^2 \leq 2u_1^2$, we obtain

$$I \geq \frac{1}{2u_1^2 \log u_1} \frac{\psi^2}{8u_1^2\psi'^2}\left(\frac{\gamma'}{\gamma}+\frac{\eta_1}{u_1\eta}\right)^2 \geq \frac{1}{4}\log u_1\left(\frac{\gamma'}{\gamma}+\frac{\eta_1}{u_1\eta}\right)^2.$$

Concerning II in (6), we have

$$\begin{aligned} II &= \frac{H}{\sqrt{1+u_1^2}}\left(\frac{\gamma'}{\gamma}-\frac{\eta_1}{\eta}u_1\right)+H_1\frac{\sqrt{1+u_1^2}}{u_1\log u_1} \\ &\geq -|H|\left(\frac{1}{M}+\frac{|\eta_1|}{\eta}\right)-|H_1| \end{aligned}$$

if $u_1 > e^2$. We also note, with the expression of a_{11},

$$\left|\left(\frac{\gamma''}{\gamma}-\frac{\gamma'^2}{\gamma^2}\right)a_{ij}u_iu_j\right| = \frac{\gamma'^2}{\gamma^2}a_{11}u_1^2 \leq \frac{1}{M^2}.$$

By substitutions in (1) and the condition $a_{ij}(\log v)_{ij}(x_0) \leq 0$, we obtain

$$\frac{1}{4}\log u_1\left(\frac{\gamma'}{\gamma}+\frac{\eta_1}{u_1\eta}\right)^2 \leq a_{ij}\left(\frac{\eta_i\eta_j}{\eta^2}-\frac{\eta_{ij}}{\eta}\right)+\frac{1}{M^2}+|H|\left(\frac{1}{M}+\frac{|\eta_1|}{\eta}\right)+|\nabla H|.$$

We now choose η such that

$$|\nabla^2\eta| + \frac{|\nabla\eta|^2}{\eta} \leq \frac{c}{R^2} \quad \text{in } B_R,$$

for some positive constant c. For example, we can take

$$\eta(x) = \left(1-\frac{|x|^2}{R^2}\right)^2.$$

Then,

$$\log u_1\left(\frac{\gamma'}{\gamma}+\frac{\eta_1}{u_1\eta}\right)^2 \leq C\left\{\frac{1}{R^2\eta}+\frac{1}{M^2}+\left(\frac{1}{M}+\frac{1}{R\sqrt{\eta}}\right)|H|+|\nabla H|\right\},$$

where C is a positive constant depending only on n. For simplicity, we set

$$\widetilde{M} = \frac{M^2}{R^2}+1+M\left(1+\frac{M}{R}\right)|H|_{L^\infty}+M^2|\nabla H|_{L^\infty}.$$

Then, if $u_1 > e^3$,

$$\log u_1\left(\frac{\gamma'}{\gamma}+\frac{\eta_1}{u_1\eta}\right)^2 \leq \frac{C\widetilde{M}}{M^2\eta}.$$

We first consider the case

$$\left|\frac{\eta_1}{u_1\eta}\right| \leq \frac{\gamma'}{2\gamma}.$$

Then,

$$\frac{\gamma'^2}{4\gamma^2}\log u_1 \le \frac{C\widetilde{M}}{M^2\eta}.$$

Since $1 \le \gamma(u) \le 2$ in B_R, we get

$$\eta\gamma\log u_1 \le C\widetilde{M}.$$

Now we consider the case

$$\frac{\gamma'}{2\gamma} \le \left|\frac{\eta_1}{u_1\eta}\right|.$$

Then,

$$\eta\gamma u_1 \le C\frac{M}{R},$$

and hence, if $u_1 > e$,

$$\eta\gamma\log u_1 \le C\frac{M}{R}.$$

In conclusion, we have, at x_0,

$$\eta\gamma\log u_1 \le C\left\{1 + \frac{M^2}{R^2} + \frac{M}{R} + M\left(1 + \frac{M}{R}\right)|H|_{L^\infty} + M^2|\nabla H|_{L^\infty}\right\}.$$

This implies the desired result since $u_1 = |\nabla u|$ at x_0. □

Here, we follow Wang [**162**] for the proof of Theorem 3.3.1. Refer to [**95**] for an alternative proof. We will prove an improved interior gradient estimate in Theorem 4.3.2.

As an application, we prove a Liouville type result for the minimal surface equation.

Theorem 3.3.2. *Suppose that u is a $C^3(\mathbb{R}^n)$-solution of*

$$\Delta u - \frac{u_i u_j}{1+|\nabla u|^2}u_{ij} = 0 \quad \textit{in } \mathbb{R}^n.$$

If u is bounded, then u is constant. If u has at most a linear growth, then u is an affine function.

Proof. By setting, for any $p \in \mathbb{R}^n$,

$$a_{ij}(p) = \delta_{ij} - \frac{p_i p_j}{1+|p|^2},$$

we write the minimal surface equation as

$$a_{ij}(\nabla u)u_{ij} = 0 \quad \text{in } \mathbb{R}^n. \tag{1}$$

It is easy to check that

$$\sup_{p\in\mathbb{R}^n}\big(|p||\nabla a_{ij}(p)|\big) \le 2.$$

Since u has at most a linear growth, then for any $R > 0$,

$$\sup_{B_R} |u| \leq C_0(1 + R),$$

for some positive constant C_0. By Theorem 3.3.1, we have

$$|\nabla u| \leq K_0 \quad \text{in } \mathbb{R}^n,$$

where K_0 is a positive constant depending only on n and C_0 above. Therefore, the equation (1) is now uniformly elliptic. If u is bounded, Theorem 2.1.3 implies that u is constant. If u has at most a linear growth, Theorem 2.4.2 implies that u is an affine function. □

3.4. Dirichlet Problems

In this section, we employ the method of continuity to solve the Dirichlet problem for the mean curvature equations. We proceed similarly as in Section 2.6.

Theorem 3.4.1. *Let $\alpha \in (0, 1)$ be a constant and $\Omega \subset \mathbb{R}^n$ be a bounded domain in $\mathbb{R}^n$ with a $C^{3,\alpha}$-boundary. Suppose that $H \in C^{1,\alpha}(\bar{\Omega})$ satisfies*

$$\|H\|_{L^n(\Omega)} < n \left(\int_{\mathbb{R}^n} (1 + |p|^2)^{-\frac{n+2}{2}} dp \right)^{\frac{1}{n}}$$

and, for any $y \in \partial\Omega$,

$$|H(y)| \leq H_{\partial\Omega}(y),$$

where $H_{\partial\Omega}$ is the mean curvature of $\partial\Omega$ corresponding to the inner unit normal vector to $\partial\Omega$. Then for any $\varphi \in C^{3,\alpha}(\bar{\Omega})$, there exists a unique solution $u \in C^{3,\alpha}(\bar{\Omega})$ of

$$\begin{aligned} \Delta u - \frac{u_i u_j}{1 + |\nabla u|^2} u_{ij} &= H\sqrt{1 + |\nabla u|^2} \quad \text{in } \Omega, \\ u &= \varphi \quad \text{on } \partial\Omega. \end{aligned} \tag{1}$$

Proof. We need only prove the existence of a $C^{2,\alpha}(\bar{\Omega})$-solution of (1). We note that the uniqueness follows from Corollary 2.1.2 and the higher regularity $u \in C^{3,\alpha}(\bar{\Omega})$ from Proposition 2.1.5. With $u \in C^3(\bar{\Omega})$, we can apply Theorems 3.2.1, 3.2.2, and 3.2.4 and obtain

$$|u|_{C^1(\bar{\Omega})} \leq C_1,$$

where C_1 is a positive constant depending only on n, $\|H\|_{L^n(\Omega)}$, $|H|_{C^1(\bar{\Omega})}$, $|\varphi|_{C^2(\bar{\Omega})}$, and Ω. Then, by applying Theorems 2.5.1 and 2.5.2, we conclude

$$|u|_{C^{2,\alpha}(\bar{\Omega})} \leq C_*, \tag{2}$$

where C_* is a positive constant depending only on n, α, $\|H\|_{L^n(\Omega)}$, $|H|_{C^1(\bar{\Omega})}$, $|\varphi|_{C^{2,\alpha}(\bar{\Omega})}$, and Ω.

In the following, we set, for any $p \in \mathbb{R}^n$,

$$a_{ij}(p) = \frac{1}{\sqrt{1+|p|^2}}\left(\delta_{ij} - \frac{p_i p_j}{1+|p|^2}\right),$$

and write (1) as

$$\begin{aligned} a_{ij}(\nabla u)u_{ij} &= H \quad \text{in } \Omega, \\ u &= \varphi \quad \text{on } \partial\Omega. \end{aligned}$$

For each $t \in [0,1]$, consider a family of Dirichlet problems

$$\begin{aligned} a_{ij}(\nabla u)u_{ij} &= tH \quad \text{in } \Omega, \\ u &= t\varphi \quad \text{on } \partial\Omega. \end{aligned} \tag{3}$$

For $t = 0$, (3) corresponds to a Dirichlet problem whose unique solution is given by $u = 0$. Any $C^{2,\alpha}$-solutions of (3) are $C^3(\bar{\Omega})$ and satisfy the estimate (2).

Letting $v = u - t\varphi$, (3) is equivalent to

$$\begin{aligned} a_{ij}(\nabla v + t\nabla\varphi)v_{ij} + ta_{ij}(\nabla v + t\nabla\varphi)\varphi_{ij} &= tH \quad \text{in } \Omega, \\ v &= 0 \quad \text{on } \partial\Omega. \end{aligned} \tag{4}$$

Moreover, any solutions of (4) satisfy the estimate (2).

Next, we set

$$\mathcal{X} = \{v \in C^{2,\alpha}(\bar{\Omega}) : \ v = 0 \text{ on } \partial\Omega\}$$

and

$$Q(v,t) = a_{ij}(\nabla v + t\nabla\varphi)v_{ij} + ta_{ij}(\nabla v + t\nabla\varphi)\varphi_{ij} - tH.$$

Solving (4) is equivalent to finding a function $v \in \mathcal{X}$ such that $Q(v,t) = 0$ in Ω.

Set

$$I = \{t \in [0,1] : \text{there exists a } v \in \mathcal{X} \text{ such that } Q(v,t) = 0\}.$$

We note that $0 \in I$. To prove $1 \in I$, we need to prove that I is both open and closed in $[0,1]$. For the openness, we note that $Q : \mathcal{X} \times [0,1] \to C^\alpha(\bar{\Omega})$ is of class C^1 and its Frèchet derivative with respect to $v \in \mathcal{X}$ is given by

$$Q_v(v,t)w = a_{ij}(\nabla v + t\nabla\varphi)w_{ij} + (v_{ij} + t\varphi_{ij})a_{ij,p_k}w_k.$$

For any fixed $v \in \mathcal{X}$ and $t \in [0,1]$, $Q_v(v,t)$ is a uniformly elliptic linear operator with C^α-coefficients. By the Schauder theory, $Q_v(v,t)$ is an invertible operator from $\mathcal{X}$ to $C^\alpha(\bar{\Omega})$. Suppose $t_0 \in I$; i.e., $Q(v^{t_0}, t_0) = 0$ for some $v^{t_0} \in \mathcal{X}$. By the implicit function theorem, for any t close to t_0, there is a unique $v^t \in \mathcal{X}$, close to v^{t_0} in the $C^{2,\alpha}$-norm, satisfying $Q(v^t, t) = 0$. Hence $t \in I$ for all such t, and therefore I is open. For the closedness, we note by (2) that any solution $v \in \mathcal{X}$ of $Q(v,t) = 0$ in Ω satisfies a uniform $C^{2,\alpha}(\bar{\Omega})$-estimate, independent of t; i.e.,

$$|v^t|_{C^{2,\alpha}(\bar{\Omega})} \le C_*, \text{ independent of } t.$$

Hence, the closedness of I follows from the compactness in $C^2(\bar{\Omega})$ of bounded sets in $C^{2,\alpha}(\bar{\Omega})$, a consequence of the Arzela-Ascoli theorem. Therefore, I is the whole unit interval. The function v^1 is then our desired solution of (4) corresponding to $t = 1$. □

In Theorem 3.4.1, we proved the existence of $C^{3,\alpha}$-solutions in $\bar{\Omega}$ if boundary values are $C^{3,\alpha}$ on $\partial\Omega$. By an approximation, we can conclude the existence of solutions under weaker conditions. In particular, when boundary values are only continuous, we have solutions which are continuous up to the boundary.

Theorem 3.4.2. *Let $\alpha \in (0,1)$ be a constant and $\Omega \subset \mathbb{R}^n$ be a bounded domain in $\mathbb{R}^n$ with a $C^{3,\alpha}$-boundary. Suppose that $H \in C^{1,\alpha}(\bar{\Omega})$ satisfies*

$$\|H\|_{L^n(\Omega)} < n\left(\int_{\mathbb{R}^n} (1+|p|^2)^{-\frac{n+2}{2}}\,dp\right)^{\frac{1}{n}}$$

and, for any $y \in \partial\Omega$,

$$|H(y)| \le H_{\partial\Omega}(y),$$

where $H_{\partial\Omega}$ is the mean curvature of $\partial\Omega$ corresponding to the inner unit normal vector to $\partial\Omega$. Then for any $\varphi \in C(\partial\Omega)$, there exists a unique solution $u \in C(\bar{\Omega}) \cap C^{2,\alpha}(\Omega)$ of

$$\text{(1)}\qquad \begin{aligned} \Delta u - \frac{u_i u_j}{1+|\nabla u|^2}u_{ij} &= H\sqrt{1+|\nabla u|^2} \quad \text{in } \Omega,\\ u &= \varphi \quad \text{on } \partial\Omega. \end{aligned}$$

Proof. Without loss of generality, we assume $\varphi \in C(\bar{\Omega})$. Let $\{\varphi_m\}$ be a sequence of $C^{3,\alpha}(\bar{\Omega})$-functions such that

$$\varphi_m \to \varphi \quad \text{uniformly in } \bar{\Omega} \text{ as } m \to \infty.$$

Now, we consider the Dirichlet problem

$$\text{(2)}\qquad \begin{aligned} \Delta u - \frac{u_i u_j}{1+|\nabla u|^2}u_{ij} &= H\sqrt{1+|\nabla u|^2} \quad \text{in } \Omega,\\ u &= \varphi_m \quad \text{on } \partial\Omega. \end{aligned}$$

By Theorem 3.4.1, there exists a unique solution $u_m \in C^{3,\alpha}(\bar{\Omega})$ of (2).

We fix an m and proceed to derive an estimate of u_m. By Theorem 3.2.1, we have

$$\text{(3)}\qquad \sup_{\Omega}|u_m| \le \max_{\partial\Omega}|u_m| + C\text{diam}(\Omega),$$

where C is a positive constant depending only on n and $\|H\|_{L^n(\Omega)}$. It is obvious that the right-hand side of (3) can be made independent of m by the uniform convergence of φ_m. Moreover, by Theorem 3.2.3, we have, for any $x \in \Omega$ and any $x_0 \in \partial\Omega$,

$$|u_m(x) - u_m(x_0)| \le \omega(|x - x_0|), \tag{4}$$

where ω is a nondecreasing function on $(0, \text{diam}(\Omega))$, with $\lim_{r\to 0} \omega(r) = 0$, depending only on n, Ω, $|H|_{C^1(\bar{\Omega})}$, $|u_m|_{L^\infty(\Omega)}$, $|\varphi_m|_{L^\infty(\Omega)}$, and the modulus of continuity of φ_m on $\partial\Omega$. Again, ω can be made independent of m.

Next, we consider an arbitrary subdomain $\Omega' \Subset \Omega$. By using Theorem 3.3.1, Theorem 2.4.1, and Theorem 2.4.3 successively, we obtain

$$|u_m|_{C^{2,\alpha}(\Omega')} \le C_*, \tag{5}$$

where C_* is a positive constant depending only on n, $|u_m|_{L^\infty(\Omega)}$, $|H|_{C^1(\bar{\Omega})}$, Ω', and Ω. Again, C_* can be made independent of m. By (3), (4), (5), and the Arzela-Ascoli theorem, there exists a subsequence of $\{u_m\}$ convergent uniformly in $\bar{\Omega}$ and also in C^2 in any subdomain $\Omega' \Subset \Omega$ to a function $u \in C(\bar{\Omega}) \cap C^{2,\alpha}(\Omega)$. Hence by passing the limit, u satisfies (1). □

In the following, we present a result of the nonsolvability of the Dirichlet problem for the mean curvature equation if the condition in Theorem 3.4.1 concerning the boundary mean curvature is not satisfied. We note that such a condition is used only in the establishment of the boundary gradient estimates, as in Theorem 3.2.2. In this sense, the nonsolvability is due to the lack of the boundary gradient estimates.

We first prove a comparison principle. For a fixed $H \in C(\bar{\Omega})$, set, for any $v \in C^2(\Omega)$,

$$Q(v) = (1 + |\nabla v|^2)\Delta v - v_i v_j v_{ij} - H(1 + |\nabla v|^2)^{\frac{3}{2}}.$$

Lemma 3.4.3. *Let $\Omega \subset \mathbb{R}^n$ be a bounded domain and Σ a relatively open C^1-portion of $\partial\Omega$. Suppose that $u \in C(\bar{\Omega}) \cap C^2(\Omega \cup \Sigma)$ and $v \in C(\bar{\Omega}) \cap C^2(\Omega)$ satisfy $Q(u) \ge Q(v)$ in Ω, $u \le v$ on $\partial\Omega \setminus \Sigma$, $\partial v/\partial\nu = -\infty$ on Σ, where ν is the inner unit normal vector to $\partial\Omega$. Then, $u \le v$ in Ω.*

Proof. Set $w = u - v$. Then, w satisfies $Lw \ge 0$ in Ω, for a linear elliptic operator L of the form $L = a_{ij}\partial_{ij} + b_i\partial_i$. By the maximum principle, we have

$$\sup_{\Omega} w \le \sup_{\partial\Omega} w.$$

Since $\partial w/\partial\nu = \infty$ on Σ, the function w cannot achieve a maximum on Σ. Hence, $w \le 0$ in Ω. □

With Lemma 3.4.3, we establish an estimate at boundary points where the condition on the boundary mean curvature in Theorem 3.4.1 is not satisfied.

Lemma 3.4.4. *Let $\Omega \subset \mathbb{R}^n$ be a bounded domain with a C^2-boundary. Suppose that $H \in C(\bar{\Omega})$ is nonnegative in $\bar{\Omega}$ and, for some $y_0 \in \partial\Omega$,*

$$H_{\partial\Omega}(y_0) < H(y_0),$$

where $H_{\partial\Omega}$ is the mean curvature of $\partial\Omega$ corresponding to the inner unit normal vector to $\partial\Omega$. Assume that $u \in C(\bar{\Omega}) \cap C^2(\Omega)$ is a solution of

$$\Delta u - \frac{u_i u_j}{1+|\nabla u|^2} u_{ij} = H\sqrt{1+|\nabla u|^2} \quad \text{in } \Omega.$$

Then, for any $a < a_0$,

$$u(y_0) \leq \max_{\partial\Omega \setminus B_a(y_0)} u + \eta(a),$$

where $a_0 \in (0, \operatorname{diam}(\Omega))$ is a constant and η is a positive continuous function on $(0, \operatorname{diam}(\Omega))$, depending only on n, $\operatorname{diam}(\Omega)$, $H(y_0) - H_{\partial\Omega}(y_0)$, the modulus of continuity of H at y_0, and the C^2-property of $\partial\Omega$ at y_0, with

$$\lim_{a\to 0} \eta(a) = 0.$$

Proof. Without loss of generality, we assume $y_0 = 0$. Set $D = \operatorname{diam}(\Omega)$ and, for any $v \in C^2(\Omega)$,

$$Q(v) = (1+|\nabla v|^2)\Delta v - v_i v_j v_{ij} - H(1+|\nabla v|^2)^{\frac{3}{2}}.$$

Take a constant $a \in (0, D)$ to be fixed. We now divide the proof into two steps.

Step 1. Set $r = |x|$ and consider

$$w(x) = c + \psi(r) \quad \text{in } \Omega \setminus B_a,$$

where c is a constant and ψ is a C^2-function in (a, D) such that

$$\text{(1)} \qquad \psi(D) = 0, \quad \psi' \leq 0, \quad \psi'(a) = -\infty.$$

A straightforward calculation yields

$$Q(w) = \psi'' + \frac{n-1}{r}\psi'(1+\psi'^2) - H(1+\psi'^2)^{\frac{3}{2}}.$$

By $H \geq 0$ in Ω and $\psi' \leq 0$ in (a, D), we have

$$Q(w) \leq \psi'' + \frac{n-1}{r}\psi'^3 \quad \text{in } \Omega \setminus B_a.$$

We now solve

$$\psi'' + \frac{n-1}{r}\psi'^3 = 0 \quad \text{in } (a, D),$$

with the condition (1). This yields

$$\psi(r) = \frac{1}{\sqrt{2(n-1)}} \int_r^D \left(\log \frac{t}{a}\right)^{-\frac{1}{2}} dt \quad \text{in } (a, D).$$

With such a ψ, we have $Q(w) \leq 0$ in $\Omega \setminus B_a$. Next, we take

$$c = \max_{\partial\Omega \setminus B_a} u.$$

Then,

$$u \leq w \text{ on } \partial\Omega \setminus B_a, \quad \frac{\partial w}{\partial \nu} = -\infty \text{ on } \Omega \cap \partial B_a,$$

where ν is the inner unit normal vector to $\partial(\Omega \setminus B_a)$. By Lemma 3.4.3, we obtain $u \leq w$ in $\Omega \setminus B_a$, and hence

$$u \leq \max_{\partial\Omega \setminus B_a} u + \psi(a) \quad \text{in } \Omega \setminus B_a. \tag{2}$$

A simple change of variables yields

$$\psi(a) = \frac{a}{\sqrt{2(n-1)}} \int_1^{\frac{D}{a}} \frac{1}{\sqrt{\log s}} \, ds.$$

We conclude easily $\psi(a) \to 0$ as $a \to 0$.

Step 2. Without loss of generality, we assume that e_n is the inner unit normal vector to $\partial\Omega$ at $y_0 = 0$ and that $\partial\Omega$ in a neighborhood of the origin is given by

$$\rho(x') = \frac{1}{2}\sum_{i=1}^{n-1} \kappa_i x_i^2 + o(|x'|^2),$$

where $\kappa_1, \ldots, \kappa_{n-1}$ are the principal curvatures of $\partial\Omega$ at 0 with respect to the inner unit normal vector to $\partial\Omega$. Consider the paraboloid $\mathcal{S}$ given by

$$\widetilde{\rho}(x') = \frac{1}{2}\sum_{i=1}^{n-1} \widetilde{\kappa}_i x_i^2,$$

for constants $\widetilde{\kappa}_i > \kappa_i$, for $i = 1, \ldots, n-1$. In the following, we will take $\widetilde{\kappa}_i$ sufficiently close to κ_i, for $i = 1, \ldots, n-1$, and then take a sufficiently small such that

$$\widetilde{\rho}(x') \geq \rho(x') \quad \text{for any } |x'| < a.$$

Take any $\varepsilon \in (0, a)$. Set $d(x) = \text{dist}(x, \mathcal{S})$ and

$$\widetilde{\Omega}_\varepsilon = \{x \in \Omega : x \in B_a, \, x_n > \widetilde{\rho}(x'), \, d(x) > \varepsilon\},$$

$$\widetilde{\Omega} = \{x \in \Omega : x \in B_a, \, x_n > \widetilde{\rho}(x')\}.$$

Then, $d(x) < 2a$ for any $x \in \widetilde{\Omega}_\varepsilon$. Consider

$$\widetilde{w}(x) = \widetilde{c} + \widetilde{\psi}(d) \quad \text{in } \widetilde{\Omega}_\varepsilon,$$

where $\widetilde{c}$ is a constant and $\widetilde{\psi}$ is a C^2-function in $(\varepsilon, 2a)$ such that

$$\widetilde{\psi}(2a) = 0, \quad \widetilde{\psi}' \leq 0, \quad \widetilde{\psi}'(\varepsilon) = -\infty. \tag{3}$$

A straightforward calculation yields

$$Q(\widetilde{w}) = \widetilde{\psi}'' + \widetilde{\psi}'(1 + \widetilde{\psi}'^2)\Delta d - H(1 + \widetilde{\psi}'^2)^{\frac{3}{2}}.$$

By $H \geq 0$ in Ω and $\widetilde{\psi}' \leq 0$ in $(\varepsilon, 2a)$, we have

$$Q(\widetilde{w}) \leq \widetilde{\psi}'' + \widetilde{\psi}'(1 + \widetilde{\psi}'^2)(\Delta d + H).$$

Since $\mathcal{S}$ is given by a quadratic polynomial, its principal curvatures are constants. By Lemma 3.1.9, we have

$$\Delta d = -\sum_{i=1}^{n-1} \frac{\widetilde{\kappa}_i}{1 - \widetilde{\kappa}_i d} = -\sum_{i=1}^{n-1} \frac{\widetilde{\kappa}_i^2 d}{1 - \widetilde{\kappa}_i d} - \sum_{i=1}^{n-1} \widetilde{\kappa}_i,$$

and hence

$$-(\Delta d + H) = \sum_{i=1}^{n-1} \frac{\widetilde{\kappa}_i^2 d}{1 - \widetilde{\kappa}_i d} + \sum_{i=1}^{n-1} (\widetilde{\kappa}_i - \kappa_i) + (H(0) - H) - (H(0) - H_{\partial\Omega}(0)).$$

By $H_{\partial\Omega}(0) < H(0)$ and the continuity of H at $y_0 = 0$, we can choose $\widetilde{\kappa}_i > \kappa_i$ sufficiently close to κ_i, for $i = 1, \ldots, n-1$, and choose a sufficiently small, such that

$$-(\Delta d + H) \leq -\delta \quad \text{in } \widetilde{\Omega}_\varepsilon,$$

for some positive constant δ. In fact, we can take $\delta = (H(0) - H_{\partial\Omega}(0))/2$. We point out that δ is a fixed constant, independent of ε. By $\widetilde{\psi}' \leq 0$ in $(\varepsilon, 2a)$ again, we have

$$Q(\widetilde{w}) \leq \widetilde{\psi}'' + \delta\widetilde{\psi}'(1 + \widetilde{\psi}'^2) \leq \widetilde{\psi}'' + \delta\widetilde{\psi}'^3 \quad \text{in } \widetilde{\Omega}_\varepsilon.$$

We now solve

$$\widetilde{\psi}'' + \delta\widetilde{\psi}'^3 = 0 \quad \text{in } (\varepsilon, 2a),$$

with the condition (3). This yields

$$\widetilde{\psi}(d) = K\left[\sqrt{2a - \varepsilon} - \sqrt{d - \varepsilon}\right] \quad \text{in } (\varepsilon, 2a),$$

for $K = \sqrt{2/\delta}$. With such a $\widetilde{\psi}$, we have $Q(\widetilde{w}) \leq 0$ in $\widetilde{\Omega}_\varepsilon$. Next, we take

$$\widetilde{c} = \sup_{\Omega \cap \partial B_a} u.$$

Then,

$$u \le \widetilde{w} \text{ on } \partial\widetilde{\Omega}_\varepsilon \cap \partial B_a, \quad \frac{\partial w}{\partial \nu} = -\infty \text{ on } \partial\widetilde{\Omega}_\varepsilon \setminus \partial B_a,$$

where ν is the inner unit normal vector to $\partial\widetilde{\Omega}_\varepsilon$. By Lemma 3.4.3, we obtain $u \le \widetilde{w}$ in $\widetilde{\Omega}_\varepsilon$, and hence

$$u \le \sup_{\Omega\cap\partial B_a} u + K\sqrt{2a-\varepsilon} \quad \text{in } \widetilde{\Omega}_\varepsilon.$$

Letting $\varepsilon \to 0$, we have

$$u \le \sup_{\Omega\cap\partial B_a} u + K\sqrt{2a} \quad \text{in } \widetilde{\Omega}$$

and, in particular,

$$u(0) \le \sup_{\Omega\cap\partial B_a} u + K\sqrt{2a}. \tag{4}$$

We point out that $\widetilde{\Omega}$ may be a proper subset of $\Omega \cap B_a$.

By combining (2) and (4), we obtain

$$u(0) \le \max_{\partial\Omega\setminus B_a} + \psi(a) + K\sqrt{2a}.$$

This is the desired result. □

Lemma 3.4.4 demonstrates that the boundary values of the solutions of the mean curvature equation satisfy a constraint and hence cannot be arbitrarily prescribed. This leads to the following nonexistence result.

Theorem 3.4.5. *Let $\Omega \subset \mathbb{R}^n$ be a bounded domain with a C^2-boundary. Suppose that $H \in C(\bar{\Omega})$ is either nonnegative or nonpositive in $\bar{\Omega}$ and, for some $y_0 \in \partial\Omega$,*

$$H_{\partial\Omega}(y_0) < |H(y_0)|,$$

where $H_{\partial\Omega}$ is the mean curvature of $\partial\Omega$ corresponding to the inner unit normal vector to $\partial\Omega$. Then, for any constant $\varepsilon > 0$, there exists a $\varphi \in C(\partial\Omega)$ with $|\varphi| \le \varepsilon$ on $\partial\Omega$, such that there exist no $C(\bar{\Omega}) \cap C^2(\Omega)$-solutions of

$$\Delta u - \frac{u_i u_j}{1+|\nabla u|^2} u_{ij} = H\sqrt{1+|\nabla u|^2} \quad \text{in } \Omega,$$
$$u = \varphi \quad \text{on } \partial\Omega.$$

Proof. We first assume $H \ge 0$ in $\bar{\Omega}$. For any given constant $\varepsilon > 0$, we take $a > 0$ such that $\eta(a) < \varepsilon$, where η is the function in Lemma 3.4.4. We construct a function $\varphi \in C(\partial\Omega)$ such that $\varphi(y_0) = \varepsilon$, $0 \le \varphi \le \varepsilon$ on $\partial\Omega$, and $\varphi = 0$ on $\partial\Omega \setminus B_a(y_0)$. Then, such a φ violates the conclusion of Lemma 3.4.4 and hence cannot be the boundary value of the solutions of the mean curvature equation. If $H \le 0$ in $\bar{\Omega}$, we consider the equation satisfied by $-u$ and repeat the argument. □

To end this section, we formulate results in this section for the minimal surface equation. For the existence of solutions of the Dirichlet problem, we have the following results.

Theorem 3.4.6. *Let $\alpha \in (0,1)$ be a constant and $\Omega \subset \mathbb{R}^n$ be a bounded domain in $\mathbb{R}^n$ with a $C^{3,\alpha}$-boundary satisfying*

$$H_{\partial\Omega} \geq 0 \quad \text{on } \partial\Omega,$$

where $H_{\partial\Omega}$ is the mean curvature of $\partial\Omega$ corresponding to the inner unit normal vector to $\partial\Omega$. Then for any $\varphi \in C^{3,\alpha}(\bar{\Omega})$, there exists a unique solution $u \in C^{3,\alpha}(\bar{\Omega}) \cap C^{\infty}(\Omega)$ of

$$\begin{aligned} \Delta u - \frac{u_i u_j}{1+|\nabla u|^2} u_{ij} &= 0 \quad \text{in } \Omega, \\ u &= \varphi \quad \text{on } \partial\Omega. \end{aligned}$$

Theorem 3.4.7. *Let $\alpha \in (0,1)$ be a constant and $\Omega \subset \mathbb{R}^n$ be a bounded domain in $\mathbb{R}^n$ with a $C^{3,\alpha}$-boundary satisfying*

$$H_{\partial\Omega} \geq 0 \quad \text{on } \partial\Omega,$$

where $H_{\partial\Omega}$ is the mean curvature of $\partial\Omega$ corresponding to the inner unit normal vector to $\partial\Omega$. Then for any $\varphi \in C(\partial\Omega)$, there exists a unique solution $u \in C(\bar{\Omega}) \cap C^{\infty}(\Omega)$ of

$$\begin{aligned} \Delta u - \frac{u_i u_j}{1+|\nabla u|^2} u_{ij} &= 0 \quad \text{in } \Omega, \\ u &= \varphi \quad \text{on } \partial\Omega. \end{aligned}$$

For the nonexistence of solutions of the Dirichlet problem, we have the following result.

Theorem 3.4.8. *Let $\Omega \subset \mathbb{R}^n$ be a bounded domain with a C^2-boundary such that for some $y_0 \in \partial\Omega$,*

$$H_{\partial\Omega}(y_0) < 0,$$

where $H_{\partial\Omega}$ is the mean curvature of $\partial\Omega$ corresponding to the inner unit normal vector to $\partial\Omega$. Then, for any constant $\varepsilon > 0$, there exists a $\varphi \in C(\partial\Omega)$ with $|\varphi| \leq \varepsilon$ on $\partial\Omega$, such that there exist no $C(\bar{\Omega}) \cap C^2(\Omega)$-solutions of

$$\begin{aligned} \Delta u - \frac{u_i u_j}{1+|\nabla u|^2} u_{ij} &= 0 \quad \text{in } \Omega, \\ u &= \varphi \quad \text{on } \partial\Omega. \end{aligned}$$

Chapter 4

Minimal Surface Equations

In this chapter, we discuss minimal surface equations. Needless to say, the minimal surface equation is a special class of mean curvature equations; namely, the mean curvature vanishes identically. It might appear that this chapter should be included in the previous one. However, there is a reason for an independent chapter. Results in this chapter are proved by analysis "upon surfaces". In other words, we treat minimal surfaces as submanifolds in the ambient Euclidean spaces and write equations on these submanifolds. In this chapter, we will derive an improved interior gradient estimate and an interior curvature estimate for solutions of the minimal surface equation.

In Section 4.1, we derive some important integral formulas. In particular, we prove a mean-value inequality for subharmonic functions on hypersurfaces and the Sobolev inequality on hypersurfaces, both of which are based on a monotonicity identity for nonnegative functions on hypersurfaces.

In Section 4.2, we discuss differential operators on hypersurfaces and derive some fundamental identities concerning the Laplace operators acting on the unit normal vector to the hypersurface and on the length of the second fundamental forms. We treat hypersurfaces as Riemannian submanifolds and employ standard methods from the Riemannian geometry. In Section 4.5, we provide alternative proofs of these identities without requiring knowledge of the Riemannian geometry.

In Section 4.3, we derive an improved interior gradient estimate for the minimal surface equation. The basis is the mean-value inequality for subharmonic functions on minimal hypersurfaces.

In Section 4.4, we derive an integral curvature estimate and a pointwise curvature estimate for minimal graphs. The pointwise curvature estimate is proved by the Moser iteration, with the help of the Sobolev inequality on hypersurfaces. These curvature estimates imply a Bernstein type result for the minimal surface equation in low dimensions.

In this chapter, we denote by $B_r(x)$ a ball in $\mathbb{R}^n$ and by $D_r(p)$ a ball in $\mathbb{R}^{n+1}$. Indices i, j, k range from 1 to n, and a, b, c range from 1 to $n+1$. For our presentation, we follow the relevant chapters in the books [**49**] and [**59**].

4.1. Integral Formulas

In this section, we derive several important integral formulas. Throughout this and subsequent sections, we adopt the notion of the n-dimensional Hausdorff measure $\mathcal{H}^n$. However, no knowledge of the geometric measure theory is required. For n-dimensional differentiable hypersurfaces Σ in $\mathbb{R}^{n+1}$, $d\mathcal{H}^n$ coincides with the area element on Σ.

Let $\Sigma \subset \mathbb{R}^{n+1}$ be a C^1-hypersurface and ν be its unit normal field. Suppose that u is a C^1-function defined in a neighborhood of Σ. We denote by Du the Euclidean gradient of u; i.e.,

$$Du = (\partial_1 u, \ldots, \partial_{n+1} u).$$

We define the *tangential gradient* $\nabla_\Sigma u$ of u on Σ by

$$\nabla_\Sigma u = Du - \langle Du, \nu \rangle \nu.$$

Hence, ∇_Σ is the Du minus the component of Du in the normal direction. In other words, $\nabla_\Sigma u$ is simply the projection of Du in the tangent space.

We note that $\nabla_\Sigma u$ depends only on the values of u on Σ; namely, if $u = 0$ on Σ, then $\nabla_\Sigma u = 0$. To check this, let $p_0 \in \Sigma$. If $Du(p_0) = 0$, then $\nabla_\Sigma u(p_0) = 0$; otherwise, we take the unit normal vector ν near p_0 in the form

$$\nu = \frac{Du}{|Du|}$$

and again conclude $\nabla_\Sigma u(p_0) = 0$. In conclusion, ∇_Σ is well-defined on functions defined only on Σ.

Next, for a C^1-vector field X on Σ (not necessarily tangential to Σ), we define the *divergence* $\operatorname{div}_\Sigma X$ of X by

$$\operatorname{div}_\Sigma X = \langle \nabla_\Sigma, X \rangle,$$

where we regard ∇_Σ as a vector in $\mathbb{R}^{n+1}$ and $\langle \cdot, \cdot \rangle$ is the standard inner product in $\mathbb{R}^{n+1}$. In a fixed coordinate system, we denote by $\mathbf{e_1}, \ldots, \mathbf{e_{n+1}}$ the standard orthonormal basis in $\mathbb{R}^{n+1}$. By writing $X = (X_1, \ldots, X_{n+1}) = X_a \mathbf{e_a}$, we have

$$\operatorname{div}_\Sigma X = \langle \nabla_\Sigma X_a, \mathbf{e_a} \rangle.$$

Here, repeated indices indicate a summation and the index a ranges from 1 to $n+1$.

If Σ is C^2, we define the *Laplace-Beltrami operator* Δ_Σ (on C^2-functions) on Σ by

$$\Delta_\Sigma u = \operatorname{div}_\Sigma(\nabla_\Sigma u).$$

We point out that $\operatorname{div}_\Sigma$ and Δ_Σ do not depend on the coordinate system in the ambient space $\mathbb{R}^{n+1}$.

First, we prove a formula of an integration by parts on hypersurfaces.

Lemma 4.1.1. *Let Σ be a C^2-hypersurface in $\mathbb{R}^{n+1}$ and ν be its unit normal field. Then, for any $u \in C_0^1(\Sigma)$,*

$$\int_\Sigma \nabla_\Sigma u \, d\mathcal{H}^n = -\int_\Sigma uH\nu \, d\mathcal{H}^n,$$

and, for any $X \in C_0^1(\Sigma, \mathbb{R}^{n+1})$,

$$\int_\Sigma \operatorname{div}_\Sigma X \, d\mathcal{H}^n = -\int_\Sigma H\langle X, \nu\rangle \, d\mathcal{H}^n,$$

where H is the mean curvature of Σ corresponding to ν.

Proof. These two identities are equivalent; we will prove the first one only. We consider the case where Σ is given by a graph of a C^2-function, say ρ, over a domain $\Omega \subset \mathbb{R}^n$. The general case then follows by a partition of unity. By our convention, indices i, j range from 1 to n, and indices a, b range from 1 to $n+1$.

For convenience, we set

$$\nabla_\Sigma = (\delta_1, \ldots, \delta_{n+1}).$$

Hence,

$$\delta_a = \partial_a - \nu_a \nu_b \partial_b.$$

In the following, we set

$$v = \sqrt{1 + |\nabla\rho|^2}.$$

At points $(x, \rho(x)) \in \Sigma$, we have

$$\begin{aligned} \nu &= \frac{1}{v}(-\partial_1\rho, \ldots, -\partial_n\rho, 1), \\ H &= \partial_i\left(\frac{\partial_i\rho}{v}\right) = -\partial_i\nu_i, \end{aligned} \tag{1}$$

and

$$d\mathcal{H}^n = v\,dx.$$

The expression of H in (1) follows from Lemma 3.1.5 or Remark 3.1.6. Define

$$\widetilde{u}(x, x_{n+1}) = u(x, \rho(x)).$$

We obviously have $\widetilde{u} = u$ on Σ and hence $\nabla_\Sigma \widetilde{u} = \nabla_\Sigma u$. Upon an integration by parts, we have, for $i = 1, \ldots, n$,

$$\int_\Sigma \delta_i u \, d\mathcal{H}^n = \int_\Sigma \delta_i \widetilde{u} \, d\mathcal{H}^n = \int_\Omega (\partial_i \widetilde{u} - \nu_i \nu_j \partial_j \widetilde{u}) v \, dx = -\int_\Omega \widetilde{u} I_i \, dx,$$

where

$$\begin{aligned} I_i &= \partial_i v - \partial_j(\nu_i \nu_j v) = \partial_i v - \nu_j \partial_j(\nu_i v) - \nu_i v \partial_j \nu_j \\ &= \partial_i v + \nu_j \partial_{ij} \rho + H \nu_i v = H \nu_i v. \end{aligned}$$

In the above calculation, we used (1) twice. Hence,

$$\int_\Sigma \delta_i u \, d\mathcal{H}^n = -\int_\Omega \widetilde{u} H \nu_i v \, dx = -\int_\Sigma u H \nu_i \, d\mathcal{H}^n.$$

Next, we have

$$\begin{aligned} \int_\Sigma \delta_{n+1} u \, d\mathcal{H}^n &= \int_\Sigma \delta_{n+1} \widetilde{u} \, d\mathcal{H}^n = -\int_\Omega \nu_{n+1} \nu_j \partial_j \widetilde{u} v \, dx \\ &= -\int_\Omega \nu_j \partial_j \widetilde{u} \, dx = \int_\Omega \widetilde{u} \partial_j \nu_j \, dx \\ &= -\int_\Omega \widetilde{u} H \, dx = -\int_\Sigma u H \nu_{n+1} \, d\mathcal{H}^n. \end{aligned}$$

We then have the desired identity. □

If X is a C^1-tangential vector field on Σ with a compact support, then

$$\int_\Sigma \operatorname{div}_\Sigma X \, d\mathcal{H}^n = 0.$$

Hence, for any $u \in C^2(\Sigma)$ and $v \in C^1(\Sigma)$, if one of them has a compact support, then

$$\int_\Sigma v \Delta_\Sigma u \, d\mathcal{H}^n = -\int_\Sigma \langle \nabla_\Sigma u, \nabla_\Sigma v \rangle \, d\mathcal{H}^n,$$

where we used the fact that $\nabla_\Sigma u$ is a tangential vector field. Later on, we will refer to this formula as an *integration by parts.*

The most basic formula in this section is the following monotonicity identity. By our convention, we denote by $D_r(p)$ balls in $\mathbb{R}^{n+1}$.

Lemma 4.1.2. *Let Σ be a C^2-hypersurface in $\mathbb{R}^{n+1}$ such that $\Sigma\cap D_R(p)\neq\emptyset$ and $\partial\Sigma\cap D_R(p)=\emptyset$ for some ball $D_R(p)\subset\mathbb{R}^{n+1}$. Then, for any function $u\in C_0^1(D_R(p))$ and any $0<\sigma<\rho<R$,*

$$\begin{aligned}\rho^{-n}\int_{\Sigma\cap D_\rho(p)} u\,d\mathcal{H}^n - \sigma^{-n}\int_{\Sigma\cap D_\sigma(p)} u\,d\mathcal{H}^n\\ = \int_{\Sigma\cap D_\rho(p)\setminus D_\sigma(p)} u|x-p|^{-n}|(D|x-p|)^\perp|^2\,d\mathcal{H}^n\\ + \int_\sigma^\rho s^{-n-1}\left(\int_{\Sigma\cap D_s(p)}\langle uH\nu+\nabla_\Sigma u, x-p\rangle\,d\mathcal{H}^n\right)ds,\end{aligned}$$

where D is the Euclidean gradient in $\mathbb{R}^{n+1}$, ν is a unit normal vector to Σ, and H is the mean curvature of Σ corresponding to ν.

Proof. For simplicity, we assume $p=0$. For an arbitrarily fixed $s>0$, we take a function $\gamma\in C^1(\mathbb{R})$ such that $\gamma(t)=1$ for $t<s/2$, $\gamma(t)=0$ for $t>s$, and $\gamma'(t)\le 0$ for all t. Set, for any $x\in D_R(p)$,

$$X(x)=u(x)\gamma(|x|)x.$$

Then,

$$\operatorname{div}_\Sigma X=\langle\nabla_\Sigma u,x\rangle\gamma(|x|)+u(x)\langle\nabla_\Sigma\gamma(|x|),x\rangle+u(x)\gamma(|x|)\operatorname{div}_\Sigma x.$$

First, we note that

$$\begin{aligned}\operatorname{div}_\Sigma x &= \langle\mathbf{e_a},\nabla_\Sigma x_a\rangle=\langle\mathbf{e_a},(\mathbf{e_a})^\top\rangle\\ &= \langle\mathbf{e_a},P\mathbf{e_a}\rangle=\operatorname{trace}P=n,\end{aligned}$$

where $P:\mathbb{R}^{n+1}\to T_x\Sigma$ is the orthogonal projection and $\mathbf{e_1},\ldots,\mathbf{e_{n+1}}$ form the standard orthonormal basis in $\mathbb{R}^{n+1}$. Next,

$$\begin{aligned}\nabla_\Sigma\gamma(|x|) &= \gamma'(r)\nabla_\Sigma|x|=\gamma'(|x|)\left(\frac{x}{|x|}\right)^\top\\ &= \gamma'(|x|)\left(\frac{x}{|x|}-\left(\frac{x}{|x|}\right)^\perp\right),\end{aligned}$$

and hence

$$\begin{aligned}\langle\nabla_\Sigma\gamma(|x|),x\rangle &= |x|\gamma'(|x|)-|x|\gamma'(|x|)\left|\frac{x^\perp}{|x|}\right|^2\\ &= |x|\gamma'(|x|)-|x|\gamma'(|x|)|(D|x|)^\perp|^2.\end{aligned}$$

Therefore,

$$\begin{aligned}\operatorname{div}_\Sigma X &= u(x)\big(|x|\gamma'(|x|)-|x|\gamma'(|x|)|(D|x|)^\perp|^2\big)\\ &\quad+nu(x)\gamma(|x|)+\gamma(|x|)\langle\nabla_\Sigma u,x\rangle.\end{aligned}$$

Lemma 4.1.1 implies

$$\int_\Sigma \operatorname{div}_\Sigma X \, d\mathcal{H}^n = -\int_\Sigma H\langle X, \nu\rangle \, d\mathcal{H}^n.$$

By a simple substitution, we have

$$\begin{aligned}\int_\Sigma u\big(|x|\gamma'(|x|) - |x|\gamma'(|x|)|(D|x|)^\perp|^2\big)\, d\mathcal{H}^n + n\int_\Sigma u\gamma(|x|)\, d\mathcal{H}^n \\ = -\int_\Sigma \gamma(|x|)\langle \nabla_\Sigma u + uH\nu, x\rangle \, d\mathcal{H}^n.\end{aligned}$$

We take a C^1-function $\phi = \phi(t)$ such that $\phi(t) = 1$ for $t \le 1/2$, $\phi(t) = 0$ for $t \ge 1$, and $\phi'(t) \le 0$ for all t. Set

$$\gamma(t) = \phi\left(\frac{t}{s}\right).$$

Then,

$$|x|\gamma'(|x|) = |x|s^{-1}\phi'\left(\frac{|x|}{s}\right) = -s\frac{\partial}{\partial s}\phi\left(\frac{|x|}{s}\right).$$

A simple substitution yields

$$\begin{aligned}n\int_\Sigma u\phi\left(\frac{|x|}{s}\right) d\mathcal{H}^n - s\frac{\partial}{\partial s}\int_\Sigma u\phi\left(\frac{|x|}{s}\right) d\mathcal{H}^n \\ = -\int_\Sigma s\frac{\partial}{\partial s}\phi\left(\frac{|x|}{s}\right) u|(D|x|)^\perp|^2 \, d\mathcal{H}^n \\ - \int_\Sigma \phi\left(\frac{|x|}{s}\right)\langle \nabla_\Sigma u + uH\nu, x\rangle \, d\mathcal{H}^n.\end{aligned}$$

By multiplying by $-s^{-n-1}$, we obtain

$$\begin{aligned}\frac{\partial}{\partial s}\left[s^{-n}\int_\Sigma u\phi\left(\frac{|x|}{s}\right) d\mathcal{H}^n\right] \\ = s^{-n}\frac{\partial}{\partial s}\int_\Sigma \phi\left(\frac{|x|}{s}\right) u|(D|x|)^\perp|^2 \, d\mathcal{H}^n \\ + s^{-n-1}\int_\Sigma \phi\left(\frac{|x|}{s}\right)\langle \nabla_\Sigma u + uH\nu, x\rangle \, d\mathcal{H}^n.\end{aligned}$$

Upon a simple integration, we have

$$\begin{aligned}
\rho^{-n}\int_\Sigma u\phi\left(\frac{|x|}{\rho}\right)d\mathcal{H}^n - \sigma^{-n}\int_\Sigma u\phi\left(\frac{|x|}{\sigma}\right)d\mathcal{H}^n\\
= \rho^{-n}\int_\Sigma \phi\left(\frac{|x|}{\rho}\right)u|(D|x|)^\perp|^2\,d\mathcal{H}^n\\
- \sigma^{-n}\int_\Sigma \phi\left(\frac{|x|}{\sigma}\right)u|(D|x|)^\perp|^2\,d\mathcal{H}^n\\
+ n\int_\sigma^\rho s^{-n-1}\int_\Sigma \phi\left(\frac{|x|}{s}\right)u|(D|x|)^\perp|^2\,d\mathcal{H}^n ds\\
+ \int_\sigma^\rho s^{-n-1}\int_\Sigma \phi\left(\frac{|x|}{s}\right)\langle\nabla_\Sigma u + uH\nu, x\rangle\,d\mathcal{H}^n ds.
\end{aligned}$$

By letting ϕ tend to the characteristic function $\chi_{(-\infty,1)}$, we obtain

$$\begin{aligned}
\rho^{-n}\int_{\Sigma\cap D_\rho} u\,d\mathcal{H}^n - \sigma^{-n}\int_{\Sigma\cap D_\sigma} u\,d\mathcal{H}^n\\
= \rho^{-n}\int_{\Sigma\cap D_\rho} u|(D|x|)^\perp|^2\,d\mathcal{H}^n - \sigma^{-n}\int_{\Sigma\cap D_\sigma} u|(D|x|)^\perp|^2\,d\mathcal{H}^n\\
+ n\int_\sigma^\rho s^{-n-1}\int_{\Sigma\cap D_s} u|(D|x|)^\perp|^2\,d\mathcal{H}^n ds\\
+ \int_\sigma^\rho s^{-n-1}\int_{\Sigma\cap D_s} \langle\nabla_\Sigma u + uH\nu, x\rangle\,d\mathcal{H}^n ds.
\end{aligned}\tag{1}$$

By exchanging the order of integration, we have

$$\begin{aligned}
n\int_\sigma^\rho s^{-n-1}\int_{\Sigma\cap D_s} u|(D|x|)^\perp|^2\,d\mathcal{H}^n ds\\
= \sigma^{-n}\int_{\Sigma\cap D_\sigma} u|(D|x|)^\perp|^2\,d\mathcal{H}^n - \rho^{-n}\int_{\Sigma\cap D_\rho} u|(D|x|)^\perp|^2\,d\mathcal{H}^n\\
+ \int_{\Sigma\cap D_\rho\setminus D_\sigma} u|x|^{-n}|(D|x|)^\perp|^2\,d\mathcal{H}^n.
\end{aligned}\tag{2}$$

A simple substitution of (2) in (1) yields the desired result. □

Lemma 4.1.2, referred to as the *monotonicity identity*, has several important applications. As the first application, we derive a mean-value inequality for subharmonic functions on hypersurfaces, which generalizes the well-known mean-value inequality for subharmonic functions in Euclidean spaces.

Theorem 4.1.3. *Let Σ be a C^2-hypersurface in $\mathbb{R}^{n+1}$ such that $\partial\Sigma \cap D_R(p) = \emptyset$ for some ball $D_R(p) \subset \mathbb{R}^{n+1}$, with $p \in \Sigma$. Suppose that u is a nonnegative subharmonic function on Σ; i.e., $\Delta_\Sigma u \geq 0$. Then, for any $0 < \rho < R$,*

$$\begin{aligned} u(p) \leq & \frac{1}{\omega_n \rho^n} \int_{\Sigma \cap D_\rho(p)} u \, d\mathcal{H}^n \\ & - \frac{1}{n\omega_n} \int_{\Sigma \cap D_\rho(p)} \left(\frac{1}{|x-p|^n} - \frac{1}{\rho^n} \right) uH \langle \nu, x-p \rangle \, d\mathcal{H}^n, \end{aligned}$$

where ω_n is the volume of the unit ball in $\mathbb{R}^n$.

Proof. Lemma 4.1.2 implies, for any $0 < \sigma < \rho < R$,

$$\begin{aligned} & \rho^{-n} \int_{\Sigma \cap D_\rho(p)} u \, d\mathcal{H}^n - \sigma^{-n} \int_{\Sigma \cap D_\sigma(p)} u \, d\mathcal{H}^n \\ & \geq \int_\sigma^\rho s^{-n-1} \left(\int_{\Sigma \cap D_s(p)} \langle uH\nu + \nabla_\Sigma u, x-p \rangle \, d\mathcal{H}^n \right) ds. \end{aligned} \tag{1}$$

For any $s \in (0, R)$, we note that

$$\langle \nabla_\Sigma u, x-p \rangle = \langle \nabla_\Sigma u, (x-p)^\top \rangle = -\frac{1}{2} \langle \nabla_\Sigma u, \nabla_\Sigma (s^2 - |x-p|^2) \rangle.$$

Since $s^2 - |x-p|^2 = 0$ on $\partial D_s(p)$, an integration by parts yields

$$\begin{aligned} \int_{\Sigma \cap D_s(p)} \langle \nabla_\Sigma u, x-p \rangle \, d\mathcal{H}^n &= \frac{1}{2} \int_{\Sigma \cap D_s(p)} \langle \nabla_\Sigma u, \nabla_\Sigma (s^2 - |x-p|^2) \rangle \, d\mathcal{H}^n \\ &= \frac{1}{2} \int_{\Sigma \cap D_s(p)} (s^2 - |x-p|^2) \Delta_\Sigma u \, d\mathcal{H}^n \geq 0. \end{aligned}$$

Next, by exchanging the order of integration, we have

$$\begin{aligned} & \int_\sigma^\rho s^{-n-1} \left(\int_{\Sigma \cap D_s(p)} uH \langle \nu, x-p \rangle d\mathcal{H}^n \right) ds \\ & = \frac{1}{n} \int_{\Sigma \cap D_\sigma(p)} \left(\frac{1}{\sigma^n} - \frac{1}{\rho^n} \right) uH \langle \nu, x-p \rangle \, d\mathcal{H}^n \\ & \quad + \frac{1}{n} \int_{\Sigma \cap D_\rho(p) \setminus D_\sigma(p)} \left(\frac{1}{|x-p|^n} - \frac{1}{\rho^n} \right) uH \langle \nu, x-p \rangle \, d\mathcal{H}^n. \end{aligned}$$

Since $p \in \Sigma$, we have

$$\lim_{\sigma \to 0} \left(\frac{1}{\omega_n \sigma^n} \int_{\Sigma \cap D_\sigma(p)} u \, d\mathcal{H}^n \right) = u(p).$$

Then, the desired result follows from simple substitutions in (1). □

As the second application of Lemma 4.1.2, we derive the Sobolev inequality on hypersurfaces, which plays an essential role in the derivation of the pointwise curvature estimates in Section 4.4. To this end, we derive two auxiliary results. The first result is an easy calculus lemma.

Lemma 4.1.4. *Let $f, h : (0,\infty) \to \mathbb{R}^+$ be bounded and nondecreasing functions satisfying, for any $0 < \sigma < \rho < \infty$ and some $n \geq 2$,*

$$\sigma^{-n} f(\sigma) \leq \rho^{-n} f(\rho) + \int_0^\rho s^{-n} h(s)\, ds. \tag{1}$$

Define $\rho_0 = 2(f(\infty))^{1/n}$ and suppose that

$$\sup_{(0,\rho_0)} \left(\rho^{-n} f(\rho)\right) \geq 1.$$

Then, there exists a $\rho \in (0, \rho_0)$ such that

$$f(5\rho) \leq \frac{5^n}{2} \rho_0 h(\rho). \tag{2}$$

Proof. We prove by contradiction. Assume that (2) is not correct for all $\rho \in (0,\rho_0)$. Then, (1) would imply

$$\begin{aligned}
\sup_{0<\sigma<\rho_0} (\sigma^{-n} f(\sigma)) &\leq \rho_0^{-n} f(\rho_0) + \frac{2}{5^n \rho_0} \int_0^{\rho_0} s^{-n} f(5s)\, ds \\
&\leq 2^{-n} + \frac{2}{5\rho_0} \left(\int_0^{\rho_0} s^{-n} f(s)\, ds + \int_{\rho_0}^{5\rho_0} s^{-n} f(s)\, ds \right) \\
&\leq 2^{-n} + \frac{2}{5} \sup_{0<s<\rho_0} (s^{-n} f(s)) + \frac{2}{5(n-1)} \rho_0^{-n} f(\infty),
\end{aligned}$$

or

$$\frac{3}{5} \leq \frac{3}{5} \sup_{0<s<\rho_0} (s^{-n} f(s)) \leq \frac{1}{4} + \frac{2}{5} \cdot \frac{1}{4} = \frac{7}{20}.$$

This would lead to a contradiction. □

The following Vitali covering lemma is an important tool in the proof of the Sobolev inequality on hypersurfaces. In the next result, we use the notation $\widehat{B} = \overline{B_{5r}(x)}$ if $B = \overline{B_r(x)}$.

Lemma 4.1.5. *Suppose that $\mathcal{B}$ is a family of closed balls in $\mathbb{R}^n$ with uniformly bounded radii. Then, there is a pairwise disjoint subcollection $\mathcal{B}' \subset \mathcal{B}$ such that*

$$\bigcup_{B\in\mathcal{B}} B \subset \bigcup_{B'\in\mathcal{B}'} \widehat{B}'.$$

Moreover, for any $B \in \mathcal{B}$, there exists an $S \in \mathcal{B}'$ such that $S \cap B \neq \emptyset$ and $B \subset \widehat{S}$.

Obviously, $\mathcal{B}'$ is a countable collection.

Proof. Set

$$R = \sup\{\text{rad}(B) : B \in \mathcal{B}\} < \infty.$$

For $j = 1, 2, \ldots,$ set

$$\mathcal{B}_j = \{B \in \mathcal{B} : R/2^j < \text{rad}(B) \leq R/2^{j-1}\}.$$

Then, $\mathcal{B} = \bigcup_{j=1}^{\infty} \mathcal{B}_j$. We proceed to define $\mathcal{B}'_j \subset \mathcal{B}_j$ as follows:

(i) let $\mathcal{B}'_1$ be any maximal pairwise disjoint subcollection of $\mathcal{B}_1$;

(ii) assuming $j \geq 2$ and $\mathcal{B}'_1, \ldots, \mathcal{B}'_{j-1}$ are defined, let $\mathcal{B}'_j$ be a maximal pairwise disjoint subcollection of

$$\Big\{B \in \mathcal{B}_j : B \cap B' = \emptyset \text{ whenever } B' \in \bigcup_{i=1}^{j-1} \mathcal{B}'_i\Big\}.$$

Then, if $j \geq 1$ and $B \in \mathcal{B}_j$, we have

$$B \cap B' \neq \emptyset \quad \text{for some } B' \in \bigcup_{i=1}^{j} \mathcal{B}'_i.$$

Otherwise, $\mathcal{B}'_j$ is not maximal. For such a pair B and B', we have $\text{rad}(B) \leq R/2^{j-1} = 2R/2^j \leq 2\text{rad}(B')$, so that $B \subset \widehat{B}'$.

To end the proof, we simply take $\mathcal{B}' = \bigcup_{i=1}^{\infty} \mathcal{B}'_i$. □

We are ready to prove the Sobolev inequality on hypersurfaces due to Michael and Simon [**113**].

Theorem 4.1.6. *Let Σ be a C^2-hypersurface in $\mathbb{R}^{n+1}$ and U be a domain in $\mathbb{R}^{n+1}$ such that $U \cap \Sigma \neq \emptyset$ and $U \cap \partial\Sigma = \emptyset$. Suppose that u is a nonnegative $C_0^1(U)$-function. Then,*

$$\left(\int_\Sigma u^{\frac{n}{n-1}}\, d\mathcal{H}^n\right)^{\frac{n-1}{n}} \leq c \int_\Sigma (|\nabla_\Sigma u| + u|H|)\, d\mathcal{H}^n,$$

where c is a positive constant depending only on n.

Proof. By our convention, we denote by $D_r(p)$ the $(n+1)$-dimensional ball in $\mathbb{R}^{n+1}$. For any $p \in \Sigma$, take R such that $D_R(p) \Subset U$. For any $0 < \sigma < \rho < R$, Lemma 4.1.2 implies

$$\sigma^{-n} \int_{\Sigma \cap D_\sigma(p)} u\, d\mathcal{H}^n \leq \rho^{-n} \int_{\Sigma \cap D_\rho(p)} u\, d\mathcal{H}^n + \int_\sigma^\rho s^{-n} \int_{\Sigma \cap D_s(p)} (u|H| + |\nabla_\Sigma u|)\, d\mathcal{H}^n ds.$$

Set

$$f(\rho) = \frac{1}{\omega_n} \int_{\Sigma \cap D_\rho(p)} u \, d\mathcal{H}^n$$

and

$$h(\rho) = \frac{1}{\omega_n} \int_{\Sigma \cap D_\rho(p)} (u|H| + |\nabla_\Sigma u|) \, d\mathcal{H}^n.$$

Since u has a compact support, we can let $\rho \to \infty$. Set $\rho_0 = 2[f(\infty)]^{1/n}$. Then,

$$\sup_{(0,\rho_0)} (\rho^{-n} f(\rho)) \geq 1,$$

for any $p \in \Sigma$ with $u(p) \geq 1$. Hence, for each $p \in \Sigma$ with $u(p) \geq 1$, Lemma 4.1.4 implies the existence of a $\rho = \rho(p) \in (0, \rho_0)$ such that

$$(1) \quad \int_{\Sigma \cap D_{5\rho}(p)} u \, d\mathcal{H}^n \leq 5^n \left(\omega_n^{-1} \int_\Sigma u \, d\mathcal{H}^n \right)^{\frac{1}{n}} \int_{\Sigma \cap D_\rho(p)} (u|H| + |\nabla_\Sigma u|) \, d\mathcal{H}^n.$$

By Lemma 4.1.5 applied to the collection of closed balls

$$\{\overline{D}_{\rho(p)}(p)\}_{p \in \Sigma \cap \{u(p) \geq 1\}},$$

there exist countably many disjoint balls $D_{\rho_j}(p_j)$, for some $p_j \in \Sigma$ with $u(p_j) \geq 1$, such that

$$\{p \in \Sigma : u(p) \geq 1\} \subset \bigcup_{j=1}^{\infty} \overline{D}_{5\rho_j}(p_j).$$

Applying (1) for each p_j and summing over j, we obtain

$$(2) \quad \int_{\{p \in \Sigma : u(p) \geq 1\}} u \, d\mathcal{H}^n \leq 5^n \left(\omega_n^{-1} \int_\Sigma u \, d\mathcal{H}^n \right)^{\frac{1}{n}} \int_\Sigma (u|H| + |\nabla_\Sigma u|) \, d\mathcal{H}^n.$$

Take a positive constant ε. Consider a nondecreasing function $\gamma \in C^1(\mathbb{R})$ with $\gamma(t) = 1$ for $t > \varepsilon > 0$ and $\gamma(t) = 0$ for $t < 0$ and apply (2) with $\gamma(u-t)$, $t \geq 0$, in place of u. We then have, for any $t > 0$,

$$\begin{aligned}
&\mathcal{H}^n(\Sigma \cap \{u > t + \varepsilon\}) \\
&\leq \int_{\Sigma \cap \{u - t > \varepsilon\}} 1 \, d\mathcal{H}^n \leq \int_{\Sigma \cap \{\gamma(u-t) \geq 1\}} \gamma(u-t) \, d\mathcal{H}^n \\
&\leq 5^n \left(\omega_n^{-1} \int_\Sigma \gamma(u-t) \, d\mathcal{H}^n \right)^{\frac{1}{n}} \int_\Sigma (\gamma(u-t)|H| + \gamma'(u-t)|\nabla_\Sigma u|) \, d\mathcal{H}^n \\
&\leq 5^n \left(\omega_n^{-1} \int_{\Sigma \cap \{u > t\}} d\mathcal{H}^n \right)^{\frac{1}{n}} \int_\Sigma (\gamma(u-t)|H| + \gamma'(u-t)|\nabla_\Sigma u|) \, d\mathcal{H}^n.
\end{aligned}$$

Multiplying this inequality by $(t+\varepsilon)^{\frac{1}{n-1}}$, we obtain

$$\begin{aligned}
&(t+\varepsilon)^{\frac{1}{n-1}}\mathcal{H}^n(\Sigma\cap\{u>t+\varepsilon\})\\
&\le 5^n\left(\omega_n^{-1}\int_{\Sigma\cap\{u>t\}}(u+\varepsilon)^{\frac{n}{n-1}}\,d\mathcal{H}^n\right)^{\frac{1}{n}}\\
&\quad\cdot\int_\Sigma(\gamma(u-t)|H|+\gamma'(u-t)|\nabla_\Sigma u|)\,d\mathcal{H}^n\\
&\le 5^n\left(\omega_n^{-1}\int_{\Sigma\cap\operatorname{supp}u}(u+\varepsilon)^{\frac{n}{n-1}}\,d\mathcal{H}^n\right)^{\frac{1}{n}}\\
&\quad\cdot\left\{-\frac{d}{dt}\int_\Sigma\gamma(u-t)|\nabla_\Sigma u|\,d\mathcal{H}^n+\int_\Sigma\gamma(u-t)|H|\,d\mathcal{H}^n\right\}.
\end{aligned}$$

Integrating this inequality over $t\in(0,\infty)$, we have

$$\begin{aligned}
&\int_0^\infty(t+\varepsilon)^{\frac{1}{n-1}}\mathcal{H}^n(\Sigma\cap\{u>t+\varepsilon\})\,dt\\
&\qquad\le 5^n\left(\omega_n^{-1}\int_{\Sigma\cap\operatorname{supp}u}(u+\varepsilon)^{\frac{n}{n-1}}\,d\mathcal{H}^n\right)^{\frac{1}{n}}\\
&\qquad\quad\cdot\left\{\int_\Sigma|\nabla_\Sigma u|d\mathcal{H}^n+\int_0^\infty\int_{\Sigma\cap\{u>t\}}|H|\,d\mathcal{H}^n dt\right\}.
\end{aligned}$$

By exchanging the order of integration, we conclude

$$\begin{aligned}
&\int_0^\infty(t+\varepsilon)^{\frac{1}{n-1}}\mathcal{H}^n(\Sigma\cap\{u>t+\varepsilon\})\,dt\\
&\qquad=\frac{n-1}{n}\int_{\Sigma\cap\{u>\varepsilon\}}\left(u^{\frac{n}{n-1}}-\varepsilon^{\frac{n}{n-1}}\right)\,d\mathcal{H}^n
\end{aligned}$$

and

$$\int_0^\infty\int_{\Sigma\cap\{u>t\}}|H|\,d\mathcal{H}^n dt=\int_\Sigma u|H|\,d\mathcal{H}^n.$$

We have the desired estimate by a simple substitution and then letting $\varepsilon\to 0$. □

The Sobolev inequality has the following more general form.

Theorem 4.1.7. *Let Σ be a C^2-hypersurface in $\mathbb{R}^{n+1}$ and U be a domain in $\mathbb{R}^{n+1}$ such that $U\cap\Sigma\neq\emptyset$ and $U\cap\partial\Sigma=\emptyset$. Suppose that u is a nonnegative $C_0^1(U)$-function and $1\le p<n$. Then,*

$$\left(\int_\Sigma u^{\frac{np}{n-p}}\,d\mathcal{H}^n\right)^{\frac{n-p}{np}}\le c\left(\int_\Sigma(|\nabla_\Sigma u|^p+u^p|H|^p)\,d\mathcal{H}^n\right)^{\frac{1}{p}},$$

where c is a positive constant depending only on n and p.

Proof. Take a constant $\gamma > 1$ to be determined. Applying Theorem 4.1.6 to u^γ and using Hölder's inequality, we conclude

$$\begin{aligned}\left(\int_\Sigma u^{\frac{\gamma n}{n-1}}\,d\mathcal{H}^n\right)^{\frac{n-1}{n}} &\leq c\int_\Sigma (u^\gamma|H| + \gamma u^{\gamma-1}|\nabla_\Sigma u|)\,d\mathcal{H}^n \\ &\leq c\left(\int_\Sigma u^{\frac{(\gamma-1)p}{p-1}}\,d\mathcal{H}^n\right)^{\frac{p-1}{p}}\left(\int_\Sigma u^p|H|^p\,d\mathcal{H}^n\right)^{\frac{1}{p}} \\ &\quad + c\gamma\left(\int_\Sigma u^{\frac{(\gamma-1)p}{p-1}}\,d\mathcal{H}^n\right)^{\frac{p-1}{p}}\left(\int_\Sigma |\nabla_\Sigma u|^p\,d\mathcal{H}^n\right)^{\frac{1}{p}}.\end{aligned}$$

By requiring

$$\frac{\gamma n}{n-1} = \frac{(\gamma-1)p}{p-1},$$

we have

$$\gamma = \frac{p(n-1)}{n-p} > 1,$$

and hence

$$\frac{\gamma n}{n-1} = \frac{(\gamma-1)p}{p-1} = \frac{np}{n-p}.$$

Then,

$$\begin{aligned}&\left(\int_\Sigma u^{\frac{np}{n-p}}\,d\mathcal{H}^n\right)^{\frac{n-1}{n}} \\ &\qquad \leq c\left(\int_\Sigma u^{\frac{np}{n-p}}\,d\mathcal{H}^n\right)^{\frac{p-1}{p}}\left(\int_\Sigma (|\nabla_\Sigma u|^p + u^p|H|^p)\,d\mathcal{H}^n\right)^{\frac{1}{p}}.\end{aligned}$$

This implies the desired estimate. □

4.2. Differential Identities

In this section, we discuss differential operators on hypersurfaces and derive some important identities concerning the Laplace-Beltrami operators. We treat hypersurfaces as Riemannian submanifolds and employ standard methods from the Riemannian geometry. Those with no background in the Riemannian geometry can skip this section and refer to Section 4.5 for alternative proofs.

Let Σ be a C^2-hypersurface in $\mathbb{R}^{n+1}$. We regard Σ as a Riemannian submanifold in $\mathbb{R}^{n+1}$. For each point $p \in \Sigma$, the tangent space $T_p\Sigma$ is an n-dimensional subspace in $\mathbb{R}^{n+1}$. A (tangential) vector field X on Σ assigns to each point $p \in \Sigma$ a vector $X(p) \in T_p\Sigma$. In local coordinates, we write

$$X = X^i\frac{\partial}{\partial x^i}.$$

In particular, $\{\frac{\partial}{\partial x^1}, \ldots, \frac{\partial}{\partial x^n}\}$ forms a basis of $T_p\Sigma$. The cotangent space $T_p^*\Sigma$ is the dual space of $T_p\Sigma$. In local coordinates, $dx^1, \ldots, dx^n$ are dual to the basis $\frac{\partial}{\partial x^1}, \ldots, \frac{\partial}{\partial x^n}$; i.e.,

$$dx^i\left(\frac{\partial}{\partial x^j}\right) = \delta^i_j.$$

More generally, we have tensor fields on Σ. In this chapter, we only need $(0, s)$- and $(1, s)$-tensor fields. A $(0, s)$-tensor field T on Σ assigns to each point $p \in \Sigma$ a multilinear mapping

$$T_p : T_p\Sigma \times \cdots \times T_p\Sigma \to \mathbb{R}.$$

In local coordinates, we write

$$T = T_{i_1\cdots i_s} dx^{i_1} \otimes \cdots \otimes dx^{i_s},$$

where

$$T_{i_1\cdots i_s} = T\left(\frac{\partial}{\partial x^{i_1}}, \ldots, \frac{\partial}{\partial x^{i_s}}\right).$$

We note that $\{dx^{i_1} \otimes \cdots \otimes dx^{i_s}\}_{i_1,\ldots,i_s=1,\ldots,n}$ forms a basis of the space of all $(0, s)$-tensors and

$$(dx^{i_1} \otimes \cdots \otimes dx^{i_s})\left(\frac{\partial}{\partial x^{j_1}}, \ldots, \frac{\partial}{\partial x^{j_s}}\right) = \delta^{i_1}_{j_1} \cdots \delta^{i_s}_{j_s}.$$

Similarly, a $(1, s)$-tensor field assigns to each point $p \in \Sigma$ a multilinear mapping

$$T_p : T_p\Sigma \times \cdots \times T_p\Sigma \to T_p\Sigma.$$

In local coordinates, we write

$$T = T^k_{i_1\cdots i_s} \frac{\partial}{\partial x^k} \otimes dx^{i_1} \otimes \cdots \otimes dx^{i_s}.$$

The tensor T is differentiable if all functions $T_{i_1\cdots i_s}$ or $T^k_{i_1\cdots i_s}$ are differentiable.

The standard Euclidean metric in $\mathbb{R}^{n+1}$ induces a metric g on Σ, which is a $(0, 2)$-tensor field on Σ. In local coordinates, we write

$$g = g_{ij} dx^i \otimes dx^j.$$

We also denote the metric g by $\langle \cdot, \cdot \rangle$. The metric g extends naturally to all tensor fields. For example, for any two $(0, 2)$-tensor fields T and S, we have

$$\langle T, S\rangle = g^{ik} g^{jl} T_{ij} S_{kl},$$

where (g^{ij}) is the inverse of the matrix (g_{ij}). In particular, the squared length of the $(0, 2)$-tensor field T is given by

$$|T|^2 = g^{ik} g^{jl} T_{ij} T_{kl}.$$

There exists a unique affine connection ∇, called the *Levi-Civita connection*, such that

(1) ∇ is g-compatible; i.e., for any vector fields X, Y, and Z,

$$\nabla_X\langle Y, Z\rangle = \langle\nabla_X Y, Z\rangle + \langle Y, \nabla_X Z\rangle;$$

(2) ∇ is torsion free; i.e., for any vector fields X and Y,

$$\nabla_X Y - \nabla_Y X = [X, Y].$$

In local coordinates, we define the *Christoffel symbols* Γ_{ij}^k by

$$\nabla_{\frac{\partial}{\partial x^i}}\frac{\partial}{\partial x^j} = \Gamma_{ij}^k\frac{\partial}{\partial x^k},$$

where

$$\Gamma_{ij}^k = \frac{1}{2}g^{kl}(\partial_j g_{il} + \partial_i g_{jl} - \partial_l g_{ij}).$$

For

$$X = X^i\frac{\partial}{\partial x^i}, \quad Y = Y^i\frac{\partial}{\partial x^i},$$

we have

$$\nabla_X Y = \left(X^k\frac{\partial Y^i}{\partial x^k} + X^k Y^l\Gamma_{kl}^i\right)\frac{\partial}{\partial x^i}.$$

We also write

$$\nabla_X Y = X^k\nabla_k Y^i\frac{\partial}{\partial x^i},$$

where

$$\nabla_k Y^i = \frac{\partial Y^i}{\partial x^k} + Y^l\Gamma_{kl}^i.$$

Moreover, the connection ∇ extends uniquely to all tensor fields that satisfies the Leibniz rule relative to the tensor product and commutes with contractions. For example, let T denote a $(0, s)$- or $(1, s)$-tensor and X be a vector field on Σ. Then for any vector fields $Y_1, \ldots, Y_s$ on Σ, we have

$$\begin{aligned}(\nabla_X T)(Y_1, \ldots, Y_s) = \nabla_X(T(Y_1, \ldots, Y_s)) &- T(\nabla_X Y_1, Y_2, \ldots, Y_s)\\ &- T(Y_1, \nabla_X Y_2, \ldots, Y_s) - \cdots\\ &- T(Y_1, \ldots, Y_{s-1}, \nabla_X Y_s).\end{aligned}$$

We denote by ∇T the $(0, s+1)$- or $(1, s+1)$-tensor given by

$$\nabla T(X, Y_1, \ldots, Y_s) = (\nabla_X T)(Y_1, \ldots, Y_s).$$

Let T be a $(0,s)$-tensor. Then, we write

$$\nabla T = \nabla_k T_{i_1\cdots i_s} dx^k \otimes dx^{i_1} \otimes \cdots \otimes dx^{i_s},$$

where

$$\begin{aligned}\nabla_k T_{i_1\cdots i_s} &= \nabla T\left(\frac{\partial}{\partial x^k}, \frac{\partial}{\partial x^{i_1}}, \ldots, \frac{\partial}{\partial x^{i_s}}\right)\\ &= \frac{\partial T_{i_1\cdots i_s}}{\partial x^k} - \Gamma^l_{ki_1} T_{li_2\cdots i_s} - \cdots - \Gamma^l_{ki_s} T_{i_1\cdots i_{s-1}l}.\end{aligned}$$

Let $u : \Sigma \to \mathbb{R}$ be a differentiable function on Σ. The *differential* du of u is a (0,1)-tensor given by

$$du = \nabla_i u dx^i.$$

Its dual is the *gradient* ∇u given by

$$\nabla u = g^{ij}\nabla_i u \frac{\partial}{\partial x^j}.$$

The covariant derivative $\nabla(du)$ is a (0,2)-tensor, which in local coordinates is given by

$$\nabla^2 u = \nabla_i \nabla_j u dx^i \otimes dx^j,$$

where

$$\nabla_i\nabla_j u = \frac{\partial^2 u}{\partial x^i \partial x^j} - \Gamma^k_{ij}\frac{\partial u}{\partial x^k}.$$

Taking the trace of $\nabla^2 u$ yields the *Laplace-Beltrami operator*

$$\Delta_\Sigma u = g^{ij}\nabla_i\nabla_j u = g^{ij}\left(\frac{\partial^2 u}{\partial x^i \partial x^j} - \Gamma^k_{ij}\frac{\partial u}{\partial x^k}\right).$$

The *Riemann curvature tensor* is a (1,3)-tensor R defined by

$$R(X,Y)Z = \nabla_X\nabla_Y Z - \nabla_Y\nabla_X Z - \nabla_{[X,Y]}Z,$$

for any vector fields X, Y, and Z. In local coordinates, we write

$$R\left(\frac{\partial}{\partial x^i}, \frac{\partial}{\partial x^j}\right)\frac{\partial}{\partial x^k} = R^l_{ijk}\frac{\partial}{\partial x^l},$$

where

$$R^l_{ijk} = \partial_i \Gamma^l_{jk} - \partial_j \Gamma^l_{ik} + \Gamma^l_{im}\Gamma^m_{jk} - \Gamma^l_{jm}\Gamma^m_{ik}.$$

We note that the Riemann curvature tensor is a quasilinear second-order operator on the metric g_{ij}. We also introduce another form of Riemann curvature tensor, a (0,4)-tensor R given by

$$R(X,Y,Z,W) = \langle R(X,Y)Z, W\rangle.$$

In local coordinates, we have

$$R_{ijkl} = \left\langle R\left(\frac{\partial}{\partial x^i}, \frac{\partial}{\partial x^j}\right)\frac{\partial}{\partial x^k}, \frac{\partial}{\partial x^l}\right\rangle,$$

or

$$R_{ijkl} = g_{lm}R^m_{ijk}.$$

It is straightforward to check that

$$R_{ijkl} = R_{klij} = -R_{jikl}.$$

Curvature tensors appear naturally if we exchange the order of the covariant differentiation. For example, let T be a (0,2)-tensor given by

$$T = T_{ij}dx^i \otimes dx^j.$$

Then,

$$\nabla_k\nabla_l T_{ij} = \nabla_l\nabla_k T_{ij} - R^m_{kli}T_{mj} - R^m_{klj}T_{im}.$$

This formula will be needed in the derivation of the Simons identity in Lemma 4.2.2.

We denote by D the Levi-Civita connection associated with the standard Euclidean metric in $\mathbb{R}^{n+1}$. Then, D is the Euclidean directional derivative. It is easy to check that

$$R^e(X,Y)Z := D_X D_Y Z - D_Y D_X Z - D_{[X,Y]}Z = 0.$$

This simply says that the Riemann curvature of $\mathbb{R}^{n+1}$ is zero. Let ν be the unit normal field of Σ. For any (tangential) vector fields X and Y of Σ, we decompose $D_X Y$ into tangential and normal components and write

$$D_X Y = \nabla_X Y + A(X,Y)\nu,$$

where A is the *second fundamental form* of Σ; i.e.,

$$A(X,Y) = \langle D_X Y, \nu\rangle = -\langle D_X\nu, Y\rangle.$$

We note that A is a $(0,2)$-tensor on Σ. In local coordinates, we write

$$A = h_{ij}dx^i \otimes dx^j,$$

where

$$h_{ij} = A\left(\frac{\partial}{\partial x^i}, \frac{\partial}{\partial x^j}\right).$$

Next, we derive the Gauss and Codazzi equations by decomposing the Riemann curvature of the Euclidean space into tangential and normal parts. Below, X, Y, Z, and W are (tangential) vector fields on Σ. First,

$$\begin{aligned}
D_X D_Y Z &= D_X(\nabla_Y Z + A(Y,Z)\nu) \\
&= D_X(\nabla_Y Z) + X\big(A(Y,Z)\big)\nu + A(Y,Z)D_X\nu \\
&= \nabla_X\nabla_Y Z + A(X,\nabla_Y Z)\nu + (\nabla_X A)(Y,Z)\nu \\
&\quad + A(\nabla_X Y, Z)\nu + A(Y,\nabla_X Z)\nu + A(Y,Z)D_X\nu.
\end{aligned}$$

With

$$D_{[X,Y]}Z = \nabla_{[X,Y]}Z + A([X,Y], Z)\nu,$$

we obtain

$$\begin{aligned} R^e(X,Y)Z = R(X,Y)Z &+ A(Y,Z)D_X\nu - A(X,Z)D_Y\nu \\ &+ (\nabla_X A)(Y,Z)\nu - (\nabla_Y A)(X,Z)\nu. \end{aligned}$$

We point out that the expression in the left-hand side is zero. Hence, its tangential part and normal part are zero respectively; i.e.,

$$R(X,Y)Z + A(Y,Z)D_X\nu - A(X,Z)D_Y\nu = 0,$$

and

$$(\nabla_X A)(Y,Z)\nu - (\nabla_Y A)(X,Z)\nu = 0.$$

For the tangential part, we then have

$$\langle R(X,Y)Z, W\rangle = -A(Y,Z)\langle D_X\nu, W\rangle + A(X,Z)\langle D_Y\nu, W\rangle,$$

and therefore,

$$R(X,Y,Z,W) = A(X,W)A(Y,Z) - A(X,Z)A(Y,W).$$

These are the *Gauss equations.* For the normal part, we have

$$(\nabla_Y A)(X,Z) = (\nabla_X A)(Y,Z).$$

These are the *Codazzi equations.* In local coordinates, Gauss and Codazzi equations have the forms

$$R_{ijkl} = h_{il}h_{jk} - h_{ik}h_{jl}$$

and

$$\nabla_j h_{ik} = \nabla_i h_{jk}.$$

Having finished the brief review, we now collect some notations which will be needed in the following. Let Σ be an n-dimensional differentiable hypersurface in $\mathbb{R}^{n+1}$. We denote by g or $\langle\cdot,\cdot\rangle$ the induced metric on Σ and by A the second fundamental form corresponding to a unit normal field ν. In local coordinates, we write

$$g = g_{ij}dx^i \otimes dx^j, \quad A = h_{ij}dx^i \otimes dx^j.$$

The mean curvature H is given by

$$H = g^{ij}h_{ij},$$

and the squared length of the second fundamental form is given by

$$|A|^2 = g^{ik}g^{jl}h_{ij}h_{kl}.$$

The Levi-Civita connection is denoted by ∇. We will write ∇_Σ, when ∇ acts on functions or $(0,s)$-tensor fields. For example, $\nabla_\Sigma A$ is the covariant derivative of A, a $(0,3)$-tensor, given by

$$\nabla_\Sigma A = \nabla_i h_{jk} dx^i \otimes dx^j \otimes dx^k,$$

with a norm

$$|\nabla_\Sigma A|^2 = g^{ik}g^{jl}g^{pq}\nabla_i h_{jp}\nabla_k h_{lq}.$$

We now derive a differential identity, which will be used in the derivation of the interior gradient estimates in Section 4.3.

Lemma 4.2.1. *Let Σ be a C^4-embedded hypersurface in $\mathbb{R}^{n+1}$ and ν be its unit normal field. Then,*

$$\Delta_\Sigma \nu + |A|^2\nu + \nabla_\Sigma H = 0 \quad \text{on } \Sigma.$$

Proof. We introduce Riemann normal coordinates at a given point $p \in \Sigma$, with orthonormal tangent vectors $\tau_1, \ldots, \tau_n$ and such that $\Gamma_{ij}^k = 0$ for all i, j, k at this particular point. We denote by $D_i = D_{\tau_i}$ the directional derivatives and observe that $D_iD_j = \nabla_i\nabla_j$ at this point. Furthermore, at p,

$$D_i\tau_j = h_{ij}\nu, \qquad D_i\nu = -h_{ij}\tau_j,$$

and

$$\Delta_\Sigma \nu = D_iD_i\nu.$$

The Codazzi equations yield, at p and for all $i, j, k = 1, \ldots, n$,

$$D_k h_{ij} = D_j h_{ik}.$$

We now compute at the point p:

$$\begin{aligned}\Delta_\Sigma \nu &= D_iD_i\nu = -D_i(h_{ij}\tau_j) = -(D_ih_{ij})\tau_j - h_{ij}D_i\tau_j \\ &= -(D_jh_{ii})\tau_j - \sum_{i,j} h_{ij}^2\nu = -\nabla_\Sigma H - |A|^2\nu.\end{aligned}$$

This is the desired identity. □

In the rest of this section, we derive a differential inequality which plays an essential role in the discussion of curvature estimates in Section 4.4. We proceed with some preparations.

The following identity, due to Simons [**140**], is of crucial importance.

Lemma 4.2.2. *Let Σ be a C^4-embedded hypersurface in $\mathbb{R}^{n+1}$. Then,*

$$\Delta_\Sigma h_{ij} = \nabla_i\nabla_j H + Hg^{kl}h_{ik}h_{lj} - |A|^2 h_{ij}.$$

Proof. By the Codazzi equation, we have

$$\nabla_l h_{ij} = \nabla_i h_{jl},$$

and hence

$$\nabla_k\nabla_l h_{ij} = \nabla_k\nabla_i h_{jl}.$$

By exchanging the order of differentiation, we get

$$\begin{aligned}\nabla_k\nabla_l h_{ij} &= \nabla_i\nabla_k h_{jl} - R_{kil}^p h_{pj} - R_{kij}^p h_{pl} \\ &= \nabla_i\nabla_j h_{kl} - g^{pq}R_{kilq}h_{pj} - g^{pq}R_{kijq}h_{pl},\end{aligned}$$

where we used the Codazzi equation again. Then,

$$\Delta_\Sigma h_{ij} = g^{kl}\nabla_k\nabla_l h_{ij} = g^{kl}\nabla_i\nabla_j h_{kl} - g^{kl}g^{pq}R_{kilq}h_{pj} - g^{kl}g^{pq}R_{kijq}h_{pl}.$$

The Gauss equation yields

$$\begin{aligned} g^{kl}g^{pq}R_{kilq}h_{pj} &+ g^{kl}g^{pq}R_{kijq}h_{pl} \\ &= g^{kl}g^{pq}(h_{kq}h_{il} - h_{kl}h_{iq})h_{pj} + g^{kl}g^{pq}(h_{kq}h_{ij} - h_{kj}h_{iq})h_{pl} \\ &= h_{ij}g^{kl}g^{pq}h_{kq}h_{lp} - Hg^{pq}h_{iq}h_{jp}. \end{aligned}$$

A simple substitution implies the desired identity. □

Next, we calculate the Laplace-Beltrami operator acting on the squared length of the second fundamental form.

Lemma 4.2.3. *Let Σ be a C^4-embedded hypersurface in $\mathbb{R}^{n+1}$. Then,*

$$\frac{1}{2}\Delta_\Sigma|A|^2 = |\nabla_\Sigma A|^2 + h^{ij}\nabla_i\nabla_j H + Hh_i^j h_j^k h_k^i - |A|^4,$$

where $h_j^i = g^{ik}h_{kj}$ and $h^{ij} = g^{ik}g^{jl}h_{kl}$.

Proof. We first recall the identity

$$(1) \qquad |A|^2 = g^{ik}g^{jl}h_{kl}h_{ij}.$$

Then,

$$\begin{aligned} \frac{1}{2}\Delta_\Sigma|A|^2 &= \frac{1}{2}\Delta_\Sigma(g^{ik}g^{jl}h_{kl}h_{ij}) \\ &= g^{ik}g^{jl}h_{kl}\Delta_\Sigma h_{ij} + g^{ik}g^{jl}g^{pq}\nabla_p h_{kl}\nabla_q h_{ij}. \end{aligned}$$

By Lemma 4.2.2 and (1) again, we obtain

$$\begin{aligned} g^{ik}g^{jl}h_{kl}\Delta_\Sigma h_{ij} &= g^{ik}g^{jl}h_{kl}(\nabla_i\nabla_j H + Hg^{pq}h_{ip}h_{qj} - |A|^2 h_{ij}) \\ &= g^{ik}g^{jl}h_{kl}\nabla_i\nabla_j H + g^{ik}g^{jl}g^{pq}h_{kl}h_{ip}h_{qj} - |A|^4. \end{aligned}$$

A simple substitution yields the desired result. □

As a consequence, we prove an estimate of the second fundamental form for minimal hypersurfaces, which plays an essential role in the discussion of integral curvature estimates in Section 4.4.

Lemma 4.2.4. *Let Σ be a C^4-minimal hypersurface in $\mathbb{R}^{n+1}$. Then,*

$$\frac{1}{2}\Delta_\Sigma|A|^2 \geq \left(1 + \frac{2}{n}\right)|\nabla_\Sigma|A||^2 - |A|^4.$$

Proof. Since $H \equiv 0$, then

$$\frac{1}{2}\Delta_\Sigma |A|^2 = |\nabla_\Sigma A|^2 - |A|^4.$$

We need to estimate $|\nabla_\Sigma A|^2$. In the following, we write $h_{ij,k}$ for the covariant derivative $\nabla_k h_{ij}$ of the $(0,2)$-tensor h_{ij}. In this notation, the Codazzi equations have the form, for any $i, j, k = 1, \ldots, n$,

$$h_{ij,k} = h_{ik,j}.$$

In the following, we fix a point $p \in \Sigma$ and introduce Riemann normal coordinates near p, so that at the point p the vectors $\tau_1, \ldots, \tau_n$ are orthonormal, $\Gamma_{ij}^k = 0$, and $h_{ij} = \kappa_i \delta_{ij}$ is diagonal. Then, we have, at p,

$$|A|^2 = \sum_i h_{ii}^2$$

and

$$|\nabla_\Sigma A|^2 = \sum_{i,j,k} h_{ij,k}^2.$$

Here and hereafter, summations are from 1 to n.

We first note that

$$|\nabla A|^2 \geq |\nabla_\Sigma |A||^2.$$

To verify this, we consider the case $|A(p)| \neq 0$. Then, at p,

$$|\nabla_\Sigma |A||^2 = \frac{1}{|A|^2} \sum_k \Big(\sum_{i,j} h_{ij} h_{ij,k} \Big)^2 = \frac{1}{|A|^2} \sum_k \Big(\sum_i h_{ii} h_{ii,k} \Big)^2.$$

Hence, the desired estimate follows from the Cauchy inequality. In fact, we have

$$|\nabla_\Sigma |A||^2 \leq \sum_{i,k} h_{ii,k}^2.$$

Now we claim

$$|\nabla_\Sigma A|^2 \geq \Big(1 + \frac{n}{2}\Big) |\nabla_\Sigma |A||^2.$$

With the normal coordinates introduced above near p, we have at p,

$$\begin{aligned}
|\nabla_\Sigma A|^2 - |\nabla_\Sigma |A||^2 &\geq \sum_{i,j,k} h_{ij,k}^2 - \sum_{i,k} h_{ii,k}^2 = \sum_{i,k} \sum_{j \neq i} h_{ij,k}^2 \\
&\geq \sum_i \sum_{j \neq i} h_{ij,i}^2 + \sum_i \sum_{j \neq i} h_{ij,j}^2 \qquad (1)\\
&= 2 \sum_i \sum_{j \neq i} h_{ij,j}^2 = 2 \sum_i \sum_{j \neq i} h_{jj,i}^2,
\end{aligned}$$

where we used the Codazzi equations. Next, since

$$H = \sum_j h_{jj} = 0,$$

we have

$$h_{ii,i} = -\sum_{j\neq i} h_{jj,i},$$

and hence

$$\begin{aligned}|\nabla_\Sigma |A||^2 &\leq \sum_{i,k} h_{ii,k}^2 = \sum_i \sum_{k\neq i} h_{ii,k}^2 + \sum_i h_{ii,i}^2 \\ &= \sum_i \sum_{k\neq i} h_{ii,k}^2 + \sum_i \Big(\sum_{j\neq i} h_{jj,i}\Big)^2 \qquad (2) \\ &\leq \sum_i \sum_{k\neq i} h_{ii,k}^2 + (n-1)\sum_i \sum_{j\neq i} h_{jj,i}^2 = n \sum_i \sum_{j\neq i} h_{jj,i}^2.\end{aligned}$$

We obtain the desired estimate by combining (1) and (2). □

4.3. Interior Gradient Estimates

The aim of this section is to derive an improved interior gradient estimate for the minimal surface equation. The basis is the mean-value inequality provided by Theorem 4.1.3.

We first derive a simple equation.

Lemma 4.3.1. *Let Ω be a domain in $\mathbb{R}^n$ and u be a $C^2(\Omega)$-solution of*

$$\operatorname{div}\left(\frac{\nabla u}{\sqrt{1+|\nabla u|^2}}\right) = 0 \quad \text{in } \Omega.$$

Then on $\Sigma = \operatorname{graph} u$, the function $w = \log\sqrt{1+|\nabla u|^2}$ satisfies

$$\Delta_\Sigma w = |A|^2 + |\nabla_\Sigma w|^2.$$

Proof. Let ν be the unit upward normal vector to Σ. Then,

$$\nu = \frac{1}{\sqrt{1+|\nabla u|^2}}(-\nabla u, 1).$$

By Lemma 4.2.1 or Lemma 4.5.3, ν satisfies

$$\Delta_\Sigma \nu + |A|^2 \nu = 0.$$

Let v be the last component of ν. Then,

$$\Delta_\Sigma v + |A|^2 v = 0.$$

Since $w = -\log v$, then,

$$\nabla_\Sigma w = -\frac{1}{v}\nabla_\Sigma v,$$

and

$$\Delta_\Sigma w = -\frac{1}{v}\Delta_\Sigma v + \frac{1}{v^2}|\nabla_\Sigma v|^2.$$

Hence,

$$\Delta_\Sigma w = |A|^2 + |\nabla_\Sigma w|^2.$$

This is the desired identity. □

We are ready to derive the following improved interior gradient estimate due to Finn [**58**] for $n = 2$ and Bombieri, de Giorgi, and Miranda [**12**] for $n \geq 3$.

Theorem 4.3.2. *Let u be an $L^\infty(B_R(x_0)) \cap C^2(B_R(x_0))$-solution of*

$$\operatorname{div}\left(\frac{\nabla u}{\sqrt{1+|\nabla u|^2}}\right) = 0 \quad \text{in } B_R(x_0).$$

Then,

$$|\nabla u(x_0)| \leq \exp\left\{C\left[1 + \frac{1}{R}\left(\sup_{B_R(x_0)} u - u(x_0)\right)\right]\right\},$$

where C is a positive constant depending only on n.

Proof. By our convention, we denote by $B_r(x_0)$ balls in $\mathbb{R}^n$ and by $D_r(p_0)$ balls in $\mathbb{R}^{n+1}$. For simplicity, we take $x_0 = 0$ and assume $u(0) = 0$. Set

$$w = \log\sqrt{1+|\nabla u|^2}.$$

Then, $w \geq 0$ and, by Lemma 4.3.1,

$$\Delta_\Sigma w = |\nabla_\Sigma w|^2 + |A|^2 \geq |\nabla_\Sigma w|^2 \geq 0. \tag{1}$$

By Theorem 4.1.3, we obtain

$$w(0) \leq \frac{c}{R^n}\int_{\Sigma\cap D_{R/2}} w\, d\mathcal{H}^n, \tag{2}$$

where c is a positive constant depending only on n. We now proceed to estimate the expression in the right-hand side.

Take a cutoff function $\eta \in C_0^1(B_{3R/4})$, with $\eta = 1$ in $B_{R/2}$ and $0 \leq \eta \leq 1$ and $|\nabla\eta| \leq 8/R$. Set

$$u_R = \begin{cases} R & \text{if } u \geq R, \\ u & \text{if } |u| < R, \\ -R & \text{if } u \leq -R. \end{cases}$$

By multiplying the equation by $\eta w(u_R + R)$ and integrating by parts, we obtain

$$\int_{B_R} \frac{\nabla u}{\sqrt{1+|\nabla u|^2}} \cdot \nabla(\eta w(u_R + R))\, dx = 0.$$

Note that $\nabla u_R = 0$ if $|u| > R$, $u_R + R = 0$ if $u < -R$, and $0 \le u_R + R \le 2R$ if $u > -R$. Then,

$$\int_{B_R\cap\{|u|<R\}} \frac{|\nabla u|^2}{\sqrt{1+|\nabla u|^2}}\eta w\, dx \le 2R \int_{B_R\cap\{u>-R\}} (w|\nabla\eta| + \eta|\nabla w|)\, dx.$$

Next, we note that

$$\begin{aligned}\eta w\sqrt{1+|\nabla u|^2} &= \frac{|\nabla u|^2}{\sqrt{1+|\nabla u|^2}}\eta w + \frac{1}{\sqrt{1+|\nabla u|^2}}\eta w\\ &= \frac{|\nabla u|^2}{\sqrt{1+|\nabla u|^2}}\eta w + e^{-w}\eta w \le \frac{|\nabla u|^2}{\sqrt{1+|\nabla u|^2}}\eta w + \eta,\end{aligned}$$

where we used the trivial estimate $e^{-w}w \le 1$. (In fact, the optimal upper bound is e^{-1}.) Hence,

$$\begin{aligned}\int_{\Sigma\cap\{|u|<R\}} \eta w\, d\mathcal{H}^n &= \int_{B_R\cap\{|u|<R\}} \eta w\sqrt{1+|\nabla u|^2}\, dx\\ &\le \int_{B_R} \eta\, dx + 2R\int_{B_R\cap\{u>-R\}} (w|\nabla\eta| + \eta|\nabla w|)\, dx.\end{aligned} \tag{3}$$

Next,

$$\begin{aligned}2R\int_{B_R\cap\{u>-R\}} w|\nabla\eta|\, dx &\le 8\int_{B_{3R/4}\cap\{u>-R\}} e^w\, dx\\ &= 8\int_{B_{3R/4}\cap\{u>-R\}} \sqrt{1+|\nabla u|^2}\, dx\\ &\le 8\mathcal{H}^n(\Sigma\cap B_{3R/4}\cap\{u>-R\}).\end{aligned} \tag{4}$$

To estimate the integral of $\eta|\nabla w|$, we take $\varphi \in C_0^1(B_R\times\mathbb{R})$. By multiplying (1) by φ^2 and then integrating by parts, we get

$$\int_\Sigma \varphi^2|\nabla_\Sigma w|^2\, d\mathcal{H}^n \le -2\int_\Sigma \varphi\langle\nabla_\Sigma w, \nabla_\Sigma\varphi\rangle\, d\mathcal{H}^n.$$

The Cauchy inequality implies

$$\int_\Sigma \varphi^2|\nabla_\Sigma w|^2\, d\mathcal{H}^n \le 4\int_\Sigma |\nabla_\Sigma\varphi|^2\, d\mathcal{H}^n.$$

In particular, we take φ such that

$$\varphi = \eta\tau(x_{n+1}),$$

where $\tau = 1$ in $(-R, \sup_{B_R} u)$, $\tau = 0$ outside $(-2R, R+\sup_{B_R} u)$, $0 \le \tau \le 1$ and $|\tau'| < 2/R$ in $\mathbb{R}$, and η as above. Hence,

$$\int_\Sigma \varphi^2|\nabla_\Sigma w|^2\, d\mathcal{H}^n \le cR^{-2}\mathcal{H}^n(\Sigma\cap B_{3R/4}\cap\{u>-2R\}).$$

Since w does not depend on x_{n+1}, we have

$$|\nabla_\Sigma w|^2 = |\nabla w|^2 - |\nabla w \cdot \nu|^2 \geq |\nabla w|^2 \left(1 - \sum_{i=1}^n \nu_i^2\right) = |\nabla w|^2 \nu_{n+1}^2,$$

and hence

$$|\nabla w| \leq |\nabla_\Sigma w| \sqrt{1 + |\nabla u|^2}.$$

Then by the Hölder inequality, we obtain

$$\begin{aligned} (5) \qquad 2R \int_{B_R \cap \{u > -R\}} \eta |\nabla w| \, dx &\leq 2R \int_{\Sigma_R \cap \{u > -R\}} \eta |\nabla_\Sigma w| \, d\mathcal{H}^n \\ &\leq c\mathcal{H}^n(\Sigma \cap B_{3R/4} \cap \{u > -2R\}). \end{aligned}$$

By combining (3), (4), and (5), we have

$$(6) \qquad \int_{\Sigma \cap \{|u| < R\}} \eta w \, d\mathcal{H}^n \leq c\big(R^n + \mathcal{H}^n(\Sigma \cap B_{3R/4} \cap \{u > -2R\})\big),$$

where c is a positive constant depending only on n. To estimate the last term in the right-hand side of (6), we take $\widetilde{\eta} \in C_0^1(B_R)$ such that $\widetilde{\eta} = 1$ in $B_{3R/4}$, $0 \leq \widetilde{\eta} \leq 1$, and $|\nabla \widetilde{\eta}| < 8/R$. Then,

$$\mathcal{H}^n(\Sigma \cap B_{3R/4} \cap \{u > -2R\}) \leq \int_{B_R \cap \{u > -2R\}} \widetilde{\eta} \sqrt{1 + |\nabla u|^2} \, dx.$$

Note that

$$\int_{B_R} \frac{\nabla u}{\sqrt{1 + |\nabla u|^2}} \cdot \nabla(\widetilde{\eta}(u + 2R)^+) \, dx = 0,$$

where $(u + 2R)^+ = \max\{u + 2R, 0\}$. We then have

$$\int_{B_R \cap \{u > -2R\}} \widetilde{\eta} \sqrt{1 + |\nabla u|^2} \, dx \leq cR^n \left(1 + R^{-1} \sup_{B_R} u\right),$$

and hence

$$\mathcal{H}^n(\Sigma \cap B_{3R/4} \cap \{u > -2R\}) \leq cR^n \left(1 + R^{-1} \sup_{B_R} u\right).$$

A simple substitution in (6) yields

$$\int_{\Sigma \cap \{|u| < R\}} \eta w \, d\mathcal{H}^n \leq cR^n \left(1 + R^{-1} \sup_{B_R} u\right).$$

We therefore have the desired estimate by combining the above estimate with (2). □

As an application of Theorem 4.3.2, we prove a type of Liouville theorem.

Theorem 4.3.3. *Suppose that u is a nonnegative $C^2(\mathbb{R}^n)$-solution of*

$$\operatorname{div}\left(\frac{\nabla u}{\sqrt{1+|\nabla u|^2}}\right)=0 \quad \text{in } \mathbb{R}^n.$$

Then, u is constant.

Proof. We first put Theorem 4.3.2 in a slightly different form. For any $x_0 \in \mathbb{R}^n$ and $R > 0$, by changing u to $-u$, we get

$$|\nabla u(x_0)| \leq \exp\left\{C\left[1+\frac{1}{R}\left(u(x_0)-\inf_{B_R(x_0)} u\right)\right]\right\},$$

where C is a positive constant depending only on n. Since u is nonnegative in $B_R(x_0)$, we have

$$|\nabla u(x_0)| \leq \exp\left\{C\left[1+\frac{1}{R}u(x_0)\right]\right\}.$$

Letting $R \to \infty$, we conclude

$$|\nabla u(x_0)| \leq \exp\{C\}.$$

This holds for any $x_0 \in \mathbb{R}^n$. With the global estimate of ∇u, the minimal surface equation is uniformly elliptic. By Theorem 3.3.2, u is an affine function. Evidently, a nonnegative affine function has to be constant. □

We also have the following type of Liouville theorem, in which only a one-sided bound is assumed.

Theorem 4.3.4. *Suppose that u is a $C^2(\mathbb{R}^n)$-solution of*

$$\operatorname{div}\left(\frac{\nabla u}{\sqrt{1+|\nabla u|^2}}\right)=0 \quad \text{in } \mathbb{R}^n.$$

Assume, for any $x \in \mathbb{R}^n$,

$$u(x) \leq K + K|x|,$$

for some constant $K > 0$. Then, u is an affine function.

Proof. For any $x_0 \in \mathbb{R}^n$ and $R > 0$, we have

$$|\nabla u(x_0)| \leq \exp\left\{C\left[1+\frac{1}{R}\left(\sup_{B_R(x_0)} u - u(x_0)\right)\right]\right\},$$

where C is a positive constant depending only on n. By the assumption on u, we get

$$\frac{1}{R}\left(\sup_{B_R(x_0)} u - u(x_0)\right) \le \frac{1}{R}(K + K|x_0| + KR + |u(x_0)|)$$
$$\le K + \frac{1}{R}(K + K|x_0| + |u(x_0)|).$$

By letting $R \to \infty$, we obtain

$$|\nabla u(x_0)| \le \exp\{C(1+K)\}.$$

This holds for any $x_0 \in \mathbb{R}^n$. With the global estimate of ∇u, the minimal surface equation is uniformly elliptic. By Theorem 3.3.2, u is an affine function. □

4.4. Interior Curvature Estimates

In this section, we prove an integral curvature estimate and a pointwise curvature estimate for solutions of the minimal surface equation. These estimates immediately imply a Bernstein type result.

We adopt the following terminology. A C^2-hypersurface Σ in $\mathbb{R}^{n+1}$ is a *minimal graph* over $\Omega \subset \mathbb{R}^n$ if $\Sigma = \operatorname{graph} u$ for some $C^2(\Omega)$-solution u of the minimal surface equation

$$\operatorname{div}\left(\frac{\nabla u}{\sqrt{1+|\nabla u|^2}}\right) = 0 \quad \text{in } \Omega.$$

We will simply say a minimal graph in $\mathbb{R}^{n+1}$ if Ω does not play any role. By our convention, we denote by $D_r(p)$ balls in $\mathbb{R}^{n+1}$ and by $B_r(x_0)$ balls in $\mathbb{R}^n$.

We first prove two preliminary results for minimal graphs.

Lemma 4.4.1. *Let Σ be a minimal graph in $\mathbb{R}^{n+1}$ satisfying $\partial\Sigma \cap D_R(p_0) = \emptyset$ for some $p_0 \in \Sigma$ and $R > 0$. Then,*

$$\mathcal{H}^n(\Sigma \cap D_{R/2}(p_0)) \le KR^n,$$

where K is a positive constant depending only on n.

Proof. Without loss of generality, we assume $p_0 = 0 \in \mathbb{R}^{n+1}$. Set $\Sigma = \operatorname{graph} u$, for some C^2-solution u of the minimal surface equation in B_R, and define the truncated function $u_{R/2}$ by

$$u_{R/2} = \begin{cases} R/2 & \text{if } u \ge R/2, \\ u & \text{if } |u| < R/2, \\ -R/2 & \text{if } u \le -R/2. \end{cases}$$

Take a cutoff function $\eta \in C_0^\infty(B_R)$ such that $\eta = 1$ in $B_{R/2}$, and $0 \le \eta \le 1$ and $|\nabla \eta| \le 4R^{-1}$ in B_R. We multiply the minimal surface equation by $\varphi = \eta u_{R/2}$ and then integrate by parts to get

$$\int_{B_R} \frac{\langle \nabla u, \nabla \varphi \rangle}{\sqrt{1+|\nabla u|^2}}\, dx = 0.$$

Since $\nabla \varphi = \nabla \eta u_{R/2} + \eta \nabla u_{R/2}$, we have

$$\begin{aligned}\int_{B_{R/2}\cap\{|u|<R/2\}} & \frac{|\nabla u|^2}{\sqrt{1+|\nabla u|^2}}\, dx \\ &\le \int_{B_{R/2}\cap\{|u|<R/2\}} \eta \frac{|\nabla u|^2}{\sqrt{1+|\nabla u|^2}}\, dx \\ &= -\int_{B_R} \frac{\langle \nabla u, \nabla \eta\rangle u_{R/2}}{\sqrt{1+|\nabla u|^2}}\, dx \le \frac{R}{2}\int_{B_R} |\nabla \eta|\, dx \le cR^n,\end{aligned}$$

where c is a positive constant depending only on n. Consequently, we obtain

$$\begin{aligned}\mathcal{H}^n(\Sigma \cap D_{R/2}) &= \int_{\{|x|^2+u^2<(R/2)^2\}} \sqrt{1+|\nabla u|^2}\, dx \\ &\le \int_{B_{R/2}\cap\{|u|<R/2\}} \sqrt{1+|\nabla u|^2}\, dx \le cR^n.\end{aligned}$$

This is the desired estimate. □

Lemma 4.4.2. *Let Σ be a minimal graph in $\mathbb{R}^{n+1}$. Then, for any $\varphi \in C_0^1(\Sigma)$,*

$$\int_\Sigma \varphi^2 |A|^2 d\mathcal{H}^n \le \int_\Sigma |\nabla_\Sigma \varphi|^2 d\mathcal{H}^n.$$

Proof. Set $\Sigma = \operatorname{graph} u$, for some C^2-solution u of the minimal surface equation, and

$$w = \log \sqrt{1+|\nabla u|^2}.$$

By Lemma 4.3.1, we have

$$\Delta_\Sigma w = |A|^2 + |\nabla_\Sigma w|^2.$$

Now multiplying the above identity by φ^2 and integrating by parts, we obtain

$$\begin{aligned}\int_\Sigma \varphi^2 |A|^2 d\mathcal{H}^n &+ \int_\Sigma \varphi^2 |\nabla_\Sigma w|^2 d\mathcal{H}^n \\ &= -2\int_\Sigma \varphi \langle \nabla_\Sigma w, \nabla_\Sigma \varphi\rangle\, d\mathcal{H}^n \\ &\le \int_\Sigma \varphi^2 |\nabla_\Sigma w|^2 d\mathcal{H}^n + \int_\Sigma |\nabla_\Sigma \varphi|^2 d\mathcal{H}^n.\end{aligned}$$

This implies the desired estimate. □

Lemma 4.4.2 is related to the second variation of minimal hypersurfaces along normal directions. In fact, a minimal hypersurface Σ (not necessarily a graph) is called to be *stable* if, for any $\varphi \in C_0^1(\Sigma)$,

$$\int_\Sigma \varphi^2 |A|^2 d\mathcal{H}^n \leq \int_\Sigma |\nabla_\Sigma \varphi|^2 d\mathcal{H}^n.$$

Lemma 4.4.2 simply asserts that minimal graphs are stable.

We also note that Lemma 4.4.2 yields an L^2-estimate of the second fundamental form. We now improve this estimate to get an L^4-estimate of the second fundamental form. The following integral curvature estimate is due to Schoen, Simon, and Yau [**132**].

Theorem 4.4.3. *Let Σ be a minimal graph in $\mathbb{R}^{n+1}$ and $0 \leq q < \sqrt{\frac{2}{n}}$. Then, for any nonnegative function $\varphi \in C_0^1(\Sigma)$,*

$$\int_\Sigma \varphi^{4+2q} |A|^{4+2q} d\mathcal{H}^n \leq c \int_\Sigma |\nabla_\Sigma \varphi|^{4+2q} d\mathcal{H}^n,$$

where c is a positive constant depending only on n and q.

Proof. Lemma 4.2.4 or Lemma 4.5.6 implies

$$-\frac{1}{2}\Delta_\Sigma |A|^2 \leq |A|^4 - \left(1 + \frac{2}{n}\right) |\nabla_\Sigma |A||^2.$$

By multiplying this inequality by $\varphi^2 |A|^{2q}$ and integrating by parts over Σ, we obtain

$$\begin{aligned} \int_\Sigma \varphi |A|^{2q} \langle \nabla_\Sigma |A|^2, \nabla_\Sigma \varphi \rangle \, d\mathcal{H}^n &+ \frac{1}{2} \int_\Sigma \varphi^2 \langle \nabla_\Sigma |A|^2, \nabla_\Sigma |A|^{2q} \rangle \, d\mathcal{H}^n \\ &\leq \int_\Sigma \varphi^2 |A|^{4+2q} d\mathcal{H}^n - \left(1 + \frac{2}{n}\right) \int_\Sigma \varphi^2 |A|^{2q} |\nabla_\Sigma |A||^2 d\mathcal{H}^n. \end{aligned}$$

This implies

$$\begin{aligned} &\left(1 + \frac{2}{n} + 2q\right) \int_\Sigma \varphi^2 |A|^{2q} |\nabla_\Sigma |A||^2 d\mathcal{H}^n \\ &\qquad \leq \int_\Sigma \varphi^2 |A|^{4+2q} d\mathcal{H}^n - 2 \int_\Sigma \varphi |A|^{1+2q} \langle \nabla_\Sigma |A|, \nabla_\Sigma \varphi \rangle \, d\mathcal{H}^n. \end{aligned} \tag{1}$$

In Lemma 4.4.2, by replacing φ by $\varphi |A|^{1+q}$ and $\nabla_\Sigma \varphi$ by

$$(1+q)\varphi |A|^q \nabla_\Sigma |A| + |A|^{1+q} \nabla_\Sigma \varphi,$$

we get

$$\begin{aligned} \int_\Sigma \varphi^2 |A|^{4+2q} d\mathcal{H}^n \leq \int_\Sigma \big\{ &(1+q)^2 \varphi^2 |A|^{2q} |\nabla_\Sigma |A||^2 + |A|^{2+2q} |\nabla_\Sigma \varphi|^2 \\ &+ 2(1+q)\varphi |A|^{1+2q} \langle \nabla_\Sigma |A|, \nabla_\Sigma \varphi \rangle \big\} \, d\mathcal{H}^n. \end{aligned} \tag{2}$$

Substituting (2) in (1), we obtain

$$\begin{aligned}\left(\frac{2}{n}-q^2\right)&\int_\Sigma \varphi^2|A|^{2q}|\nabla_\Sigma|A||^2 d\mathcal{H}^n\\ &\le \int_\Sigma \left\{|A|^{2+2q}|\nabla_\Sigma\varphi|^2+2q\varphi|A|^{1+2q}\langle\nabla_\Sigma|A|,\nabla_\Sigma\varphi\rangle\right\}d\mathcal{H}^n.\end{aligned}$$

Using the Cauchy inequality, for some constant $\varepsilon>0$ to be determined,

$$2q\varphi|A|^{1+2q}\langle\nabla_\Sigma|A|,\nabla_\Sigma\varphi\rangle\le \varepsilon q^2\varphi^2|A|^{2q}|\nabla_\Sigma|A||^2+\varepsilon^{-1}|A|^{2+2q}|\nabla_\Sigma\varphi|^2,$$

we get

$$\begin{aligned}\left(\frac{2}{n}-(1+\varepsilon)q^2\right)&\int_\Sigma \varphi^2|A|^{2q}|\nabla_\Sigma|A||^2 d\mathcal{H}^n\\ &\le (1+\varepsilon^{-1})\int_\Sigma |A|^{2+2q}|\nabla_\Sigma\varphi|^2 d\mathcal{H}^n.\end{aligned}$$

If $0\le q<\sqrt{\frac{2}{n}}$, we take $\varepsilon>0$ such that

$$\frac{2}{n}-(1+\varepsilon)q^2>0.$$

With such a choice of ε, we obtain

$$\int_\Sigma \varphi^2|A|^{2q}|\nabla_\Sigma|A||^2 d\mathcal{H}^n\le c\int_\Sigma |A|^{2+2q}|\nabla_\Sigma\varphi|^2 d\mathcal{H}^n, \tag{3}$$

where c is a positive constant depending only on n and q. Next, in (2), we use the Cauchy inequality in the form

$$\varphi|A|^{1+2q}\langle\nabla_\Sigma|A|,\nabla_\Sigma\varphi\rangle\le\frac{1}{2}\varphi^2|A|^{2q}|\nabla_\Sigma|A||^2+\frac{1}{2}|A|^{2+2q}|\nabla_\Sigma\varphi|^2$$

and get by (2) and (3)

$$\int_\Sigma \varphi^2|A|^{4+2q}d\mathcal{H}^n\le c\int_\Sigma |A|^{2+2q}|\nabla_\Sigma\varphi|^2 d\mathcal{H}^n.$$

By replacing φ by φ^{2+q}, we obtain

$$\int_\Sigma \varphi^{4+2q}|A|^{4+2q}d\mathcal{H}^n\le c\int_\Sigma \varphi^{2+2q}|A|^{2+2q}|\nabla_\Sigma\varphi|^2 d\mathcal{H}^n.$$

Finally, for each constant $\delta>0$, we conclude from the Hölder inequality that

$$(\varphi|A|)^{2+2q}|\nabla_\Sigma\varphi|^2\le\delta(\varphi|A|)^{4+2q}+c(\delta,q)|\nabla_\Sigma\varphi|^{4+2q}.$$

By choosing δ sufficiently small, we obtain the desired result. □

Now we are ready to prove the Bernstein theorem for minimal graphs over $\mathbb{R}^n$.

Theorem 4.4.4. *Let Σ be a minimal graph over $\mathbb{R}^n$. If $2\le n\le 5$, then Σ is a hyperplane.*

Proof. For simplicity, we assume $0 \in \Sigma$. Take an arbitrary $R > 0$. Consider a C^1-function γ on $\mathbb{R}$ such that $\gamma(t) = 1$ for $t \leq R/2$, $\gamma(t) = 0$ for $t \geq R$, and $0 \leq \gamma \leq 1$ and $|\gamma'| \leq 4/R$ on $\mathbb{R}$. Set $\varphi(x) = \gamma(|x|)$. Then, $|\nabla_\Sigma \varphi| \leq 4/R$ on Σ. By Theorem 4.4.3, we have, for any $q \in [0, \sqrt{\frac{2}{n}})$,

$$\int_{\Sigma \cap D_{R/2}} |A|^{4+2q} d\mathcal{H}^n \leq cR^{-4-2q}\mathcal{H}^n(\Sigma \cap D_R),$$

where c is a positive constant depending only on n and q. By Lemma 4.4.1, we obtain

$$\int_{\Sigma \cap D_{R/2}} |A|^{4+2q} d\mathcal{H}^n \leq cR^{n-4-2q}.$$

If

$$n < 4 + 2\sqrt{\frac{2}{n}}, \tag{1}$$

we can take a $q \in [0, \sqrt{\frac{2}{n}})$ such that $n - 4 - 2q < 0$. Now we let $R \to \infty$ and conclude $|A| \equiv 0$. Hence, Σ must be a hyperplane. Last, we note that (1) holds if $n \leq 5$. Therefore, we have the desired result for $n \leq 5$. □

We need to point out that Theorem 4.4.4 holds for $2 \leq n \leq 7$. This is a result due to Simons [**140**]. On the other hand, Bombieri, de Giorgi, and Giusti [**11**] constructed minimal graphs over $\mathbb{R}^n$ which are not hyperplanes for $n \geq 8$.

In the rest of this section, we derive an interior pointwise curvature estimate for minimal hypersurface by a Moser type iteration "on the surface", based on the integral estimate we just derived in Theorem 4.4.3. An essential tool is the Sobolev inequality on hypersurfaces provided by Theorem 4.1.7. The following pointwise curvature estimate is due to Schoen, Simon, and Yau [**132**].

Theorem 4.4.5. *Let Σ be a minimal graph in $\mathbb{R}^{n+1}$ satisfying $\partial\Sigma \cap D_R(p_0) = \emptyset$ for some $p_0 \in \Sigma$ and $R > 0$. If $2 \leq n \leq 5$, then,*

$$\sup_{\Sigma \cap D_{R/2}(p_0)} |A| \leq \frac{C}{R},$$

where C is a positive constant depending only on n.

Proof. For simplicity, we take $p_0 = 0$. Lemma 4.2.4 or Lemma 4.5.6 implies

$$\Delta_\Sigma |A|^2 \geq 2|\nabla_\Sigma |A||^2 - 2|A|^4,$$

and hence

$$|A|\Delta_\Sigma |A| \geq -|A|^4.$$

Next, for $p \geq 2$,

$$\Delta_\Sigma |A|^p = \operatorname{div}_\Sigma(p|A|^{p-1}\nabla_\Sigma|A|)$$
$$= p(p-1)|A|^{p-2}|\nabla_\Sigma|A||^2 + p|A|^{p-1}\Delta_\Sigma|A| \geq -p|A|^{p+2}.$$

Take a nonnegative function $\varphi \in C_0^1(D_{3R/4})$, and multiply the inequality above by $\varphi^2|A|^p$. Then,

$$\int_\Sigma \varphi^2|A|^p\Delta_\Sigma|A|^p d\mathcal{H}^n \geq -p\int_\Sigma \varphi^2|A|^{2p+2}d\mathcal{H}^n.$$

An integration by parts yields

$$\int_\Sigma \varphi^2|\nabla_\Sigma|A|^p|^2 d\mathcal{H}^n \leq \int_\Sigma \{p\varphi^2|A|^{2p+2} - 2\varphi|A|^p\langle\nabla_\Sigma\varphi, \nabla_\Sigma|A|^p\rangle\}d\mathcal{H}^n.$$

By the Cauchy inequality, we have

$$-2\varphi|A|^p\langle\nabla_\Sigma\varphi, \nabla_\Sigma|A|^p\rangle \leq \frac{1}{2}\varphi^2|\nabla_\Sigma|A|^p|^2 + 2|A|^{2p}|\nabla_\Sigma\varphi|^2,$$

and hence

$$\text{(1)} \qquad \int_\Sigma \varphi^2|\nabla_\Sigma|A|^p|^2 d\mathcal{H}^n \leq \int_\Sigma \{2p\varphi^2|A|^{2p+2} + 4|A|^{2p}|\nabla_\Sigma\varphi|^2\}d\mathcal{H}^n.$$

We first consider $n \geq 3$ and take $\chi = \frac{n}{n-2}$. By applying Theorem 4.1.7 with $p = 2$, we obtain

$$\left(\int_\Sigma (\varphi|A|^p)^{2\chi}d\mathcal{H}^n\right)^{\frac{1}{\chi}} \leq c\int_\Sigma |\nabla_\Sigma(\varphi|A|^p)|^2 d\mathcal{H}^n$$
$$\leq 2c\int_\Sigma \{\varphi^2|\nabla_\Sigma|A|^p|^2 + |A|^{2p}|\nabla_\Sigma\varphi|^2\}d\mathcal{H}^n.$$

By combining with (1), we get

$$\text{(2)} \qquad \left(\int_\Sigma \varphi^{2\chi}|A|^{2p\chi}d\mathcal{H}^n\right)^{\frac{1}{\chi}} \leq c\int_\Sigma \{p\varphi^2|A|^{2p+2} + |A|^{2p}|\nabla_\Sigma\varphi|^2\}d\mathcal{H}^n.$$

We now estimate the first term in the right-hand side. For positive constants ε and τ to be determined, we have by the Hölder inequality

$$R^2|A|^2 \leq \tau(1+\tau)^{-(1+\frac{1}{\tau})}\varepsilon^{-\frac{1}{\tau}} + \varepsilon(R^2|A|^2)^{1+\tau}$$
$$\leq \varepsilon^{-\frac{1}{\tau}} + \varepsilon(R^2|A|^2)^{1+\tau}.$$

By writing $\varphi^2|A|^{2p+2} = R^{-2}\varphi^2|A|^{2p} \cdot R^2|A|^2$, we obtain

$$\text{(3)} \qquad \int_\Sigma \varphi^2|A|^{2p+2}d\mathcal{H}^n \leq \int_\Sigma \{\varepsilon^{-\frac{1}{\tau}}R^{-2}\varphi^2|A|^{2p} + \varepsilon R^{2\tau}\varphi^2|A|^{2p+2+2\tau}\}d\mathcal{H}^n.$$

Moreover, the Hölder inequality implies

$$\int_\Sigma \varphi^2|A|^{2p+2+2\tau}d\mathcal{H}^n \le \left(\int_\Sigma \varphi^{2\chi}|A|^{2p\chi}d\mathcal{H}^n\right)^{\frac{1}{\chi}}\left(\int_{\Sigma\cap D_{3R/4}}|A|^{n(1+\tau)}d\mathcal{H}^n\right)^{\frac{2}{n}},$$

where we used the fact $\operatorname{supp}\varphi \subset D_{3R/4}$. Now we choose τ such that

$$n+\tau n < 4+2\sqrt{\frac{2}{n}}.$$

This can be achieved if $3 \le n \le 5$. Let $\gamma = \gamma(t)$ be a C^1-function on $\mathbb{R}$ such that $\gamma(t) = 1$ for $t \le 3R/4$, $\gamma(t) = 0$ for $t \ge R$, and $|\gamma'| \le 8/R$ on $\mathbb{R}$. Set $\psi(x) = \gamma(|x|)$. Then, $|\nabla_\Sigma\psi| \le 8/R$ on Σ. Using the integral curvature estimate provided by Theorem 4.4.3, we obtain

$$\begin{aligned}\int_{\Sigma\cap D_{3R/4}} & |A|^{n+\tau n}d\mathcal{H}^n \\ & \le \int_{\Sigma\cap D_R}\psi^{n+\tau n}|A|^{n+\tau n}d\mathcal{H}^n \le c\int_{\Sigma\cap D_R}|\nabla_\Sigma\psi|^{n+\tau n}d\mathcal{H}^n,\end{aligned}$$

where c is a positive constant depending only on n. By Lemma 4.4.1, we get

$$\int_{\Sigma\cap D_{R/2}}|A|^{n+\tau n}d\mathcal{H}^n \le KR^{-\tau n},$$

where K is a positive constant depending only on n. Therefore, we have

$$\int_\Sigma \varphi^2|A|^{2p+2+2\tau}d\mathcal{H}^n \le K^{\frac{2}{n}}R^{-2\tau}\left(\int_\Sigma \varphi^{2\chi}|A|^{2p\chi}d\mathcal{H}^n\right)^{\frac{1}{\chi}}. \tag{4}$$

By combining (3) and (4), we get

$$\begin{aligned}\int_\Sigma \varphi^2|A|^{2p+2}d\mathcal{H}^n \le \varepsilon^{-\frac{1}{\tau}}R^{-2}&\int_\Sigma \varphi^2|A|^{2p}d\mathcal{H}^n \\ &+\varepsilon K^{\frac{2}{n}}\left(\int_\Sigma \varphi^{2\chi}|A|^{2p\chi}d\mathcal{H}^n\right)^{\frac{1}{\chi}}.\end{aligned}\tag{5}$$

By combining with (2), we obtain

$$\begin{aligned}\left(\int_\Sigma \varphi^{2\chi}|A|^{2p\chi}d\mathcal{H}^n\right)^{\frac{1}{\chi}} &\le c\int_\Sigma \{\varepsilon^{-\frac{1}{\tau}}pR^{-2}\varphi^2|A|^{2p}+|A|^{2p}|\nabla_\Sigma\varphi|^2\}d\mathcal{H}^n \\ &\quad + c\varepsilon pK^{\frac{2}{n}}\left(\int_\Sigma \varphi^{2\chi}|A|^{2p\chi}d\mathcal{H}^n\right)^{\frac{1}{\chi}}.\end{aligned}$$

By taking ε with $c\varepsilon pK^{\frac{2}{n}} = 1/2$, we have

$$\left(\int_\Sigma \varphi^{2\chi}|A|^{2p\chi}d\mathcal{H}^n\right)^{\frac{1}{\chi}} \le c\int_\Sigma \{p^{1+\tau^{-1}}R^{-2}\varphi^2|A|^{2p}+|A|^{2p}|\nabla_\Sigma\varphi|^2\}d\mathcal{H}^n,$$

where c is a positive constant depending only on n and K, and hence only on n. Take arbitrary numbers $0 < \sigma < \rho < 3R/4$ and let φ be a nonnegative cutoff function in $\mathbb{R}^{n+1}$ such that $\varphi = 1$ in $D_{\rho-\sigma}$, $\varphi = 0$ in $\mathbb{R}^{n+1} \setminus D_\rho$, and $|\nabla_\Sigma \varphi| \leq 2\sigma^{-1}$. We then conclude, for $3 \leq n \leq 5$, $\chi = \frac{n}{n-2}$, and any $p \geq 2$,

$$\left(\int_{\Sigma \cap D_{\rho-\sigma}} |A|^{2p\chi} d\mathcal{H}^n \right)^{\frac{1}{\chi}} \leq cp^{1+\tau^{-1}} \sigma^{-2} \int_{\Sigma \cap D_\rho} |A|^{2p} d\mathcal{H}^n,$$

where c is a positive constant depending only on n. By renaming p, we have, for any $p \geq 1$,

$$(6) \quad \left(\frac{1}{R^n} \int_{\Sigma \cap D_{\rho-\sigma}} |A|^{4p\chi} d\mathcal{H}^n \right)^{\frac{1}{\chi}} \leq cp^{1+\tau^{-1}} \left(\frac{R}{\sigma} \right)^2 \frac{1}{R^n} \int_{\Sigma \cap D_\rho} |A|^{4p} d\mathcal{H}^n.$$

We now consider $n = 2$. By applying Theorem 4.1.6 with $p = 1$, we obtain, for any constant $\sigma \in (0, R)$,

$$\begin{aligned} &\left(\int_\Sigma (\varphi^2 |A|^{2p})^2 d\mathcal{H}^2 \right)^{\frac{1}{2}} \\ &\quad \leq c \int_\Sigma |\nabla_\Sigma (\varphi^2 |A|^{2p})| \, d\mathcal{H}^2 \\ &\quad \leq \int_\Sigma \{ \sigma \varphi^2 |\nabla_\Sigma |A|^p|^2 + \sigma |A|^{2p} |\nabla_\Sigma \varphi|^2 + 2c^2 \sigma^{-1} \varphi^2 |A|^{2p} \} d\mathcal{H}^2. \end{aligned}$$

By combining with (1), we get

$$(7) \quad \begin{aligned} &\left(\int_\Sigma \varphi^4 |A|^{4p} d\mathcal{H}^2 \right)^{\frac{1}{2}} \\ &\quad \leq \int_\Sigma \{ 2\sigma p \varphi^2 |A|^{2p+2} + 5\sigma |A|^{2p} |\nabla_\Sigma \varphi|^2 + 2c^2 \sigma^{-1} \varphi^2 |A|^{2p} \} d\mathcal{H}^2. \end{aligned}$$

We note that (3) holds for $n = 2$. By the Cauchy inequality, we have

$$\begin{aligned} &\int_\Sigma \varphi^2 |A|^{2p+2+2\tau} d\mathcal{H}^2 \\ &\quad \leq \left(\int_\Sigma \varphi^4 |A|^{4p} d\mathcal{H}^2 \right)^{\frac{1}{2}} \left(\int_{\Sigma \cap D_{3R/4}} |A|^{4(1+\tau)} d\mathcal{H}^2 \right)^{\frac{1}{2}}, \end{aligned}$$

where we used the fact that $\operatorname{supp} \varphi \subset D_{3R/4}$. Now we choose τ such that $4 + 4\tau < 6$. Let $\gamma = \gamma(t)$ be a C^1-function on $\mathbb{R}$ such that $\gamma(t) = 1$ for $t \leq 3R/4$, $\gamma(t) = 0$ for $t \geq R$, and $0 \leq \gamma \leq 1$ and $|\gamma'| \leq 8/R$ on $\mathbb{R}$. Set $\psi(x) = \gamma(|x|)$. Then, $|\nabla_\Sigma \psi| \leq 8/R$ on Σ. Using the integral curvature estimate provided by Theorem 4.4.3, we obtain

$$\begin{aligned}\int_{\Sigma\cap D_{3R/4}}|A|^{4+4\tau}d\mathcal{H}^2 &\le \int_{\Sigma\cap D_R}\psi^{4+4\tau}|A|^{4+4\tau}d\mathcal{H}^2\\ &\le c\int_{\Sigma\cap D_R}|\nabla_\Sigma\psi|^{4+4\tau}d\mathcal{H}^2.\end{aligned}$$

By Lemma 4.4.1, we get

$$\int_{\Sigma\cap D_{3R/4}}|A|^{4+4\tau}d\mathcal{H}^2 \le KR^{-(2+4\tau)}.$$

Therefore, we have, instead of (4),

$$\int_\Sigma \varphi^2|A|^{2p+2+2\tau}d\mathcal{H}^2 \le K^{\frac{1}{2}}R^{-(1+2\tau)}\left(\int_\Sigma \varphi^4|A|^{4p}d\mathcal{H}^2\right)^{\frac{1}{2}}. \tag{8}$$

By combining (3) and (8), we get

$$\begin{aligned}\int_\Sigma \varphi^2|A|^{2p+2}d\mathcal{H}^2 &\le \varepsilon^{-\frac{1}{\tau}}R^{-2}\int_\Sigma \varphi^2|A|^{2p}d\mathcal{H}^2\\ &\quad + \varepsilon R^{-1}K^{\frac{1}{2}}\left(\int_\Sigma \varphi^4|A|^{4p}d\mathcal{H}^2\right)^{\frac{1}{2}}.\end{aligned} \tag{9}$$

By combining with (7), we obtain

$$\begin{aligned}&\left(\int_\Sigma \varphi^4|A|^{4p}d\mathcal{H}^2\right)^{\frac{1}{2}}\\ &\quad\le \int_\Sigma\{(2\sigma p\varepsilon^{-\frac{1}{\tau}}R^{-2}+2c^2\sigma^{-1})\varphi^2|A|^{2p}+5\sigma|A|^{2p}|\nabla_\Sigma\varphi|^2\}d\mathcal{H}^2\\ &\quad + 2\sigma p\varepsilon R^{-1}K^{\frac{1}{2}}\left(\int_\Sigma \varphi^4|A|^{4p}d\mathcal{H}^2\right)^{\frac{1}{2}}.\end{aligned}$$

Note that $\sigma/R \le 1$. By taking ε with $2p\varepsilon K^{1/2} = 1/2$, we obtain

$$\left(\int_\Sigma \varphi^4|A|^{4p}d\mathcal{H}^2\right)^{\frac{1}{2}} \le c\int_\Sigma\{p^{1+\tau^{-1}}\sigma^{-1}\varphi^2|A|^{2p}+\sigma|A|^{2p}|\nabla_\Sigma\varphi|^2\}d\mathcal{H}^2,$$

where c is a universal positive constant. Take arbitrary numbers $0<\sigma<\rho<3R/4$ and let φ be a nonnegative cutoff function in $\mathbb{R}^3$ such that $\varphi=1$ in $D_{\rho-\sigma}$, $\varphi=0$ in $\mathbb{R}^3\setminus D_\rho$, and $|\nabla_\Sigma\varphi|\le 2\sigma^{-1}$. Hence, for any $p\ge 2$,

$$\left(\int_{\Sigma\cap D_{\rho-\sigma}}|A|^{4p}d\mathcal{H}^2\right)^{\frac{1}{2}} \le cp^{1+\tau^{-1}}\sigma^{-1}\int_{\Sigma\cap D_\rho}|A|^{2p}d\mathcal{H}^2,$$

where c is a universal positive constant. By renaming p, we have, for any $p \geq 1$,

$$\left(\frac{1}{R^2}\int_{\Sigma\cap D_{\rho-\sigma}}|A|^{8p}d\mathcal{H}^2\right)^{\frac{1}{2}} \leq cp^{1+\tau^{-1}}\frac{R}{\sigma}\cdot\frac{1}{R^2}\int_{\Sigma\cap D_\rho}|A|^{4p}d\mathcal{H}^2. \tag{10}$$

In conclusion, by combining (6) and (10), we obtain, for any $p \geq 1$,

$$\left(\frac{1}{R^n}\int_{\Sigma\cap D_{\rho-\sigma}}|A|^{4p\chi}d\mathcal{H}^n\right)^{\frac{1}{\chi}} \leq cp^{1+\tau^{-1}}\left(\frac{R}{\sigma}\right)^m\frac{1}{R^n}\int_{\Sigma\cap D_\rho}|A|^{4p}d\mathcal{H}^n, \tag{11}$$

where c and τ are positive constants depending only on n, $\chi = \frac{n}{n-2}$ and $m = 2$ for $3 \leq n \leq 5$, and $\chi = 2$ and $m = 1$ for $n = 2$.

We now carry out iterations. Set, for $k = 0, 1, 2, \ldots,$

$$\rho_0 = \frac{3R}{4}, \quad \sigma_k = \frac{R}{2^{k+3}}, \quad \rho_{k+1} = \rho_k - \sigma_k.$$

Then,

$$\rho_k \to \frac{R}{2} \quad \text{as } k \to \infty.$$

For each $k \geq 1$, with $p = \chi^{k-1}$, $\rho = \rho_{k-1}$, and $\sigma = \sigma_{k-1}$ in (11), we obtain

$$\left(\frac{1}{R^n}\int_{\Sigma\cap D_{\rho_k}}|A|^{4\chi^k}d\mathcal{H}^n\right)^{\frac{1}{\chi}} \leq c^k\frac{1}{R^n}\int_{\Sigma\cap D_{\rho_{k-1}}}|A|^{4\chi^{k-1}}d\mathcal{H}^n. \tag{12}$$

Set

$$I_k = \left(\frac{1}{R^n}\int_{\Sigma\cap D_{\rho_k}}|A|^{4\chi^k}d\mathcal{H}^n\right)^{\frac{1}{4\chi^k}}.$$

Then, for any $k \geq 1$,

$$I_k \leq c^{\frac{k}{\chi^k}} I_{k-1}.$$

A simple iteration yields

$$I_k \leq c^{\sum_{i=1}^k \frac{i}{\chi^i}} I_0.$$

Since the series

$$\sum_{i=1}^{\infty}\frac{i}{\chi^i}$$

is convergent, we then have

$$\left(\frac{1}{R^n}\int_{\Sigma\cap D_{R/2}}|A|^{4\chi^k}d\mathcal{H}^n\right)^{\frac{1}{4\chi^k}} \leq C\left(\frac{1}{R^n}\int_{\Sigma\cap D_{3R/4}}|A|^4d\mathcal{H}^n\right)^{\frac{1}{4}}.$$

By taking $k \to \infty$ in the left-hand side, we have

$$\sup_{\Sigma\cap D_{R/2}} |A| \leq C\left(\frac{1}{R^n}\int_{\Sigma\cap D_{3R/4}} |A|^4 d\mathcal{H}^n\right)^{\frac{1}{4}}.$$

Finally by the L^4-estimates on the second fundamental form provided by Theorem 4.4.3 (with $q = 0$), we have

$$\sup_{\Sigma\cap D_{R/2}} |A| \leq C\left(\frac{1}{R^{n+4}}\mathcal{H}^n(\Sigma\cap D_R)\right)^{\frac{1}{4}}.$$

With Lemma 4.4.1 again, we obtain

$$\sup_{\Sigma\cap D_{R/2}} |A| \leq \frac{C}{R}.$$

This is the desired estimate. □

4.5. Differential Identities: An Alternative Approach

In this section, we provide alternative proofs of Lemma 4.2.1 and Lemma 4.2.4. This section does not require any knowledge of Riemannian geometry.

In Section 4.1, we introduced gradient operators and Laplace-Beltrami operators on hypersurfaces. We now briefly review these concepts.

Let Σ be a C^1-hypersurface in $\mathbb{R}^{n+1}$ and ν be its unit normal field. Suppose that u is a C^1-function defined in a neighborhood of Σ. We denote by Du the Euclidean gradient of u; i.e.,

$$Du = (\partial_1 u, \ldots, \partial_{n+1}u).$$

We define the *tangential gradient* $\nabla_\Sigma u$ of u on Σ by

$$\nabla_\Sigma u = Du - \langle Du, \nu\rangle\nu.$$

Hence, ∇_Σ is the Du minus the component of Du in the normal direction. In other words, $\nabla_\Sigma u$ is simply the projection of Du in the tangent space.

It was proved in Section 4.1 that $\nabla_\Sigma u$ depends only on the values of u on Σ; namely, if $u = 0$ on Σ, then $\nabla_\Sigma u = 0$. We now repeat this simple argument. Let $p_0 \in \Sigma$. If $Du(p_0) = 0$, then $\nabla_\Sigma u(p_0) = 0$; otherwise, we take the unit normal vector ν near p_0 in the form

$$\nu = \frac{Du}{|Du|}$$

and again conclude $\nabla_\Sigma u(p_0) = 0$. In conclusion, ∇_Σ is well-defined on functions defined only on Σ.

We set

$$\nabla_\Sigma = (\delta_1, \ldots, \delta_{n+1}).$$

It is convenient to adopt the summation convention. In any expression, repeated indices indicate a summation from 1 to $n+1$. With this notation, we write

$$\delta_a = \partial_a - \nu_a\nu_b\partial_b.$$

It seems more natural to denote by ∇_a the components of ∇_Σ. However, ∇_a is reserved for the components of the covariant derivative in Riemannian geometry. This leaves us with no obvious choices for notations for the components of ∇_Σ in $\mathbb{R}^{n+1}$. We need to emphasize that our main results in this section, Lemma 4.5.3 and Lemma 4.5.6, are expressed in the standard notations in Riemannian geometry.

By our convention, indices i, j, k range from 1 to n, and a, b, c range from 1 to $n+1$.

We now derive several formulas.

Lemma 4.5.1. *Let Σ be a C^2-hypersurface in $\mathbb{R}^{n+1}$ and ν be its unit normal field. Then, for any $a, b = 1, \ldots, n+1$,*

(1) $\nu_c\delta_c = 0$ *and* $\nu_c\delta_b\nu_c = 0$;

(2) $\delta_a\nu_b = \delta_b\nu_a$;

(3) $\delta_a\delta_b = \delta_b\delta_a + (\nu_a\delta_b\nu_c - \nu_b\delta_a\nu_c)\delta_c$.

Proof. (1) Both identities are trivial.

(2) Set $\Sigma = \{F = 0\}$, for some C^2-function F defined in a neighborhood of Σ with $DF \neq 0$ on Σ. Then, we take

$$\nu = \frac{DF}{|DF|}.$$

Hence,

$$\begin{aligned}\partial_a\nu_b = \partial_a\left(\frac{\partial_b F}{|DF|}\right) &= \frac{1}{|DF|}\partial_{ab}F - \frac{1}{|DF|^3}\partial_b F\partial_c F\partial_{ac}F \\ &= \frac{1}{|DF|}\big(\partial_{ab}F - \nu_b\nu_c\partial_{ac}F\big).\end{aligned}$$

This implies, with (1),

$$\begin{aligned}\delta_a\nu_b &= \partial_a\nu_b - \nu_a\nu_d\partial_d\nu_b \\ &= \frac{1}{|DF|}\big(\partial_{ab}F - \nu_b\nu_c\partial_{ac}F\big) - \frac{1}{|DF|}\nu_a\nu_d\big(\partial_{db}F - \nu_b\nu_c\partial_{dc}F\big) \\ &= \frac{1}{|DF|}\big(\partial_{ab}F - \nu_b\nu_c\partial_{ac}F - \nu_a\nu_c\partial_{cb}F + \nu_a\nu_b\nu_c\nu_d\partial_{dc}F\big).\end{aligned}$$

This is symmetric in a, b.

(3) Performing a straightforward calculation, we have

$$\begin{aligned}\delta_a\delta_b &= \delta_a(\partial_b - \nu_b\nu_c\partial_c) = \delta_a\partial_b - \nu_b\nu_c\delta_a\partial_c - \delta_a(\nu_b\nu_c)\partial_c\\ &= (\partial_a - \nu_a\nu_c\partial_c)\partial_b - \nu_b\nu_c(\partial_a - \nu_a\nu_d\partial_d)\partial_c - \delta_a(\nu_b\nu_c)\partial_c\\ &= \partial_a\partial_b - \nu_a\nu_c\partial_c\partial_b - \nu_b\nu_c\partial_a\partial_c + \nu_a\nu_b\nu_c\nu_d\partial_d\partial_c - (\nu_b\delta_a\nu_c + \nu_c\delta_a\nu_b)\partial_c.\end{aligned}$$

By taking a difference and canceling those terms symmetric in a and b, we have

$$\delta_a\delta_b - \delta_b\delta_a = (\nu_a\delta_b\nu_c - \nu_b\delta_a\nu_c)\partial_c,$$

where we also used (2) that we just proved. Note that

$$\nu_c(\nu_a\delta_b\nu_c - \nu_b\delta_a\nu_c) = 0.$$

This implies

$$(\nu_a\delta_b\nu_c - \nu_b\delta_a\nu_c)\partial_c = (\nu_a\delta_b\nu_c - \nu_b\delta_a\nu_c)\delta_c.$$

We then have the desired identity. □

Now we briefly review the mean curvature of hypersurfaces introduced in Section 3.1. Let Σ be a hypersurface in $\mathbb{R}^{n+1}$ and $\kappa_1, \ldots, \kappa_n$ be the principal curvatures of Σ corresponding to a unit normal vector ν. The *mean curvature* H of Σ corresponding to ν is the sum of the principal curvatures corresponding to ν; namely,

$$H = \sum_{i=1}^{n} \kappa_i.$$

We now introduce another important quantity. The *squared length of the second fundamental form* of Σ is the sum of the squared principal curvatures; namely,

$$|A|^2 = \sum_{i=1}^{n} \kappa_i^2.$$

We now give a representation of the mean curvature and the squared length of the second fundamental form.

Lemma 4.5.2. *Let Σ be a C^2-hypersurface in $\mathbb{R}^{n+1}$ and $\nu = (\nu_1, \ldots, \nu_{n+1})$ be its unit normal field.*

(i) *The mean curvature H of Σ corresponding to ν is given by*

$$H = -\delta_a\nu_a.$$

(ii) *The squared length of the second fundamental form $|A|^2$ of Σ is given by*

$$|A|^2 = (\delta_a\nu_b)(\delta_b\nu_a).$$

Proof. Without loss of generality, we assume ν is extended to a neighborhood of Σ.

Step 1. We first note that $\nu_a \partial_c \nu_a = \partial_c |\nu|^2/2 = 0$. Then,

$$\delta_a \nu_a = \partial_a \nu_a - \nu_a \nu_c \partial_c \nu_a = \partial_a \nu_a,$$

and

$$\begin{aligned}(\delta_a \nu_b)(\delta_b \nu_a) &= (\partial_a \nu_b - \nu_a \nu_c \partial_c \nu_b)(\partial_b \nu_a - \nu_b \nu_d \partial_d \nu_a) \\ &= (\partial_a \nu_b)(\partial_b \nu_a) - \nu_a \nu_c (\partial_c \nu_b)(\partial_b \nu_a) - \nu_b \nu_d (\partial_d \nu_a)(\partial_a \nu_b) \\ &\quad + \nu_a \nu_b \nu_c \nu_d (\partial_c \nu_b)(\partial_d \nu_a) \\ &= (\partial_a \nu_b)(\partial_b \nu_a).\end{aligned}$$

Let $\mathbf{e_1}, \ldots, \mathbf{e_{n+1}}$ be the standard orthonormal basis of $\mathbb{R}^{n+1}$. We now consider

$$\partial_a \nu_a = \partial_{x_a} \langle \nu, \mathbf{e_a} \rangle \tag{1}$$

and

$$(\partial_a \nu_b)(\partial_b \nu_a) = \partial_{x_a} \langle \nu, \mathbf{e_b} \rangle \partial_{x_b} \langle \nu, \mathbf{e_a} \rangle. \tag{2}$$

Step 2. We claim that the expressions in (1) and (2) are independent of the orthonormal basis $\{\mathbf{e_1}, \ldots, \mathbf{e_{n+1}}\}$ of $\mathbb{R}^{n+1}$. To check this, consider the change of coordinates $x = P\widetilde{x}$ for an orthogonal matrix P. Straightforward calculations yield

$$\sum_{a=1}^{n+1} \partial_{x_a} \langle \nu, \mathbf{e_a} \rangle = \sum_{a=1}^{n+1} \partial_{\widetilde{x}_a} \langle \nu, \widetilde{\mathbf{e}}_\mathbf{a} \rangle$$

and

$$\sum_{a,b=1}^{n+1} \partial_{x_a} \langle \nu, \mathbf{e_b} \rangle \partial_{x_b} \langle \nu, \mathbf{e_a} \rangle = \sum_{a,b=1}^{n+1} \partial_{\widetilde{x}_a} \langle \nu, \widetilde{\mathbf{e}}_\mathbf{b} \rangle \partial_{\widetilde{x}_b} \langle \nu, \widetilde{\mathbf{e}}_\mathbf{a} \rangle.$$

Step 3. We fix a point p on Σ. Without loss of generality, we assume p is the origin in $\mathbb{R}^{n+1}$. We choose a special coordinate in $\mathbb{R}^{n+1}$ such that $\mathbf{e_{n+1}} = \nu(p)$ and assume Σ is expressed by $x_{n+1} = \rho(x)$ in a neighborhood of p. Then, $\rho(0) = 0$ and $\nabla\rho(0) = 0$. The hypersurface Σ can be represented by

$$\mathbf{r} = (x, \rho(x)).$$

Then,

$$\mathbf{r}_i = (\mathbf{e_i}, \rho_i).$$

We take

$$\nu = \frac{1}{\sqrt{1 + |\nabla\rho|^2}}(-\nabla\rho, 1).$$

Then at p, $g_{ij} = \delta_{ij}$ and $h_{ij} = \rho_{ij}(0)$. By a rotation in $\mathbb{R}^n \times \{0\}$, we assume $\nabla^2\rho(0)$ is diagonal. Then,

$$\nabla^2\rho(0) = \mathrm{diag}(\kappa_1, \dots, \kappa_n),$$

where $\kappa_1, \dots, \kappa_n$ are principal curvatures of Σ at p. A simple calculation yields

$$(\partial_a \nu_b(p))_{1\le a,b\le n+1} = \mathrm{diag}(-\kappa_1, \dots, -\kappa_n, 0).$$

We then conclude

$$\sum_{i=1}^n \kappa_i = -\sum_{a=1}^{n+1} \partial_a \nu_a(p)$$

and

$$\sum_{i=1}^n \kappa_i^2 = \sum_{a,b=1}^{n+1} \partial_a \nu_b(p) \partial_b \nu_a(p).$$

By combining Steps 1–3, we have the desired identities. □

Next, for a C^1-vector field X on Σ (not necessarily tangential to Σ), by writing $X = (X_1, \dots, X_{n+1})$, we define the *divergence* $\mathrm{div}_\Sigma X$ of X by

$$\mathrm{div}_\Sigma X = \delta_a X_a.$$

We also define the *Laplace-Beltrami operator* Δ_Σ (on C^2-functions) on Σ by

$$\Delta_\Sigma = \delta_a \delta_a.$$

We point out that div_Σ and Δ_Σ do not depend on the choice of coordinate systems in the ambient space $\mathbb{R}^{n+1}$ although each δ_a does. Note that we can write Δ_Σ as

$$\Delta_\Sigma u = \mathrm{div}_\Sigma(\nabla_\Sigma u).$$

With the divergence operator, the mean curvature in Lemma 4.5.2 can be expressed as

$$H = -\mathrm{div}_\Sigma \nu.$$

We now provide an alternative proof of Lemma 4.2.1.

Lemma 4.5.3. *Let Σ be a C^2-hypersurface in $\mathbb{R}^{n+1}$ and ν be its unit normal field. Then,*

$$\Delta_\Sigma \nu + |A|^2 \nu + \nabla_\Sigma H = 0 \quad \text{on } \Sigma,$$

where H is the mean curvature of Σ corresponding to ν and A is its second fundamental form.

Proof. By applying Lemma 4.5.1, we have, for any $c = 1, \dots, n+1$,

$$\begin{aligned}\Delta_\Sigma \nu_c &= \delta_a \delta_a \nu_c = \delta_a \delta_c \nu_a \\ &= \delta_c \delta_a \nu_a + (\nu_a \delta_c \nu_d - \nu_c \delta_a \nu_d)\delta_d \nu_a \\ &= \delta_c \delta_a \nu_a - \nu_c (\delta_a \nu_d)(\delta_d \nu_a).\end{aligned}$$

By Lemma 4.5.2, we obtain

$$\Delta_\Sigma \nu_c + |A|^2\nu_c + \delta_c H = 0.$$

This implies the desired result. □

In the rest of this section, we provide an alternative proof of Lemma 4.2.4. We proceed with some preparations.

Lemma 4.5.4. *Let Σ be a C^3-hypersurface in $\mathbb{R}^{n+1}$ and ν be its unit normal field. Then, for any $c = 1, \ldots, n+1$,*

$$\begin{aligned}\Delta_\Sigma \delta_c = \delta_c \Delta_\Sigma &- 2\nu_c(\delta_a\nu_b)\delta_a\delta_b - 2(\delta_c\nu_a)(\delta_a\nu_b)\delta_b \\ &+ \nu_c(\delta_a H)\delta_a - H(\delta_c\nu_a)\delta_a.\end{aligned}$$

Proof. By Lemma 4.5.1(3), we have

$$\begin{aligned}\Delta_\Sigma \delta_c &= \delta_a\delta_a\delta_c = \delta_a\big[\delta_c\delta_a + (\nu_a\delta_c\nu_b - \nu_c\delta_a\nu_b)\delta_b\big] \\ &= \delta_a\delta_c\delta_a + (\nu_a\delta_c\nu_b - \nu_c\delta_a\nu_b)\delta_a\delta_b + \delta_a(\nu_a\delta_c\nu_b - \nu_c\delta_a\nu_b)\delta_b.\end{aligned}$$

By Lemma 4.5.1(3) again, we get

$$\begin{aligned}\delta_a\delta_c\delta_a &= \delta_c\delta_a\delta_a + (\nu_a\delta_c\nu_b - \nu_c\delta_a\nu_b)\delta_b\delta_a \\ &= \delta_c\Delta_\Sigma + (\nu_a\delta_c\nu_b - \nu_c\delta_a\nu_b)[\delta_a\delta_b + (\nu_b\delta_a\nu_d - \nu_a\delta_b\nu_d)\delta_d].\end{aligned}$$

A simple substitution yields

$$\begin{aligned}\Delta_\Sigma \delta_c = \delta_c\Delta_\Sigma &+ 2(\nu_a\delta_c\nu_b - \nu_c\delta_a\nu_b)\delta_a\delta_b \\ &+ (\nu_a\delta_c\nu_b - \nu_c\delta_a\nu_b)(\nu_b\delta_a\nu_d - \nu_a\delta_b\nu_d)\delta_d \\ &+ (\nu_a\delta_a\delta_c\nu_b - \nu_c\delta_a\delta_a\nu_b)\delta_b + [(\delta_a\nu_a)(\delta_c\nu_b) - (\delta_a\nu_c)(\delta_a\nu_b)]\delta_b.\end{aligned}$$

By Lemma 4.5.1(1), we then obtain

$$\begin{aligned}\Delta_\Sigma \delta_c = \delta_c\Delta_\Sigma &- 2\nu_c(\delta_a\nu_b)\delta_a\delta_b - \nu_a^2(\delta_c\nu_b)(\delta_b\nu_d)\delta_d - \nu_c(\Delta_\Sigma\nu_b)\delta_b \\ &+ (\delta_a\nu_a)(\delta_c\nu_b)\delta_b - (\delta_a\nu_c)(\delta_a\nu_b)\delta_b.\end{aligned}$$

Then, Lemma 4.5.3 implies the desired identity. □

Next, we calculate the Laplace-Beltrami operator acting on the squared length of the second fundamental form.

Lemma 4.5.5. *Let Σ be a C^4-hypersurface in $\mathbb{R}^{n+1}$ and ν be its unit normal field. Then,*

$$\begin{aligned}\frac{1}{2}\Delta_\Sigma|A|^2 = (\delta_a\delta_b\nu_c)(\delta_a\delta_b\nu_c) &- 2\nu_a\nu_b(\delta_c\delta_a\nu_d)(\delta_d\delta_b\nu_c) - |A|^4 \\ &- (\delta_a\nu_b)(\delta_a\delta_b H) - H(\delta_a\nu_b)(\delta_b\nu_c)(\delta_c\nu_a).\end{aligned}$$

Proof. First, we note that

$$\frac{1}{2}\Delta_\Sigma|A|^2 = \frac{1}{2}\Delta_\Sigma[(\delta_c\nu_d)(\delta_c\nu_d)] = (\delta_c\nu_d)\Delta_\Sigma(\delta_c\nu_d) + (\delta_a\delta_c\nu_d)(\delta_a\delta_c\nu_d).$$

In the following, we calculate the first term in the right-hand side. By Lemma 4.5.4, we have

$$\begin{aligned}\Delta_\Sigma(\delta_c\nu_d) = \delta_c\Delta_\Sigma\nu_d &- 2\nu_c(\delta_a\nu_b)(\delta_a\delta_b\nu_d) - 2(\delta_c\nu_a)(\delta_a\nu_b)(\delta_b\nu_d)\\ &+ \nu_c(\delta_b H)(\delta_b\nu_d) - H(\delta_c\nu_b)(\delta_b\nu_d).\end{aligned}$$

Lemma 4.5.3 yields

$$\delta_c\Delta_\Sigma\nu_d = \delta_c(-|A|^2\nu_d - \delta_d H) = -|A|^2\delta_c\nu_d - \nu_d\delta_c|A|^2 - \delta_c\delta_d H.$$

By Lemma 4.5.1(1), we have

$$\begin{aligned}(\delta_c\nu_d)\Delta_\Sigma(\delta_c\nu_d) = &-2(\delta_c\nu_d)(\delta_c\nu_a)(\delta_a\nu_b)(\delta_b\nu_d) - |A|^2(\delta_c\nu_d)(\delta_c\nu_d)\\ &- (\delta_c\nu_d)(\delta_c\delta_d H) - H(\delta_c\nu_d)(\delta_c\nu_b)(\delta_b\nu_d).\end{aligned}$$

Lemma 4.5.2 implies that the second term in the right-hand side is $|A|^4$. For the first term, we note that, by Lemma 4.5.1(3),

$$\begin{aligned}\nu_a\delta_c\delta_a\nu_b &= \nu_a[\delta_a\delta_c\nu_b + (\nu_c\delta_a\nu_d - \nu_a\delta_c\nu_d)\delta_d\nu_b]\\ &= -\nu_a^2(\delta_c\nu_d)(\delta_d\nu_b) = -(\delta_c\nu_d)(\delta_d\nu_b).\end{aligned}$$

Similarly, we have

$$\nu_d\delta_b\delta_d\nu_c = -(\delta_b\nu_a)(\delta_a\nu_c).$$

Hence,

$$\begin{aligned}(\delta_c\nu_d)\Delta_\Sigma(\delta_c\nu_d) = &-2\nu_a\nu_d(\delta_c\delta_a\nu_b)(\delta_b\delta_d\nu_c) - |A|^4\\ &- (\delta_c\nu_d)(\delta_c\delta_d H) - H(\delta_c\nu_d)(\delta_c\nu_b)(\delta_b\nu_d).\end{aligned}$$

This implies the desired result. □

We are ready to provide an alternative proof of Lemma 4.2.4.

Lemma 4.5.6. *Let Σ be a C^4-minimal hypersurface in $\mathbb{R}^{n+1}$. Then,*

$$\frac{1}{2}\Delta_\Sigma|A|^2 \geq \left(1 + \frac{2}{n}\right)|\nabla_\Sigma|A||^2 - |A|^4.$$

Proof. Since $H \equiv 0$, then Lemma 4.5.5 implies

$$\frac{1}{2}\Delta_\Sigma|A|^2 = (\delta_a\delta_b\nu_c)(\delta_a\delta_b\nu_c) - 2\nu_a\nu_b(\delta_c\delta_a\nu_d)(\delta_d\delta_b\nu_c) - |A|^4.$$

We need to estimate the difference from the first two terms in the right-hand side. We point out that this difference is independent of the choice of coordinates in the ambient Euclidean space. By our convention, indices a, b, c range from 1 to $n+1$ and i, j, k range from 1 to n.

Take a point $p_0 \in \Sigma$ and choose the x_{n+1}-axis to be the same direction as $\nu(p_0)$. Then, at p_0,

$$\nu_i = 0, \quad \nu_{n+1} = 1$$

and

$$\delta_i = \partial_i, \quad \delta_{n+1} = 0.$$

By Lemma 4.5.1(3), we have

$$\delta_i\delta_j = \delta_j\delta_i \quad \text{at } p_0.$$

Next, we note

$$\delta_a\delta_{n+1}\nu_{n+1} = 0 \quad \text{at } p_0. \tag{1}$$

To verify this, we apply Lemma 4.5.1(3) and get

$$\begin{aligned}\delta_a\delta_{n+1}\nu_{n+1} &= \delta_{n+1}\delta_a\nu_{n+1} + (\nu_a\delta_{n+1}\nu_c - \nu_{n+1}\delta_a\nu_c)\delta_c\nu_{n+1}\\ &= -\nu_{n+1}(\delta_a\nu_c)(\delta_c\nu_{n+1}) = 0\end{aligned}$$

since

$$\nu_{n+1}\delta_c\nu_{n+1} = \nu_{n+1}\delta_c\nu_{n+1} + \nu_i\delta_c\nu_i = \nu_b\delta_c\nu_b = 0.$$

Hence,

$$\begin{aligned}&(\delta_a\delta_b\nu_c)(\delta_a\delta_b\nu_c) - 2\nu_a\nu_b(\delta_c\delta_a\nu_d)(\delta_d\delta_b\nu_c)\\ &\quad = (\delta_i\delta_j\nu_k)(\delta_i\delta_j\nu_k) + 2(\delta_i\delta_j\nu_{n+1})(\delta_i\delta_j\nu_{n+1}) - 2(\delta_i\delta_j\nu_{n+1})(\delta_i\delta_j\nu_{n+1})\\ &\quad = (\delta_i\delta_j\nu_k)(\delta_i\delta_j\nu_k).\end{aligned}$$

Therefore,

$$\frac{1}{2}\Delta_\Sigma|A|^2 = \sum_{i,j,k}(\delta_i\delta_j\nu_k)^2 - |A|^4 \quad \text{at } p_0. \tag{2}$$

Next, by

$$|A| = \left(\sum_{a,b}(\delta_a\nu_b)^2\right)^{\frac{1}{2}},$$

we have

$$|\nabla_\Sigma|A||^2 = \sum_c(\delta_c|A|)^2 = \frac{1}{|A|^2}\sum_c\left(\sum_{a,b}(\delta_a\nu_b)(\delta_c\delta_a\nu_b)\right)^2.$$

With $\delta_{n+1} = 0$, we get

$$|\nabla_\Sigma|A||^2 = \frac{1}{|A|^2}\sum_i\left(\sum_{j,k}(\delta_j\nu_k)(\delta_i\delta_j\nu_k)\right)^2. \tag{3}$$

Using

$$|A|^2 = \sum_{j,k} (\delta_j \nu_k)^2,$$

the Cauchy inequality implies

$$(4) \qquad |\nabla_\Sigma |A||^2 \le \frac{1}{|A|^2} \sum_i \Big(\sum_{j,k} (\delta_j \nu_k)^2 \Big) \Big(\sum_{j,k} (\delta_i \delta_j \nu_k)^2 \Big) \le \sum_{i,j,k} (\delta_i \delta_j \nu_k)^2.$$

By combining (2) and (4), we obtain

$$\frac{1}{2} \Delta_\Sigma |A|^2 \ge |\nabla_\Sigma |A||^2 - |A|^4.$$

In the proof of the above inequality, we used the condition $H = 0$ simply to drop terms involving H in the expression of $\Delta_\Sigma |A|^2$. Next, we will make use of such a condition in an essential way to improve this inequality.

We denote by x points in $\mathbb{R}^n$ and set $p_0 = (x^0, x^0_{n+1})$. Suppose Σ is given by

$$x_{n+1} = \rho(x)$$

in a neighborhood of p_0, with

$$\partial_{x_i} \rho(x^0) = 0.$$

By rotating coordinates in $\mathbb{R}^n$ if necessary, we assume

$$\partial_{x_i x_j} \rho(x^0) = -\kappa_i \delta_{ij},$$

where $\kappa_1, \ldots, \kappa_n$ are the principal curvatures of Σ at p_0. The components of the normal vector ν are given by

$$\nu_i = -\frac{\partial_i \rho}{\sqrt{1 + |\nabla \rho|^2}}.$$

Hence, at p_0,

$$\delta_i \nu_j = \kappa_i \delta_{ij}.$$

In the following, calculations will be made at p_0. By (3), we have

$$(5) \qquad |\nabla_\Sigma |A||^2 = \frac{1}{|A|^2} \sum_i \Big(\sum_j \kappa_j \delta_i \delta_j \nu_j \Big)^2 \le \sum_{i,j} (\delta_i \delta_j \nu_j)^2.$$

Then,

$$\begin{aligned} \sum_{i,j,k} (\delta_i \delta_j \nu_k)^2 - \sum_{i,j} (\delta_i \delta_j \nu_j)^2 &= \sum_{i,j} \sum_{k \ne j} (\delta_i \delta_j \nu_k)^2 \\ (6) \qquad &\ge \sum_j \sum_{k \ne j} (\delta_j \delta_j \nu_k)^2 + \sum_j \sum_{k \ne j} (\delta_k \delta_j \nu_k)^2 = 2 \sum_j \sum_{k \ne j} (\delta_j \delta_j \nu_k)^2 \\ &= 2 \sum_j \sum_{k \ne j} (\delta_j \delta_k \nu_j)^2 = 2 \sum_j \sum_{k \ne j} (\delta_k \delta_j \nu_j)^2. \end{aligned}$$

Next, by $H = 0$, we have

$$\delta_i \nu_i = -\sum_{a \neq i} \delta_a \nu_a.$$

We point out that there is no summation in the left-hand side. Then,

$$\delta_i \delta_i \nu_i = -\sum_{a \neq i} \delta_i \delta_a \nu_a = -\sum_{j \neq i} \delta_i \delta_j \nu_j,$$

where we used (1). Hence,

$$\begin{aligned}
\sum_{i,j} (\delta_i \delta_j \nu_j)^2 &= \sum_i (\delta_i \delta_i \nu_i)^2 + \sum_i \sum_{j \neq i} (\delta_i \delta_j \nu_j)^2 \\
&\le \sum_i \Big(\sum_{j \neq i} \delta_i \delta_j \nu_j \Big)^2 + \sum_i \sum_{j \neq i} (\delta_i \delta_j \nu_j)^2 \\
&\le (n-1) \sum_i \sum_{j \neq i} (\delta_i \delta_j \nu_j)^2 + \sum_i \sum_{j \neq i} (\delta_i \delta_j \nu_j)^2 \\
&= n \sum_i \sum_{j \neq i} (\delta_i \delta_j \nu_j)^2.
\end{aligned} \tag{7}$$

Therefore, by combining (6), (7), and then (5), we obtain

$$\sum_{i,j,k} (\delta_i \delta_j \nu_k)^2 \ge \left(1 + \frac{2}{n}\right) \sum_{i,j} (\delta_i \delta_j \nu_j)^2 \ge \left(1 + \frac{2}{n}\right) |\nabla_\Sigma |A||^2.$$

Hence, the desired result follows from (2). □

Part 2

Fully Nonlinear Elliptic Equations

Chapter 5

Fully Nonlinear Uniformly Elliptic Equations

In this chapter, we discuss fully nonlinear uniformly elliptic differential equations. We derive various a priori estimates for their solutions and solve the Dirichlet boundary-value problems under a key assumption that the associated functional is concave with respect to Hessian matrices.

In Section 5.1, we study basic properties of fully nonlinear uniformly elliptic differential equations. We prove a comparison principle and the higher regularity of solutions.

In Sections 5.2 and 5.3, we discuss estimates of derivatives up to the second order of solutions of fully nonlinear uniformly elliptic differential equations. Section 5.2 concerns interior estimates and Section 5.3 global estimates for the Dirichlet boundary-value problems. These two sections are based on the maximum principle. The concavity plays an essential role in the estimate of derivatives of the second order.

In Sections 5.4 and 5.5, we discuss estimates of the Hölder semi-norms of derivatives of the second order. Section 5.4 concerns interior estimates and Section 5.5 global estimates. These estimates are based on Krylov-Safonov's weak Harnack inequality. The concavity plays an essential role in this section.

In Section 5.6, we solve the Dirichlet boundary-value problems for fully nonlinear uniformly elliptic differential equations by the method of continuity. The estimates of the Hölder semi-norms of the second derivatives play an important role.

The three topics reviewed in Chapter 1 play different roles in this chapter. The maximum principle will be used to derive estimates of derivatives up to the second order, the Harnack inequality will be used to derive estimates of the Hölder semi-norms of derivatives of the second order, and the Schauder theory will be used to solve the linearized equations.

To illustrate ideas and techniques more clearly, we study a special class of fully nonlinear elliptic equations in this chapter.

5.1. Basic Properties

Denote by $\mathcal{S}$ the space of $n\times n$ symmetric matrices and consider a continuous function

$$F : \mathcal{S} \longrightarrow \mathbb{R}.$$

In some cases, F is defined only in a subset of $\mathcal{S}$. Let Ω be a domain in $\mathbb{R}^n$. For a given function f in Ω, we discuss a differential equation of the form

$$F(\nabla^2 u) = f \quad \text{in } \Omega.$$

This is a fully nonlinear differential equation if F is nonlinear in $\nabla^2 u$. We often say that F is a fully nonlinear differential operator on $\nabla^2 u$. The ellipticity is defined through its linearized equations.

For any $M = (m_{ij}) \in \mathcal{S}$, we denote

$$\begin{aligned} F_{ij}(M) &= \partial_{m_{ij}} F(M), \\ F_{ij,kl}(M) &= \partial_{m_{ij}m_{kl}} F(M), \end{aligned}$$

when these derivatives exist.

Let F be a C^1-function in $\mathcal{S}$. For an arbitrarily fixed $u \in C^2(\Omega)$, the linearized operator of F at $\nabla^2 u$ is given by

$$Lv = \frac{d}{dt}\bigg|_{t=0} F(\nabla^2 u + t\nabla^2 v),$$

for any $v \in C^2(\Omega)$. It is easy to check that

$$Lv = F_{ij}(\nabla^2 u) v_{ij} \quad \text{in } \Omega.$$

When it is clear from the context, we often write

$$L = F_{ij}\partial_{ij}.$$

Then, F is elliptic at the range of $\nabla^2 u$ if L is elliptic in Ω, or if the matrix (F_{ij}) is positive definite at $\nabla^2 u(x)$ for any $x \in \Omega$.

In general, the function F is *elliptic* in a subset $\mathcal{S}' \subset \mathcal{S}$ if the symmetric matrix $\big(F_{ij}(M)\big)$ is positive definite for any $M \in \mathcal{S}'$. The function F is *uniformly elliptic* in $\mathcal{S}'$ if, for any $M \in \mathcal{S}'$ and any $\xi \in \mathbb{R}^n$,

$$\lambda|\xi|^2 \leq F_{ij}(M)\xi_i\xi_j \leq \Lambda|\xi|^2,$$

for some positive constants $\lambda \leq \Lambda$. Throughout this chapter, the uniform ellipticity is always assumed. The constants λ and Λ are called the *ellipticity constants.*

For some $F \in C^1(\mathcal{S})$, suppose u is a $C^3(\Omega)$-solution of

$$F(\nabla^2 u) = f \quad \text{in } \Omega,$$

for some $f \in C^1(\Omega)$. For any unit vector $\gamma \in \mathbb{R}^n$, we write $u_\gamma = \nabla u \cdot \gamma$ and refer to this as the derivative of u with respect to x_γ.

By differentiating the equation with respect to x_γ, we get

$$F_{ij}u_{ij\gamma} = f_\gamma,$$

where F_{ij} is evaluated at $(\nabla^2 u(x))$. With the linearized operator L of F at $\nabla^2 u$ introduced earlier, we have

$$Lu_\gamma = f_\gamma.$$

In other words, u_γ satisfies a linear elliptic differential equation whose coefficients depend on the second derivatives of u.

Now for $F \in C^2(\mathcal{S})$, $u \in C^4(\Omega)$, and $f \in C^2(\Omega)$, we differentiate with respect to x_γ again to get

$$F_{ij}u_{ij\gamma\gamma} + F_{ij,kl}u_{ij\gamma}u_{kl\gamma} = f_{\gamma\gamma},$$

or

$$Lu_{\gamma\gamma} + F_{ij,kl}u_{ij\gamma}u_{kl\gamma} = f_{\gamma\gamma}.$$

We note that the third-order derivatives of u appear quadratically in the left-hand side. In general, $u_{\gamma\gamma}$ does not satisfy a linear differential equation. We need to introduce further assumptions on F in order for $u_{\gamma\gamma}$ to be a subsolution.

We say that the function F is *concave* at some $M \in \mathcal{S}$ if, for any $T = (t_{ij}) \in \mathcal{S}$,

$$F_{ij,kl}(M)t_{ij}t_{kl} \leq 0.$$

Similarly, F is concave in a subset $\mathcal{S}' \subset \mathcal{S}$ if the above condition holds for any $M \in \mathcal{S}'$.

If F is concave at the range of $\nabla^2 u$, then $u_{\gamma\gamma}$ satisfies

$$Lu_{\gamma\gamma} \geq f_{\gamma\gamma}.$$

In other words, $u_{\gamma\gamma}$ is a subsolution of some linear elliptic differential equation.

Now we collect some of the results we have shown in the next lemma.

Lemma 5.1.1. *Let Ω be a domain in $\mathbb{R}^n$ and F be a C^2-function in $\mathcal{S}$. Suppose that u is a $C^4(\Omega)$-solution of*

$$F(\nabla^2 u) = f \quad in\ \Omega,$$

for some $f \in C^2(\Omega)$, and that L is the linearized operator of F at $\nabla^2 u$; i.e.,

$$L = F_{ij}(\nabla^2 u)\partial_{ij}.$$

(1) *If $F\big(t\nabla^2 u(x)\big)$ is a concave function of $t \in [0, 1]$ for any $x \in \Omega$, then*

$$Lu \leq f - F(0) \quad in\ \Omega.$$

(2) *For any unit vector γ in $\mathbb{R}^n$,*

$$Lu_\gamma = f_\gamma \quad in\ \Omega.$$

(3) *If F is concave at $\nabla^2 u(x)$ for any $x \in \Omega$, then, for any unit vector γ in $\mathbb{R}^n$,*

$$Lu_{\gamma\gamma} \geq f_{\gamma\gamma} \quad in\ \Omega.$$

Proof. We need only prove (1). For any fixed $x \in \Omega$, set

$$h(t) = F\big(t\nabla^2 u(x)\big) \quad \text{for any } t \in [0, 1].$$

By the concavity of h, we have

$$h(0) \leq h(1) + h'(1)(0 - 1).$$

A simple calculation yields

$$F(0) \leq F\big(\nabla^2 u(x)\big) - F_{ij}\big(\nabla^2 u(x)\big)u_{ij}(x).$$

This implies the desired result. □

It turns out that u also satisfies a linear differential equation with coefficients depending on the second derivatives of u.

Lemma 5.1.2. *Let Ω be a domain in $\mathbb{R}^n$ and F be a C^1-function in $\mathcal{S}$. Suppose that u is a $C^2(\Omega)$-solution of*

$$F(\nabla^2 u) = f \quad in\ \Omega,$$

for some $f \in C(\Omega)$. Then,

$$a_{ij}u_{ij} = f - F(0) \quad in\ \Omega,$$

where a_{ij} is given by

$$a_{ij} = \int_0^1 F_{ij}(t\nabla^2 u)dt.$$

If F is uniformly elliptic at $t\nabla^2 u(x)$ for any $t \in [0,1]$ and any $x \in \Omega$, then $a_{ij}\partial_{ij}$ is a uniformly elliptic linear operator with the same ellipticity constants.

Proof. We write

$$F(\nabla^2 u) - F(0) = f - F(0).$$

For the left-hand side, we have

$$F(\nabla^2 u) - F(0) = \int_0^1 \frac{d}{dt} F(t\nabla^2 u)dt = \int_0^1 F_{ij}(t\nabla^2 u)dt \cdot u_{ij}.$$

This implies the desired result. □

As a consequence of Lemma 5.1.2, we have a Liouville type theorem. Compare this with Theorem 1.2.15.

Theorem 5.1.3. *Let F be a uniformly elliptic C^1-function in $\mathcal{S}$ with $F(0) = 0$. Suppose that u is a $C^2(\mathbb{R}^n)$-solution of*

$$F(\nabla^2 u) = 0 \quad \text{in } \mathbb{R}^n.$$

If u is bounded, then u is constant.

Proof. By Lemma 5.1.2, u is a solution of

$$a_{ij}u_{ij} = 0 \quad \text{in } \mathbb{R}^n,$$

where (a_{ij}) satisfies, for any $x \in \mathbb{R}^n$ and any $\xi \in \mathbb{R}^n$,

$$\lambda|\xi|^2 \leq a_{ij}(x)\xi_i\xi_j \leq \Lambda|\xi|^2.$$

By Theorem 1.2.15, we conclude that u is constant. □

We note that the concavity is not assumed in Theorem 5.1.3.

Lemma 5.1.2, although simple, has an important consequence. As we pointed out earlier, pure second derivatives $u_{\gamma\gamma}$ usually do not satisfy a linear differential equation. If F is concave, then $u_{\gamma\gamma}$ is a subsolution of some linear differential equation as shown in Lemma 5.1.1(3). Conceptually, an upper bound can be derived for $u_{\gamma\gamma}$ and it is impossible to derive a lower bound simply from the fact that $u_{\gamma\gamma}$ is a subsolution. The linear equation in Lemma 5.1.2 establishes a lower bound of $u_{\gamma\gamma}$ once an upper bound of $u_{\gamma\gamma}$ is known for any unit vector γ.

Corollary 5.1.4. *Let Ω be a domain in $\mathbb{R}^n$ and F be a uniformly elliptic C^1-function in $\mathcal{S}$. Suppose that u is a $C^2(\Omega)$-solution of*

$$F(\nabla^2 u) = f \quad \text{in } \Omega,$$

for some $f \in C(\Omega)$. *Then, for any* $x \in \Omega$,

$$|\nabla^2 u(x)| \leq C\Big\{ \sup_{\gamma\in\mathbb{R}^n, |\gamma|=1} u^+_{\gamma\gamma}(x) + |f(x)| + |F(0)|\Big\},$$

where C *is a positive constant depending only on* n, λ, *and* Λ.

Proof. We fix an $x \in \Omega$ and diagonalize $\nabla^2 u(x)$. By Lemma 5.1.2, we have

$$\sum_{i=1}^n a_{ii}u_{ii}(x) = f(x) - F(0).$$

We write this as

$$\sum_{u_{ii}(x)>0} a_{ii}u^+_{ii}(x) - \sum_{u_{ii}(x)<0} a_{ii}u^-_{ii}(x) = f(x) - F(0).$$

By the uniform ellipticity, we have, for any $i = 1, \ldots, n$,

$$\lambda \leq a_{ii} \leq \Lambda.$$

Hence,

$$\lambda \sup_{u_{ii}(x)<0} u^-_{ii}(x) \leq n\Lambda \sup_{u_{ii}(x)>0} u^+_{ii}(x) + |f(x)| + |F(0)|.$$

This implies the desired estimate. □

By a similar method as in the proof of Lemma 5.1.2, we can derive an identity for $u_{ij}(y) - u_{ij}(x)$, for any $x, y \in \Omega$.

Lemma 5.1.5. *Let* Ω *be a domain in* $\mathbb{R}^n$ *and* F *be a* C^1*-function in* $\mathcal{S}$. *Suppose that* u *is a* $C^2(\Omega)$*-solution of*

$$F(\nabla^2 u) = f \quad \text{in } \Omega,$$

for some $f \in C(\Omega)$. *Then, for any* $x, y \in \Omega$,

$$\widetilde{a}_{ij} \cdot \big(u_{ij}(y) - u_{ij}(x)\big) = f(y) - f(x),$$

where $\widetilde{a}_{ij}$ *is given by*

$$\widetilde{a}_{ij} = \int_0^1 F_{ij}\big(t\nabla^2 u(y) + (1-t)\nabla^2 u(x)\big)dt.$$

Proof. For any $x, y \in \Omega$, we have

$$F\big(\nabla^2 u(y)\big) - F\big(\nabla^2 u(x)\big) = f(y) - f(x).$$

We write

$$\begin{aligned}
&F\big(\nabla^2 u(y)\big) - F\big(\nabla^2 u(x)\big)\\
&\quad = \int_0^1 \frac{d}{dt} F\big(t\nabla^2 u(y) + (1-t)\nabla^2 u(x)\big)dt\\
&\quad = \int_0^1 F_{ij}\big(t\nabla^2 u(y) + (1-t)\nabla^2 u(x)\big)dt \cdot \big(u_{ij}(y) - u_{ij}(x)\big).
\end{aligned}$$

This yields the desired result. □

For any fixed $z \in \mathbb{R}^n$, by taking $y = x + z$, we conclude that $u(\cdot + z) - u$ satisfies a linear differential equation in an appropriate subdomain of Ω.

Similar to Corollary 5.1.4, we have the following result.

Corollary 5.1.6. *Let Ω be a domain in $\mathbb{R}^n$ and F be a uniformly elliptic C^1-function in $\mathcal{S}$. Suppose that u is a $C^2(\Omega)$-solution of*

$$F(\nabla^2 u) = f \quad \text{in } \Omega,$$

for some $f \in C(\Omega)$. Then, for any $x, y \in \Omega$,

$$|\nabla^2 u(x) - \nabla^2 u(y)| \le C\Big\{\sup_{\gamma\in\mathbb{R}^n, |\gamma|=1} \big(u_{\gamma\gamma}(x) - u_{\gamma\gamma}(y)\big)^+ + |f(x) - f(y)|\Big\},$$

where C is a positive constant depending only on n, λ, and Λ.

Next, we prove a comparison principle.

Theorem 5.1.7. *Let Ω be a bounded domain in $\mathbb{R}^n$ and F be a uniformly elliptic C^1-function in $\mathcal{S}$. Suppose that $u, v \in C(\bar{\Omega}) \cap C^2(\Omega)$ satisfy $F(\nabla^2 u) \ge F(\nabla^2 v)$ in Ω and $u \le v$ on $\partial\Omega$. Then, $u \le v$ in Ω.*

Proof. By

$$\begin{aligned}
F(\nabla^2 u) - F(\nabla^2 v) &= \int_0^1 \frac{d}{dt} F\big(t\nabla^2 u + (1-t)\nabla^2 v\big)dt\\
&= \int_0^1 F_{ij}\big(t\nabla^2 u + (1-t)\nabla^2 v\big)dt \cdot \big(u - v\big)_{ij},
\end{aligned}$$

we have

$$\begin{aligned}
\widehat{a}_{ij}(u-v)_{ij} &\ge 0 \quad \text{in } \Omega,\\
u - v &\le 0 \quad \text{on } \partial\Omega,
\end{aligned}$$

where $\widehat{a}_{ij}$ satisfies, for any $x \in \Omega$ and $\xi \in \mathbb{R}^n$,

$$\lambda|\xi|^2 \le \widehat{a}_{ij}(x)\xi_i\xi_j \le \Lambda|\xi|^2.$$

By the maximum principle (for linear elliptic equations), we have $u \le v$ in Ω. □

A simple consequence is the uniqueness of solutions of the Dirichlet problem.

Corollary 5.1.8. *Let Ω be a bounded domain in $\mathbb{R}^n$ and F be a uniformly elliptic C^1-function in $\mathcal{S}$. Suppose that $u, v \in C(\bar{\Omega}) \cap C^2(\Omega)$ satisfy $F(\nabla^2 u) = F(\nabla^2 v)$ in Ω and $u = v$ on $\partial\Omega$. Then, $u = v$ in Ω.*

Now we discuss two scalings associated with fully nonlinear uniformly elliptic equations.

The first scaling concerns domains. Consider

$$F(\nabla^2 u) = f \quad \text{in } B_r.$$

For $x \in B_r$ and $y \in B_1$, set

$$y = \frac{x}{r}$$

and

$$v(y) = \frac{1}{r^2}u(x).$$

A simple calculation yields

$$F(\nabla^2 v) = \widetilde{f} \quad \text{in } B_1,$$

where

$$\widetilde{f}(y) = f(ry) \quad \text{for any } y \in B_1.$$

Due to such a scaling, it is convenient to work in the unit ball B_1.

The second scaling concerns solutions. Consider

$$F(\nabla^2 u) = f \quad \text{in } \Omega.$$

For any constant $t > 0$, we write the equation above as

$$\frac{1}{t}F\left(t \cdot \frac{\nabla^2 u}{t}\right) = \frac{f}{t}.$$

Then, $w = u/t$ satisfies

$$\widetilde{F}(\nabla^2 w) = \frac{f}{t} \quad \text{in } \Omega,$$

where $\widetilde{F}$ is given by

$$\widetilde{F}(M) = \frac{1}{t}F(tM) \quad \text{for any } M \in \mathcal{S}.$$

It follows easily that $\widetilde{F}$ is uniformly elliptic with the same ellipticity constants as F.

To end this section, we discuss the higher regularity of solutions. Let u be a $C^{2,\alpha}(\Omega)$-solution of

$$F(\nabla^2 u) = f \quad \text{in } \Omega.$$

If u is also C^3, then Lemma 5.1.1(2) shows, for any $k = 1, \ldots, n$,

$$F_{ij}(\nabla^2 u)u_{ijk} = f_k \quad \text{in } \Omega.$$

Since $\nabla^2 u$ is Hölder continuous, this is a uniformly elliptic differential equation for u_k with Hölder continuous coefficients. If $f \in C^{1,\alpha}$, the Schauder theory implies $u_k \in C^{2,\alpha}$, for any $k = 1, \ldots, n$, and hence $u \in C^{3,\alpha}$. A bootstrap argument infers the higher regularity of u. In this simple argument, we required that u is C^3 in order to have a meaningful equation for u_k. This is in fact not necessary.

Proposition 5.1.9. *Let $\alpha \in (0,1)$ be a constant, Ω be a domain in $\mathbb{R}^n$, and F be a uniformly elliptic C^1-function in $\mathcal{S}$. Suppose that u is a $C^{2,\alpha}(\Omega)$-solution of*

$$F(\nabla^2 u) = f \quad \text{in } \Omega,$$

for some $f \in C^\alpha(\Omega)$. For any integer $m \geq 1$, if $F \in C^{m,1}(\mathcal{S})$ and $f \in C^{m,\alpha}(\Omega)$, then $u \in C^{m+2,\alpha}(\Omega)$. In particular, if $F \in C^\infty(\mathcal{S})$ and $f \in C^\infty(\Omega)$, then $u \in C^\infty(\Omega)$.

Proof. We first consider $m = 1$. Fix arbitrary subdomains Ω', Ω'', and Ω''', with $\Omega' \Subset \Omega'' \Subset \Omega''' \Subset \Omega$. Take any unit vector γ in $\mathbb{R}^n$ and any small $h > 0$ such that $x + h\gamma \in \Omega''$ for any $x \in \Omega'$ and $x + h\gamma \in \Omega'''$ for any $x \in \Omega''$. We set, for any $x \in \Omega''$,

$$u^h(x) = \frac{u(x+h\gamma) - u(x)}{h},$$
$$f^h(x) = \frac{f(x+h\gamma) - f(x)}{h}.$$

By Lemma 5.1.5 for $y = x + h\gamma$, we have

$$\widetilde{a}_{ij}\partial_{ij}u^h = f^h \quad \text{in } \Omega'',$$

where $\widetilde{a}_{ij}$ is given by

$$\widetilde{a}_{ij}(x) = \int_0^1 F_{ij}\big(t\nabla^2 u(x+h\gamma) + (1-t)\nabla^2 u(x)\big)dt.$$

The uniform ellipticity of F implies, for any $x \in \Omega''$ and $\xi \in \mathbb{R}^n$,

$$\lambda|\xi|^2 \leq \widetilde{a}_{ij}(x)\xi_i\xi_j \leq \Lambda|\xi|^2.$$

Since $u \in C^{2,\alpha}(\Omega)$ and $\nabla_M F \in C^{0,1}(\mathcal{S})$, then $\widetilde{a}_{ij} \in C^\alpha(\Omega'')$ with $|\widetilde{a}_{ij}|_{C^\alpha(\Omega'')} \leq C$, independent of h. Moreover,

$$|f^h|_{C^\alpha(\Omega'')} \leq C|f|_{C^{1,\alpha}(\Omega''')},$$

independent of h. Hence, by interior Schauder estimates, we obtain

$$\begin{aligned} |u^h|_{C^{2,\alpha}(\Omega')} &\leq C\left\{|u^h|_{L^\infty(\Omega'')} + |f^h|_{C^\alpha(\Omega'')}\right\} \\ &\leq C\left\{|u|_{C^1(\Omega''')} + |f|_{C^{1,\alpha}(\Omega''')}\right\}, \end{aligned}$$

where C is a positive constant depending only on λ, Λ, Ω', Ω, the $C^{0,1}$-norm of $\nabla_M F$, and the $C^{2,\alpha}$-norm of u. This is uniform for h and, hence, we have $u \in C^{3,\alpha}(\Omega')$ and

$$|u|_{C^{3,\alpha}(\Omega')} \leq C\left\{|u|_{C^1(\Omega''')} + |f|_{C^{1,\alpha}(\Omega''')}\right\}.$$

Therefore, we conclude $u \in C^{3,\alpha}(\Omega)$.

By Lemma 5.1.1(2), we have, for any $k = 1, \ldots, n$,

$$Lu_k = f_k \quad \text{in } \Omega.$$

The coefficients of L are $C^{1,\alpha}$ in Ω if $u \in C^{3,\alpha}(\Omega)$ and $\nabla_M F \in C^{1,1}(\mathcal{S})$. If $f \in C^{2,\alpha}(\Omega)$, then by the interior Schauder regularity, we have $u_k \in C^{3,\alpha}(\Omega)$ and hence $u \in C^{4,\alpha}(\Omega)$. We can repeat this argument as long as the regularity of the coefficients of L and f allows and hence obtain the desired result. □

A global version of the higher regularity also holds if we employ the global Schauder estimates.

Proposition 5.1.10. *Let $\alpha \in (0,1)$ be a constant, Ω be a bounded domain in $\mathbb{R}^n$ with a $C^{2,\alpha}$-boundary, and F be a uniformly elliptic C^1-function in $\mathcal{S}$. Suppose that u is a $C^{2,\alpha}(\bar{\Omega})$-solution of*

$$\begin{aligned} F(\nabla^2 u) &= f \quad \text{in } \Omega, \\ u &= \varphi \quad \text{on } \partial\Omega, \end{aligned}$$

for some $f \in C^{\alpha}(\bar{\Omega})$ and $\varphi \in C^{2,\alpha}(\bar{\Omega})$. For any integer $m \geq 1$, if $F \in C^{m,1}(\mathcal{S})$, $\partial\Omega \in C^{m+2,\alpha}$, $f \in C^{m,\alpha}(\bar{\Omega})$, and $\varphi \in C^{m+2,\alpha}(\bar{\Omega})$, then $u \in C^{m+2,\alpha}(\bar{\Omega})$. In particular, if $\partial\Omega \in C^{\infty}$, $F \in C^{\infty}(\mathcal{S})$, $f \in C^{\infty}(\bar{\Omega})$, and $\varphi \in C^{\infty}(\bar{\Omega})$, then $u \in C^{\infty}(\bar{\Omega})$.

We leave the proof as an exercise.

Proposition 5.1.9 and Proposition 5.1.10 illustrate the importance of the $C^{2,\alpha}$-regularity of solutions of fully nonlinear elliptic equations. In the next four sections, we will derive a priori estimates of the $C^{2,\alpha}$-norms of solutions.

5.2. Interior C^2-Estimates

In this section, we derive interior estimates of derivatives up to the second order of solutions of fully nonlinear uniformly elliptic differential equations. Most of the results in this section are based on the maximum principle. The concavity plays an essential role in the estimate of the second derivatives.

We first derive a $C^{1,\alpha}$-estimate of solutions of uniformly elliptic equations. A simple calculus lemma is needed.

Lemma 5.2.1. *Let $\alpha \in (0,1)$, $\beta \in (0,1]$, and $K > 0$ be constants. For a given $u \in L^\infty((-1,1))$, define, for $h \in [-1,1] \setminus \{0\}$,*

$$u_{\beta,h}(x) = \frac{1}{|h|^\beta}\big(u(x+h) - u(x)\big) \quad \text{for any } x \in I_h,$$

where $I_h = [-1, 1-h]$ if $h > 0$ and $I_h = [-1-h, 1]$ if $h < 0$. Assume $u_{\beta,h} \in C^\alpha(I_h)$ and $|u_{\beta,h}|_{C^\alpha(I_h)} \leq K$ for any $h \in [-1,1] \setminus \{0\}$.

(1) *If $\alpha + \beta < 1$, then $u \in C^{\alpha+\beta}((-1,1))$ and, for any $a \in (0,1)$,*

$$|u|_{C^{\alpha+\beta}([-a,a])} \leq C\left\{K + |u|_{L^\infty((-1,1))}\right\};$$

(2) *if $\alpha + \beta > 1$, then $u \in C^{0,1}((-1,1))$ and, for any $a \in (0,1)$,*

$$|u|_{C^{0,1}([-a,a])} \leq C\left\{K + |u|_{L^\infty((-1,1))}\right\},$$

where C in (1) and (2) is a positive constant depending only on $\alpha + \beta$ and a.

Proof. We fix an $a \in (0,1)$. Take any $x \in [-1,a]$ and $\varepsilon > 0$ such that $x + \varepsilon \leq 1$. Let $m \geq 0$ be the integer satisfying

$$x + 2^m\varepsilon \leq 1 < x + 2^{m+1}\varepsilon.$$

For any $\tau > 0$ with $x + \tau \leq 1$, we set

$$w(\tau) = u(x+\tau) - u(x).$$

Then,

$$\begin{aligned} w(\tau) - \frac{1}{2}w(2\tau) &= \frac{1}{2}\big(2u(x+\tau) - u(x+2\tau) - u(x)\big) \\ &= -\frac{1}{2}\big[\big(u(x+2\tau) - u(x+\tau)\big) - \big(u(x+\tau) - u(x)\big)\big] \\ &= -\frac{1}{2}\tau^\beta\big(u_{\beta,\tau}(x+\tau) - u_{\beta,\tau}(x)\big), \end{aligned}$$

and hence

$$\left|w(\tau) - \frac{1}{2}w(2\tau)\right| \leq \frac{1}{2}K\tau^{\alpha+\beta}.$$

By writing

$$\begin{aligned} w(\varepsilon) &= \left(w(\varepsilon) - \frac{1}{2}w(2\varepsilon)\right) + \left(\frac{1}{2}w(2\varepsilon) - \frac{1}{2^2}w(2^2\varepsilon)\right) \\ &\quad + \cdots + \left(\frac{1}{2^{m-1}}w(2^{m-1}\varepsilon) - \frac{1}{2^m}w(2^m\varepsilon)\right) + \frac{1}{2^m}w(2^m\varepsilon), \end{aligned}$$

we have

$$|w(\varepsilon)| \leq \frac{1}{2}K\varepsilon^{\alpha+\beta}\sum_{i=0}^{m-1}2^{i(\alpha+\beta-1)} + \frac{2}{2^m}|u|_{L^\infty}.$$

It is easy to check that

$$\frac{1}{2}(1-a) \leq 2^m\varepsilon \leq 2.$$

Hence,

$$|u(x+\varepsilon)-u(x)| \leq \frac{1}{2}K\varepsilon^{\alpha+\beta}\sum_{i=0}^{m-1}2^{i(\alpha+\beta-1)} + \frac{4\varepsilon}{1-a}|u|_{L^\infty}.$$

If $\alpha+\beta<1$, then

$$|u(x+\varepsilon)-u(x)| \leq \frac{1}{2}\cdot\frac{1}{1-2^{\alpha+\beta-1}}K\varepsilon^{\alpha+\beta} + \frac{4\varepsilon}{1-a}|u|_{L^\infty}.$$

If $\alpha+\beta>1$, then

$$\begin{aligned}|u(x+\varepsilon)-u(x)| &\leq \frac{1}{2}\cdot\frac{1}{2^{\alpha+\beta-1}-1}K\varepsilon^{\alpha+\beta}2^{m(\alpha+\beta-1)} + \frac{4\varepsilon}{1-a}|u|_{L^\infty} \\ &\leq \frac{1}{2}\cdot\frac{2^{\alpha+\beta-1}}{2^{\alpha+\beta-1}-1}K\varepsilon + \frac{4\varepsilon}{1-a}|u|_{L^\infty}.\end{aligned}$$

Similar results hold for $x\in[-a,1]$ and $\varepsilon<0$ with $x+\varepsilon\geq -1$. □

Now we are ready to prove interior $C^{1,\alpha}$-estimates. We note that the concavity is not assumed for F.

Theorem 5.2.2. *Let B_R be a ball in $\mathbb{R}^n$ and F be a uniformly elliptic C^1-function in $\mathcal{S}$. Suppose that u is an $L^\infty(B_R)\cap C^3(B_R)$-solution of*

$$F(\nabla^2 u) = f \quad \text{in } B_R,$$

for some $f\in C^1(B_R)$, with $|f|_{C^1(B_R)}<\infty$. Then, for some $\alpha\in(0,1)$,

$$\begin{aligned}&R|\nabla u|_{L^\infty(B_{R/2})} + R^{1+\alpha}[\nabla u]_{C^\alpha(B_{R/2})} \\ &\qquad \leq C\left\{|u|_{L^\infty(B_R)} + R^2|f|_{L^\infty(B_R)} + R^3|\nabla f|_{L^\infty(B_R)} + R^2|F(0)|\right\},\end{aligned}$$

where α and C are positive constants depending only on n, λ, and Λ.

Instead of the uniform ellipticity in $\mathcal{S}$, it suffices to assume, for any $x\in B_R$ and $t\in[0,1]$,

$$\lambda|\xi|^2 \leq F_{ij}\big(t\nabla^2 u(x)\big)\xi_i\xi_j \leq \Lambda|\xi|^2,$$

and, for any $x,y\in B_R$ with $x+y\in B_R$ and any $t\in[0,1]$,

$$\lambda|\xi|^2 \leq F_{ij}\big(t\nabla^2 u(x+y)+(1-t)\nabla^2 u(x)\big)\xi_i\xi_j \leq \Lambda|\xi|^2,$$

for some positive constants $\lambda\leq\Lambda$.

Proof. We will prove this for $R = 1$, and the general case follows from a simple scaling. By Lemma 5.1.2, u satisfies a linear equation of the form

$$a_{ij}u_{ij} = f - F(0) \quad \text{in } B_1,$$

where a_{ij} satisfies, for any $x \in B_1$ and any $\xi \in \mathbb{R}^n$,

$$\lambda|\xi|^2 \leq a_{ij}(x)\xi_i\xi_j \leq \Lambda|\xi|^2.$$

By Theorem 1.2.14, we conclude, for some $\alpha \in (0, 1)$,

$$(1) \qquad |u|_{C^\alpha(B_{7/8})} \leq C\left\{|u|_{L^\infty(B_1)} + |f|_{L^\infty(B_1)} + |F(0)|\right\},$$

where α and C are positive constants depending only on n, λ, and Λ. By making α slightly smaller, we take an integer m such that $m\alpha < 1 < (m+1)\alpha$.

Next, we take a constant $\beta \in (0, 1)$ to be determined and fix a unit vector $\gamma \in \mathbb{R}^n$. For any $h \in (0, 1/8)$, we set, for any $x \in B_{7/8}$,

$$u_{\beta,h}(x) = \frac{1}{h^\beta}\big(u(x + h\gamma) - u(x)\big), \quad f_{\beta,h}(x) = \frac{1}{h^\beta}\big(f(x + h\gamma) - f(x)\big).$$

By Lemma 5.1.5, we have

$$\widetilde{a}_{ij}\partial_{ij}u_{\beta,h} = f_{\beta,h} \quad \text{in } B_{7/8},$$

where $\widetilde{a}_{ij}$ satisfies, for any $x \in B_{7/8}$ and any $\xi \in \mathbb{R}^n$,

$$\lambda|\xi|^2 \leq \widetilde{a}_{ij}(x)\xi_i\xi_j \leq \Lambda|\xi|^2.$$

For any $0 < r < s \leq 7/8$ and $0 < h < (s - r)/2$, we have, by Theorem 1.2.14 again,

$$|u_{\beta,h}|_{C^\alpha(B_r)} \leq C\left\{|u_{\beta,h}|_{L^\infty(B_{(s+r)/2})} + |f_{\beta,h}|_{L^\infty(B_{(s+r)/2})}\right\},$$

and hence

$$(2) \qquad |u_{\beta,h}|_{C^\alpha(B_r)} \leq C\left\{|u|_{C^\beta(B_s)} + |f|_{C^1(B_1)}\right\}.$$

By taking $\beta = \alpha$ and $r = r_1 < s = 7/8$ and substituting (1) in (2), we have, for any $0 < h < (7/8 - r_1)/2$,

$$|u_{\alpha,h}|_{C^\alpha(B_{r_1})} \leq C\left\{|u|_{L^\infty(B_1)} + |f|_{C^1(B_1)} + |F(0)|\right\},$$

where C is a positive constant depending only on n, λ, Λ, and r_1.

If $2\alpha < 1$, we apply Lemma 5.2.1(1) for any γ as above on segments parallel to γ and get, for any $r_2 < r_1$,

$$|u|_{C^{2\alpha}(B_{r_2})} \leq C\left\{|u|_{L^\infty(B_1)} + |f|_{C^1(B_1)} + |F(0)|\right\}.$$

If $3\alpha < 1$, we apply (2) and Lemma 5.2.1(1) for $\beta = 2\alpha$ to get a similar estimate for 3α. We can repeat this process and finally get, by Lemma 5.2.1(2),

$$(3) \qquad |u|_{C^{0,1}(B_{3/4})} \leq C\left\{|u|_{L^\infty(B_1)} + |f|_{C^1(B_1)} + |F(0)|\right\}.$$

By Lemma 5.1.1(2), we have

$$Lu_\gamma = f_\gamma.$$

By Theorem 1.2.14, we conclude

$$\begin{aligned}[\nabla u]_{C^\alpha(B_{1/2})} &\le C\left\{|\nabla u|_{L^\infty(B_{3/4})} + |f|_{C^1(B_1)} + |F(0)|\right\} \\ &\le C\left\{|u|_{L^\infty(B_1)} + |f|_{C^1(B_1)} + |F(0)|\right\}.\end{aligned} \tag{4}$$

We obtain the desired result by combining (3) and (4). □

An important application of interior estimates is to classify solutions in the entire space with a specific growth. In the next result, we discuss solutions of a linear growth.

Theorem 5.2.3. *Let F be a uniformly elliptic C^1-function in $\mathcal{S}$ with $F(0) = 0$. Suppose that u is a $C^3(\mathbb{R}^n)$-solution of*

$$F(\nabla^2 u) = 0 \quad \text{in } \mathbb{R}^n.$$

If u has a linear growth in $\mathbb{R}^n$, then u is an affine function.

Proof. We first note that, for any $R > 0$,

$$\sup_{B_R} |u| \le C(1 + R),$$

for some positive constant C. By Theorem 5.2.2, we obtain

$$R^{1+\alpha}[\nabla u]_{C^\alpha(B_{R/2})} \le C(1 + R),$$

where $\alpha \in (0, 1)$ is a constant depending only on n, λ, and Λ. By letting $R \to \infty$, we conclude that ∇u is constant in $\mathbb{R}^n$. □

In the proof of $C^{1,\alpha}$-estimates in Theorem 5.2.2, we used the Harnack inequality to derive C^1-estimates, also referred to as the gradient estimates, as an intermediate step. As we pointed out before the statement of Theorem 5.2.2, the concavity is not required. The method here is adapted from the book [**20**]. If we assume the concavity, we can also derive C^1-estimates by the classical *Bernstein method.* The basic idea is to derive a differential equation for $|\nabla u|^2$ and then apply the maximum principle. There are two classes of gradient estimates, global gradient estimates and interior gradient estimates. The global gradient estimates yield estimates of gradients ∇u in Ω in terms of ∇u on $\partial\Omega$, as well as u in Ω, while the interior gradient estimates yield estimates of ∇u in compact subsets of Ω in terms of u in Ω.

In the next result, we derive an interior gradient estimate by Bernstein's method under the assumption that F is concave. The proof is also adapted from the book [**20**].

Theorem 5.2.4. *Let B_R be a ball in $\mathbb{R}^n$ and F be a uniformly elliptic and concave C^1-function in $\mathcal{S}$. Suppose that u is an $L^\infty(B_R)\cap C^3(B_R)$-solution of*

$$F(\nabla^2 u) = f \quad \text{in } B_R,$$

for some $f \in C^1(B_R)$, with $|f|_{C^1(B_R)} < \infty$. Then,

$$R|\nabla u|_{L^\infty(B_{R/2})} \le C\left\{|u|_{L^\infty(B_R)} + R^2|F(0)| + R^2|f|_{L^\infty(B_R)} + R^3|\nabla f|_{L^\infty(B_R)}\right\},$$

where C is a positive constant depending only on n, λ, and Λ.

It suffices to assume that F is uniformly elliptic at $\nabla^2 u(x)$ for any $x \in B_R$ and that $F(t\nabla^2 u(x))$ is a concave function of $t \in [0,1]$ for any $x \in B_R$.

Proof. We fix a cutoff function $\varphi \in C_0^\infty(B_R)$ such that $0 \le \varphi \le 1$ in B_R, $\varphi = 1$ in $B_{R/2}$, and

$$|\nabla\varphi|^2 + |\nabla^2\varphi| \le \frac{c}{R^2} \quad \text{in } B_R,$$

for some positive constant c depending only on n. Set

$$L = F_{ij}\partial_{ij}.$$

A straightforward calculation yields

$$\begin{aligned} L(\varphi^2|\nabla u|^2) &= \varphi^2 L(|\nabla u|^2) + 2F_{ij}(\varphi^2)_i(|\nabla u|^2)_j + |\nabla u|^2 L(\varphi^2) \\ &= 2\varphi^2 F_{ij}u_{ki}u_{kj} + 8F_{ij}\varphi\varphi_i u_k u_{kj} + |\nabla u|^2 L(\varphi^2) + 2\varphi^2 u_k L u_k. \end{aligned}$$

By the uniform ellipticity and $Lu_k = f_k$, we have

$$L(\varphi^2|\nabla u|^2) \ge 2\lambda\varphi^2|\nabla^2 u|^2 + 8F_{ij}\varphi\varphi_i u_k u_{kj} + |\nabla u|^2 L(\varphi^2) + 2\varphi^2 u_k f_k.$$

By the Cauchy inequality, we obtain

$$|8F_{ij}\varphi\varphi_i u_k u_{kj}| \le 2\lambda\varphi^2|\nabla^2 u|^2 + \frac{8\Lambda^2}{\lambda}|\nabla\varphi|^2|\nabla u|^2$$

and

$$|2\varphi^2 u_k f_k| \le \frac{1}{R^2}\varphi^2|\nabla u|^2 + R^2\varphi^2|\nabla f|^2.$$

Hence,

$$L(\varphi^2|\nabla u|^2) \ge \left(-\frac{8\Lambda^2}{\lambda}|\nabla\varphi|^2 + L(\varphi^2) - \frac{1}{R^2}\varphi^2\right)|\nabla u|^2 - R^2\varphi^2|\nabla f|^2.$$

To control ∇u in the right-hand side, we set $K = \sup_{B_R} u$ and note that

$$L\big((K-u)^2\big) = 2F_{ij}u_i u_j - 2(K-u)Lu.$$

Since F is concave, Lemma 5.1.1(1) implies $Lu \le f - F(0)$. By the uniform ellipticity again, we have

$$L\big((K-u)^2\big) \ge 2\lambda|\nabla u|^2 - 2(K-u)\big(|f| + |F(0)|\big).$$

For some positive constant A to be determined, we have

$$\begin{aligned} L&\left(\varphi^2|\nabla u|^2 + \frac{A}{R^2}(K-u)^2\right)\\ &\geq \left(\frac{2\lambda A}{R^2} - \frac{8\Lambda^2}{\lambda}|\nabla\varphi|^2 + L(\varphi^2) - \frac{1}{R^2}\varphi^2\right)|\nabla u|^2\\ &\quad - \frac{2A}{R^2}(K-u)\big(|f| + |F(0)|\big) - R^2\varphi^2|\nabla f|^2. \end{aligned}$$

By taking A sufficiently large depending only on n, λ, and Λ, we get

$$\begin{aligned} L&\left(\varphi^2|\nabla u|^2 + \frac{A}{R^2}(K-u)^2\right)\\ &\geq -\frac{2A}{R^2}(K-u)\big(|f| + |F(0)|\big) - R^2\varphi^2|\nabla f|^2\\ &\geq -\frac{A^2}{R^4}(K-u)^2 - \big(|f| + |F(0)|\big)^2 - R^2\varphi^2|\nabla f|^2. \end{aligned}$$

Last, we note that

$$L(x_1^2) = 2F_{11} \geq 2\lambda.$$

Hence,

$$L\left(\varphi^2|\nabla u|^2 + \frac{A}{R^2}(K-u)^2 + Bx_1^2\right) \geq 0 \quad \text{in } B_R$$

if we take B such that

$$2B\lambda = \frac{A^2}{R^4}\left(\operatorname*{osc}_{B_R} u\right)^2 + \big(|f|_{L^\infty(B_R)} + |F(0)|\big)^2 + R^2|\nabla f|^2_{L^\infty(B_R)}.$$

By the maximum principle, we conclude

$$\begin{aligned} \sup_{B_R}&\left(\varphi^2|\nabla u|^2 + \frac{A}{R^2}(K-u)^2 + Bx_1^2\right)\\ &\leq \max_{\partial B_R}\left(\varphi^2|\nabla u|^2 + \frac{A}{R^2}(K-u)^2 + Bx_1^2\right), \end{aligned}$$

and hence, by $\varphi = 1$ in $B_{R/2}$ and $\varphi = 0$ on ∂B_R,

$$\sup_{B_{R/2}}|\nabla u|^2 \leq \max_{\partial B_R}\left(\frac{A}{R^2}(K-u)^2 + Bx_1^2\right).$$

By the definition of B and the simple fact that $x_1^2 \leq R^2$ on ∂B_R, we obtain

$$\sup_{B_{R/2}}|\nabla u| \leq C\left\{\frac{1}{R}|u|_{L^\infty(B_R)} + R|F(0)| + R|f|_{L^\infty(B_R)} + R^2|\nabla f|_{L^\infty(B_R)}\right\}.$$

This implies the desired result. □

Next, we derive an interior C^2-estimate by Bernstein's method under the assumption that F is concave. The proof is similar to that of Theorem 5.2.4 and is also adapted from the book [**20**].

Theorem 5.2.5. *Let B_R be a ball in $\mathbb{R}^n$ and F be a uniformly elliptic and concave C^2-function in $\mathcal{S}$. Suppose that u is a $C^4(B_R)$-solution of*

$$F(\nabla^2 u) = f \quad \text{in } B_R,$$

for some $f \in C^2(B_R)$, with $|f|_{C^2(B_R)} < \infty$ and $|\nabla u|_{L^\infty(B_R)} < \infty$. Then,

$$\begin{aligned} R^2|\nabla^2 u|_{L^\infty(B_{R/2})} \le C\big\{&R|\nabla u|_{L^\infty(B_R)} + R^2|F(0)| + R^2|f|_{L^\infty(B_R)} \\ &+R^3|\nabla f|_{L^\infty(B_R)} + R^4|\nabla^2 f|_{L^\infty(B_R)}\big\}, \end{aligned}$$

where C is a positive constant depending only on n, λ, and Λ.

It suffices to assume that F is uniformly elliptic at $t\nabla^2 u(x)$, for any $t \in [0,1]$ and any $x \in B_R$, and that F is concave at $\nabla^2 u(x)$, for any $x \in B_R$.

Proof. We fix a unit vector $\gamma \in \mathbb{R}^n$ and a cutoff function $\varphi \in C_0^\infty(B_R)$ such that $0 \le \varphi \le 1$ in B_R, $\varphi = 1$ in $B_{R/2}$, and

$$|\nabla\varphi|^2 + |\nabla^2\varphi| \le \frac{c}{R^2} \quad \text{in } B_R,$$

for some positive constant c depending only on n. As in the proof of Theorem 5.2.4, we set

$$L = F_{ij}\partial_{ij}.$$

A straightforward calculation yields

$$\begin{aligned} L(\varphi^2 u_{\gamma\gamma}^2) &= \varphi^2 L(u_{\gamma\gamma}^2) + 2F_{ij}(\varphi^2)_i(u_{\gamma\gamma}^2)_j + u_{\gamma\gamma}^2 L(\varphi^2) \\ &= 2\varphi^2 F_{ij}u_{\gamma\gamma i}u_{\gamma\gamma j} + 8F_{ij}\varphi\varphi_i u_{\gamma\gamma}u_{\gamma\gamma j} \\ &\quad + u_{\gamma\gamma}^2 L(\varphi^2) + 2\varphi^2 u_{\gamma\gamma} L u_{\gamma\gamma}. \end{aligned}$$

Since F is concave, Lemma 5.1.1(3) implies $Lu_{\gamma\gamma} \ge f_{\gamma\gamma}$. Set

$$\Omega = \{x \in B_R : u_{\gamma\gamma}(x) \ge 0\}.$$

By the uniform ellipticity, we have

$$\begin{aligned} L(\varphi^2 u_{\gamma\gamma}^2) \ge 2\lambda\sum_{i=1}^n \varphi^2 u_{\gamma\gamma i}^2 &+ 8F_{ij}\varphi\varphi_i u_{\gamma\gamma}u_{\gamma\gamma j} \\ &+ u_{\gamma\gamma}^2 L(\varphi^2) + 2\varphi^2 u_{\gamma\gamma} f_{\gamma\gamma} \quad \text{in } \Omega. \end{aligned}$$

By the Cauchy inequality as in the proof of Theorem 5.2.4, we obtain

$$L(\varphi^2 u_{\gamma\gamma}^2) \ge \left(-\frac{8\Lambda^2}{\lambda}|\nabla\varphi|^2 + L(\varphi^2) - \frac{1}{R^2}\varphi^2\right)u_{\gamma\gamma}^2 - R^2\varphi^2 f_{\gamma\gamma}^2 \quad \text{in } \Omega.$$

To control $u_{\gamma\gamma}^2$ in the right-hand side, we note that

$$L(u_\gamma^2) = 2F_{ij}u_{\gamma i}u_{\gamma j} + 2u_\gamma Lu_\gamma.$$

By the uniform ellipticity again, we have

$$L(u_\gamma^2) \geq 2\lambda\sum_{i=1}^n u_{\gamma i}^2 + 2u_\gamma f_\gamma \geq 2\lambda u_{\gamma\gamma}^2 + 2u_\gamma f_\gamma.$$

By choosing A sufficiently large, depending only on n, λ, and Λ, we obtain

$$L\left(\varphi^2 u_{\gamma\gamma}^2 + \frac{A}{R^2}u_\gamma^2\right) \geq -R^2 f_{\gamma\gamma}^2 + \frac{2A}{R^2}u_\gamma f_\gamma \quad \text{in } \Omega.$$

Hence,

$$L\left(\varphi^2 u_{\gamma\gamma}^2 + \frac{A}{R^2}u_\gamma^2 + Bx_1^2\right) \geq 0 \quad \text{in } \Omega$$

if we take B such that

$$2B\lambda = \sup_{B_R}\left(R^2 f_{\gamma\gamma}^2 + Af_\gamma^2 + \frac{A}{R^4}u_\gamma^2\right).$$

By the maximum principle, we conclude

$$\sup_\Omega\left(\varphi^2 u_{\gamma\gamma}^2 + \frac{A}{R^2}u_\gamma^2 + Bx_1^2\right) \leq \max_{\partial\Omega}\left(\varphi^2 u_{\gamma\gamma}^2 + \frac{A}{R^2}u_\gamma^2 + Bx_1^2\right).$$

Since $\varphi u_{\gamma\gamma} = 0$ on $\partial\Omega$, we obtain

$$\sup_{B_{R/2}}(u_{\gamma\gamma}^+)^2 \leq \sup_{B_R}\left(\frac{A}{R^2}u_\gamma^2 + Bx_1^2\right).$$

By the definition of B, we have

$$\sup_{B_{R/2}} u_{\gamma\gamma}^+ \leq C\left\{\frac{1}{R}|\nabla u|_{L^\infty(B_R)} + R|\nabla f|_{L^\infty(B_R)} + R^2|\nabla^2 f|_{L^\infty(B_R)}\right\}.$$

This establishes an upper bound of $u_{\gamma\gamma}$ in $B_{R/2}$ for any unit vector γ. Corollary 5.1.4 yields the desired estimate of $\nabla^2 u$ in $B_{R/2}$. □

Next, we prove a decomposition concerning symmetric matrices. Such a result, although simple, plays an important role in deriving estimates of second derivatives for fully nonlinear elliptic equations. We first introduce a notation. For any vector $\gamma = (\gamma_1, \ldots, \gamma_n) \in \mathbb{R}^n$, we denote by $\gamma \otimes \gamma$ the matrix given by $(\gamma_i\gamma_j)$. This is a symmetric matrix whose eigenvalues are given by $|\gamma|^2, 0, \ldots, 0$.

Lemma 5.2.6. *Let $\mathcal{S}(\lambda, \Lambda)$ denote the set of $n \times n$ positive definite matrices with eigenvalues in the interval $[\lambda, \Lambda]$, for some positive constants $\lambda < \Lambda$. Then, there exist finitely many unit vectors $\gamma_1, \ldots, \gamma_N \in \mathbb{R}^n$ and positive* *numbers* $\lambda^* < \Lambda^*$, *depending only on* n, λ, *and* Λ, *such that any matrix* $A = (a_{ij}) \in \mathcal{S}(\lambda, \Lambda)$ *has the form*

$$A = \sum_{k=1}^{N} \beta_k \gamma_k \otimes \gamma_k,$$

for some constants $\beta_k \in [\lambda^*, \Lambda^*]$, $k = 1, \dots, N$. *Furthermore,* $\{\gamma_1, \dots, \gamma_N\}$ *can be chosen to include the coordinate directions* $e_i, i = 1, \dots, n$, *together with the directions* $(e_i \pm e_j)/\sqrt{2}$, $i, j = 1, \dots, n$, *with* $i < j$.

If we write $\gamma_k = (\gamma_{k1}, \dots, \gamma_{kn})$ for $k = 1, \dots, N$, then, for any $i = 1, \dots, n$,

$$a_{ij} = \sum_{k=1}^{N} \beta_k \gamma_{ki} \gamma_{kj}.$$

Proof. Let $\mathcal{S}_+$ denote the cone of positive definite matrices in $\mathcal{S}$ and $n' = n(n+1)/2$. We take a matrix A_0 whose diagonal and nondiagonal elements are n and 1, respectively. It is easy to check that $A_0 \in \mathcal{S}_+$ and

$$A_0 = \sum_{i=1}^{n} e_i \otimes e_i + \sum_{i<j}^{n} (e_i + e_j) \otimes (e_i + e_j).$$

There are n' terms in this decomposition. We note that any two matrices in $\mathcal{S}_+$ are similar, and hence, in particular, each $A \in \mathcal{S}_+$ is similar to A_0. In other words, for any $A \in \mathcal{S}_+$, there exists a nonsingular matrix P such that $A = P^T A_0 P$. Therefore, we can represent any matrix $A \in \mathcal{S}_+$ in the form

$$A = \sum_{k=1}^{n'} \beta_k \gamma_k \otimes \gamma_k,$$

for some positive constants $\beta_1, \dots, \beta_{n'}$ and some unit vectors $\gamma_1, \dots, \gamma_{n'} \in \mathbb{R}^n$ with the property that $\gamma_1 \otimes \gamma_1, \dots, \gamma_{n'} \otimes \gamma_{n'}$ are linearly independent. Consequently, the family of sets of the form

$$U(\gamma_1, \dots, \gamma_{n'}) = \left\{ \sum_{k=1}^{n'} \beta_k \gamma_k \otimes \gamma_k : \beta_k > 0, \ k = 1, \dots, n' \right\},$$

where $\gamma_1, \dots, \gamma_{n'}$ are unit vectors such that $\gamma_1 \otimes \gamma_1, \dots, \gamma_{n'} \otimes \gamma_{n'}$ are linearly independent, forms an open cover of $\mathcal{S}(\lambda, \Lambda) \subset \mathcal{S}_+$. Since $\mathcal{S}(\lambda, \Lambda)$ is compact, there exists a finite subcover. Accordingly, there exist finitely many unit vectors $\gamma_1, \dots, \gamma_N$, depending only on n, λ, and Λ, such that any matrix $A \in \mathcal{S}(\lambda, \Lambda)$ may be written as

$$A = \sum_{k=1}^{N} \beta_k \gamma_k \otimes \gamma_k,$$

with $0 \leq \beta_k \leq \Lambda$, for any $k = 1, \ldots, N$. Note that, in this step, any particular finite set of unit vectors may be included in $\{\gamma_1, \ldots, \gamma_N\}$.

To create a common positive lower bound, we proceed as follows. We apply the previous process to $\mathcal{S}(\lambda/2, \Lambda)$ to get unit vectors $\gamma_1, \ldots, \gamma_N$. For any $A \in \mathcal{S}(\lambda, \Lambda)$, consider

$$A - \lambda^* \sum_{k=1}^{N} \gamma_k \otimes \gamma_k \in \mathcal{S}\left(\frac{\lambda}{2}, \Lambda\right),$$

for sufficiently small λ^*, say $\lambda^* = \lambda/(2N)$. By what we have proved, we obtain

$$A - \lambda^* \sum_{k=1}^{N} \gamma_k \otimes \gamma_k = \sum_{k=1}^{N} \beta_k \gamma_k \otimes \gamma_k,$$

with $0 \leq \beta_k \leq \Lambda$, for any $k = 1, \ldots, N$. Hence,

$$A = \sum_{k=1}^{N} (\beta_k + \lambda^*) \gamma_k \otimes \gamma_k.$$

Since $\lambda^* \leq \beta_k + \lambda^* \leq \Lambda + \lambda^*$, we have the desired result by renaming β_k. □

Applying Lemma 5.2.6 to uniformly elliptic operators, we get a relation among *pure* second derivatives of solutions. Suppose L is a linear operator of the form $L = a_{ij}\partial_{ij}$ in $\Omega \subset \mathbb{R}^n$, satisfying, for any $x \in \Omega$ and $\xi \in \mathbb{R}^n$,

$$\lambda|\xi|^2 \leq a_{ij}(x)\xi_i\xi_j \leq \Lambda|\xi|^2,$$

for some positive constants λ and Λ. Then, L can be written as

$$L = \sum_{k=1}^{N} \beta_k(x)\partial_{\gamma_k\gamma_k},$$

where functions $\beta_1, \ldots, \beta_N$ satisfy

$$\lambda^* \leq \beta_k(x) \leq \Lambda^*,$$

for some positive constants λ^* and Λ^* as in Lemma 5.2.6.

In the rest of this section, we derive interior estimates for solutions of more general fully nonlinear elliptic equations. Let Ω be a bounded domain in $\mathbb{R}^n$ and let F be a function defined in $\Omega \times \mathcal{S}$. We assume that F is *uniformly elliptic* (also uniformly for $x \in \Omega$); i.e., for any $x \in \Omega$, $M \in \mathcal{S}$, and $\xi \in \mathbb{R}^n$,

$$\lambda|\xi|^2 \leq F_{ij}(x, M)\xi_i\xi_j \leq \Lambda|\xi|^2,$$

for some positive constants λ and Λ. The upper bound of F_{ij} implies, for any $x \in \Omega$ and $M \in \mathcal{S}$,

$$|F(x, M)| \leq |F(x, 0)| + \Lambda|M|.$$

In other words, $F(x, M)$ has a linear growth in M, uniformly for x. It is natural to assume that the x-derivatives of F also have a linear growth in M and that the M-derivatives of F are bounded.

First, we derive an interior estimate of the gradients by modifying the proof of Theorem 5.2.4.

Theorem 5.2.7. *Let B_R be a ball in $\mathbb{R}^n$ and F be a $C^1(B_R \times \mathcal{S})$-function, uniformly elliptic and concave in $M \in \mathcal{S}$, and satisfy, for any $x \in B_R$ and $M \in \mathcal{S}$,*

$$|\nabla_x F(x, M)| \le F_0 + F_1|M|,$$

for some positive constants F_0 and F_1. Suppose that u is a $C^3(B_R)$-solution of

$$F(x, \nabla^2 u) = 0 \quad \text{in } B_R. \tag{1}$$

Then,

$$R|\nabla u|_{L^\infty(B_{R/2})} \le C\big\{(1 + RF_1)|u|_{L^\infty(B_R)} + R^2|F(\cdot, 0)|_{L^\infty(B_R)} + R^3 F_0\big\},$$

where C is a positive constant depending only on n, λ, and Λ.

Proof. By differentiating (1) with respect to x_k, we have

$$F_{ij}u_{ijk} + F_{x_k} = 0.$$

Set

$$L = F_{ij}(x, \nabla^2 u(x))\partial_{ij}.$$

Then,

$$Lu_k + F_{x_k} = 0,$$

and hence

$$L(|\nabla u|^2) = 2F_{ij}u_{ki}u_{kj} - 2u_k F_{x_k}.$$

We take a cutoff function $\varphi \in C_0^\infty(B_R)$ such that $0 \le \varphi \le 1$ in B_R, $\varphi = 1$ in $B_{R/2}$, and

$$|\nabla\varphi|^2 + |\nabla^2\varphi| \le \frac{c}{R^2} \quad \text{in } B_R,$$

for some positive constant c depending only on n. Then,

$$\begin{aligned} L(\varphi^2|\nabla u|^2) &= \varphi^2 L(|\nabla u|^2) + 2F_{ij}(\varphi^2)_i(|\nabla u|^2)_j + |\nabla u|^2 L(\varphi^2) \\ &= 2\varphi^2 F_{ij}u_{ki}u_{kj} + 8F_{ij}\varphi\varphi_i u_k u_{kj} - 2\varphi^2 u_k F_{x_k} + |\nabla u|^2 L(\varphi^2). \end{aligned}$$

By the uniform ellipticity, we have

$$L(\varphi^2|\nabla u|^2) \ge 2\lambda\varphi^2|\nabla^2 u|^2 + 8F_{ij}\varphi\varphi_i u_k u_{kj} - 2\varphi^2 u_k F_{x_k} + |\nabla u|^2 L(\varphi^2).$$

By the assumption on $\nabla_x F$ and the Cauchy inequality, we obtain

$$\begin{aligned}|2u_kF_{x_k}| &\leq 2|\nabla u|(F_0 + F_1|\nabla^2 u|) = 2F_0|\nabla u| + 2F_1|\nabla u||\nabla^2 u| \\ &\leq \frac{1}{R^2}|\nabla u|^2 + R^2F_0^2 + \lambda|\nabla^2 u|^2 + \frac{F_1^2}{\lambda}|\nabla u|^2\end{aligned}$$

and

$$|8F_{ij}\varphi\varphi_i u_k u_{kj}| \leq \lambda\varphi^2|\nabla^2 u|^2 + \frac{16\Lambda^2}{\lambda}|\nabla\varphi|^2|\nabla u|^2.$$

Hence,

$$L(\varphi^2|\nabla u|^2) \geq \left(L(\varphi^2) - \frac{16\Lambda^2}{\lambda}|\nabla\varphi|^2 - \frac{1}{R^2} - \frac{1}{\lambda}F_1^2\right)|\nabla u|^2 - \varphi^2R^2F_0^2.$$

To control ∇u in the right-hand side, we set $K = \sup_{B_R} u$ and note that

$$L\big((K-u)^2\big) = 2F_{ij}u_iu_j - 2(K-u)Lu.$$

Since F is concave in $M \in \mathcal{S}$, we have, for any $x \in B_R$,

$$F(x,0) \leq F(x, \nabla^2 u(x)) + F_{ij}(x, \nabla^2 u(x))(0 - u_{ij}),$$

and hence

$$Lu = F_{ij}u_{ij} \leq -F(x,0) \leq |F(x,0)|.$$

By the uniform ellipticity again, we have

$$L\big((K-u)^2\big) \geq 2\lambda|\nabla u|^2 - 2(K-u)|F(x,0)|.$$

For any positive constant A to be determined, we have

$$\begin{aligned}&L\left(\varphi^2|\nabla u|^2 + \frac{A}{R^2}(K-u)^2\right) \\ &\quad\geq \left(\frac{2\lambda A}{R^2} + L(\varphi^2) - \frac{16\Lambda^2}{\lambda}|\nabla\varphi|^2 - \frac{1}{R^2} - \frac{1}{\lambda}F_1^2\right)|\nabla u|^2 \\ &\qquad - \frac{2A}{R^2}(K-u)|F(x,0)| - \varphi^2R^2F_0^2.\end{aligned}$$

Now we take A to be of the form

$$A = C_*\left(1 + R^2F_1^2\right), \tag{2}$$

for some constant C_* sufficiently large, depending only on n, λ, and Λ. Then,

$$\begin{aligned}&L\left(\varphi^2|\nabla u|^2 + \frac{A}{R^2}(K-u)^2\right) \\ &\quad\geq \frac{c_0A}{R^2}|\nabla u|^2 - \frac{2A}{R^2}(K-u)|F(x,0)| - \varphi^2R^2F_0^2,\end{aligned}$$

where c_0 is a positive constant depending only on n, λ, and Λ. Assume $\varphi^2|\nabla u|^2 + A(K-u)^2/R^2$ attains its maximum at $x_0 \in \bar{B}_R$. If $x_0 \in B_R$, then at x_0,

$$\frac{c_0A}{R^2}|\nabla u|^2 \leq \frac{2A}{R^2}(K-u)|F(x,0)| + \varphi^2R^2F_0^2,$$

and hence

$$\begin{aligned}|\nabla u|^2 &\le C\left\{2(K-u)|F(\cdot,0)|_{L^\infty(B_R)} + \frac{R^4}{A}F_0^2\right\}\\ &\le C\left\{\frac{A}{R^2}(K-u)^2 + \frac{R^2}{A}|F(\cdot,0)|^2_{L^\infty(B_R)} + \frac{R^4}{A}F_0^2\right\}.\end{aligned}$$

By multiplying by φ^2 and then adding $A(K-u)^2/R^2$, we obtain at x_0

$$\varphi^2|\nabla u|^2 + \frac{A}{R^2}(K-u)^2 \le C\left\{\frac{A}{R^2}(K-u)^2 + \frac{R^2}{A}|F(\cdot,0)|^2_{L^\infty(B_R)} + \frac{R^4}{A}F_0^2\right\}.$$

This estimate obviously holds if $x_0 \in \partial B_R$. Since the left-hand side attains its maximum at x_0, we have

$$\varphi^2|\nabla u|^2 \le C\left\{\frac{A}{R^2}|u|^2_{L^\infty(B_R)} + \frac{R^2}{A}|F(\cdot,0)|^2_{L^\infty(B_R)} + \frac{R^4}{A}F_0^2\right\} \quad \text{in } B_R.$$

By the explicit expression of A in (2) and taking the square root, we obtain

$$\begin{aligned}R\varphi|\nabla u| \le C\Big\{&(1+RF_1)|u|_{L^\infty(B_R)}\\ &+ (1+RF_1)^{-1}\left(R^2|F(\cdot,0)|_{L^\infty(B_R)} + R^3F_0\right)\Big\}.\end{aligned}$$

This implies the desired result. □

In the interior gradient estimate in Theorem 5.2.7, the factors of R appear naturally due to scaling. To see this, we set

$$x = Ry \quad \text{and} \quad v(y) = u(x).$$

Then, $y \in B_1$ for $x \in B_R$ and $\nabla_y^2 v = R^2\nabla_x^2 u$. Furthermore, we set, for any $M \in \mathcal{S}$,

$$N = R^2M \quad \text{and} \quad G(y,N) = R^2F(x,M).$$

Then, $\nabla_N G = \nabla_M F$. It follows that $G(y,\nabla^2 v) = 0$ in B_1 if $F(x,\nabla^2 u) = 0$ in B_R and that G and F have the same ellipticity constants. Now, we assume, for any $x \in B_R$ and $M \in \mathcal{S}$,

$$|\nabla_x F(x,M)| \le F_0 + F_1|M|,$$

for some positive constants F_0 and F_1. Then,

$$\begin{aligned}|\nabla_y G(y,N)| &= R^3|\nabla_x F(x,M)|\\ &\le R^3(F_0 + F_1|M|) = R^3F_0 + RF_1|N|.\end{aligned}$$

By writing

$$|\nabla_y G(y, N)| \le G_0 + G_1|N|,$$

we can take

$$G_0 = R^3 F_0, \quad G_1 = RF_1.$$

Now we derive interior estimates of the second derivatives by modifying the proof of Theorem 5.2.5. The proof here is much more involved than that of Theorem 5.2.5.

Theorem 5.2.8. *Let B_R be a ball in $\mathbb{R}^n$ and F be a $C^2(B_R \times \mathcal{S})$-function, uniformly elliptic and concave in $M \in \mathcal{S}$, and satisfy, for any $x \in B_R$ and $M \in \mathcal{S}$,*

$$\begin{aligned} |\nabla_x F(x, M)| &\le F_{10} + F_{11}|M|, \\ |\nabla^2_{xx} F(x, M)| &\le F_{20} + F_{21}|M|, \\ |\nabla^2_{xM} F(x, M)| &\le F_{22}, \end{aligned}$$

for some positive constants F_{10}, F_{11}, F_{20}, F_{21}, and F_{22}. Suppose that u is a $C^4(B_R)$-solution of

$$F(x, \nabla^2 u) = 0 \quad \text{in } B_R. \tag{1}$$

Then,

$$\begin{aligned} R^2|\nabla^2 u|_{L^\infty(B_{R/2})} \le C\big\{ &R(1 + RF_{11} + R^2F_{21} + RF_{22})|\nabla u|_{L^\infty(B_R)} \\ &+ R^2|F(\cdot, 0)|_{L^\infty(B_R)} + R^3F_{10} + R^4F_{20}\big\}, \end{aligned}$$

where C is a positive constant depending only on n, λ, and Λ.

Proof. Let γ be an arbitrary unit vector in $\mathbb{R}^n$. By differentiating (1) with respect to x_γ twice, we have

$$F_{ij}u_{ij\gamma} + F_{x_\gamma} = 0$$

and

$$F_{ij}u_{ij\gamma\gamma} + F_{ij,kl}u_{ij\gamma}u_{kl\gamma} + 2F_{ij,x_\gamma}u_{ij\gamma} + F_{x_\gamma x_\gamma} = 0.$$

Set

$$L = F_{ij}(x, \nabla^2 u(x))\partial_{ij}.$$

Then,

$$Lu_{\gamma\gamma} = -F_{ij,kl}u_{ij\gamma}u_{kl\gamma} - 2F_{ij,x_\gamma}u_{ij\gamma} - F_{x_\gamma x_\gamma}.$$

Let b and K be two positive constants to be determined. We take a cutoff function $\varphi \in C_0^\infty(B_R)$ such that $0 \le \varphi \le 1$ in B_R, $\varphi = 1$ in $B_{R/2}$, and

$$|\nabla\varphi|^2 + |\nabla^2\varphi| \le \frac{c}{R^2} \quad \text{in } B_R,$$

for some positive constant c depending only on n. We first consider $(b\varphi u_{\gamma\gamma}+K)^2$ in B_R. A straightforward calculation yields

$$\begin{aligned}
&L\big((b\varphi u_{\gamma\gamma}+K)^2\big)\\
&\quad= 2(b\varphi u_{\gamma\gamma}+K)L(b\varphi u_{\gamma\gamma}) + 2F_{ij}(b\varphi u_{\gamma\gamma})_i(b\varphi u_{\gamma\gamma})_j\\
&\quad= 2(b\varphi u_{\gamma\gamma}+K)\big(\varphi L(bu_{\gamma\gamma}) + (L\varphi)bu_{\gamma\gamma} + 2F_{ij}\varphi_i(bu_{\gamma\gamma})_j\big)\\
&\qquad+ 2F_{ij}(b\varphi u_{\gamma\gamma})_i(b\varphi u_{\gamma\gamma})_j\\
&\quad= 2b\varphi(b\varphi u_{\gamma\gamma}+K)\big(-F_{ij,kl}u_{ij\gamma}u_{kl\gamma} - 2F_{ij,x_\gamma}u_{ij\gamma} - F_{x_\gamma x_\gamma}\big)\\
&\qquad+ 2(b\varphi u_{\gamma\gamma}+K)\big((L\varphi)bu_{\gamma\gamma} + 2F_{ij}\varphi_i(bu_{\gamma\gamma})_j\big)\\
&\qquad+ 2F_{ij}(b\varphi u_{\gamma\gamma})_i(b\varphi u_{\gamma\gamma})_j.
\end{aligned}$$

We choose K such that $b\varphi u_{\gamma\gamma}+K\geq 0$ in B_R. By the concavity and the uniform ellipticity, we have

$$F_{ij,kl}u_{ij\gamma}u_{kl\gamma}\leq 0$$

and

$$F_{ij}(b\varphi u_{\gamma\gamma})_i(b\varphi u_{\gamma\gamma})_j \geq \lambda\sum_{i=1}^n \big((b\varphi u_{\gamma\gamma})_i\big)^2.$$

Hence,

$$L\big((b\varphi u_{\gamma\gamma}+K)^2\big)\geq I+II, \tag{2}$$

where

$$\begin{aligned}
I = {}& 2\lambda\sum_{i=1}^n\big((b\varphi u_{\gamma\gamma})_i\big)^2 + 4F_{ij}\varphi_i(b\varphi u_{\gamma\gamma}+K)(bu_{\gamma\gamma})_j\\
&- 4b\varphi(b\varphi u_{\gamma\gamma}+K)F_{ij,x_\gamma}u_{ij\gamma}
\end{aligned}$$

and

$$II = 2(b\varphi u_{\gamma\gamma}+K)\big((L\varphi)bu_{\gamma\gamma} - b\varphi F_{x_\gamma x_\gamma}\big).$$

We first estimate I. By the Cauchy inequality, we have

$$2\lambda\sum_{i=1}^n\big((b\varphi u_{\gamma\gamma})_i\big)^2 \geq \lambda b^2\varphi^2\sum_{i=1}^n u_{i\gamma\gamma}^2 - 2\lambda b^2|\nabla\varphi|^2u_{\gamma\gamma}^2$$

and

$$|4F_{ij}\varphi_i(b\varphi u_{\gamma\gamma}+K)(bu_{\gamma\gamma})_j| \leq \frac{1}{2}\lambda b^2\varphi^2\sum_{i=1}^n u_{i\gamma\gamma}^2 + \frac{8\Lambda^2}{\lambda}\frac{|\nabla\varphi|^2}{\varphi^2}(b\varphi u_{\gamma\gamma}+K)^2.$$

Moreover, for $\varepsilon>0$ to be determined,

$$|4b\varphi(b\varphi u_{\gamma\gamma}+K)F_{ij,x_\gamma}u_{ij\gamma}| \leq \varepsilon b^2\varphi^2\sum_{i,j=1}^n u_{ij\gamma}^2 + \frac{4}{\varepsilon}F_{22}^2(b\varphi u_{\gamma\gamma}+K)^2.$$

Note that

$$(b\varphi u_{\gamma\gamma}+K)^2\leq 2(b\varphi u_{\gamma\gamma})^2 + 2K^2.$$

Then,

$$(3)\quad \begin{aligned} I \geq & \frac{1}{2}\lambda b^2\varphi^2 \sum_{i=1}^{n} u_{i\gamma\gamma}^2 - \varepsilon b^2\varphi^2 \sum_{i,j=1}^{n} u_{ij\gamma}^2 - 2\lambda b^2|\nabla\varphi|^2 u_{\gamma\gamma}^2 \\ & - 2\left(\frac{8\Lambda^2}{\lambda}\frac{|\nabla\varphi|^2}{\varphi^2} + \frac{4}{\varepsilon}F_{22}^2\right)\left((b\varphi u_{\gamma\gamma})^2 + K^2\right). \end{aligned}$$

Now we estimate II by using the Cauchy inequality. For the first term in II, we have

$$|2(b\varphi u_{\gamma\gamma} + K)(L\varphi)bu_{\gamma\gamma}| \leq |L\varphi|(bu_{\gamma\gamma})^2 + |L\varphi|(b\varphi u_{\gamma\gamma} + K)^2.$$

For the last term in II, we get

$$\begin{aligned} 2b\varphi(b\varphi u_{\gamma\gamma} + K)|F_{x_\gamma x_\gamma}| & \leq 2b\varphi(b\varphi u_{\gamma\gamma} + K)(F_{20} + F_{21}|\nabla^2 u|) \\ & \leq \varphi\left(\frac{1}{R^2} + \frac{R^2 F_{21}^2}{\lambda}\right)(b\varphi u_{\gamma\gamma} + K)^2 + \frac{\lambda b^2}{R^2}\varphi|\nabla^2 u|^2 + b^2R^2F_{20}^2\varphi. \end{aligned}$$

Hence,

$$(4)\quad \begin{aligned} II \geq & -\left(\frac{|L\varphi|}{\varphi^2} + 2|L\varphi| + \frac{2}{R^2} + \frac{2R^2F_{21}^2}{\lambda}\right)(b\varphi u_{\gamma\gamma})^2 \\ & - \left(2|L\varphi| + \frac{2}{R^2} + \frac{2R^2F_{21}^2}{\lambda}\right)K^2 - \frac{\lambda b^2}{R^2}\varphi|\nabla^2 u|^2 - b^2R^2F_{20}^2\varphi. \end{aligned}$$

By combining (2), (3), and (4), we obtain

$$(5)\quad \begin{aligned} L\big((b\varphi u_{\gamma\gamma} + K)^2\big) \geq & \frac{1}{2}\lambda b^2\varphi^2 \sum_{i=1}^{n} u_{i\gamma\gamma}^2 - \varepsilon b^2\varphi^2 \sum_{i,j=1}^{n} u_{ij\gamma}^2 \\ & - \frac{B_1}{\varphi^2}(b\varphi u_{\gamma\gamma})^2 - \frac{B_1}{\varphi^2}K^2 - \frac{\lambda b^2}{R^2}|\nabla^2 u|^2 - b^2R^2F_{20}^2, \end{aligned}$$

where

$$(6)\quad B_1 = \sup_{B_R}\left(3|L\varphi| + \frac{16\Lambda^2}{\lambda}|\nabla\varphi|^2 + 2\lambda|\nabla\varphi|^2\right) + \frac{2}{R^2} + \frac{2}{\lambda}R^2F_{21}^2 + \frac{8}{\varepsilon}F_{22}^2.$$

We note that the first term in the right-hand side in (5) is quadratic in terms of *special* third derivatives $u_{i\gamma\gamma}$ and that the second term involves *arbitrary* third derivatives $u_{ij\gamma}$. We need to choose γ appropriately so that the first term controls the second. By writing (1) as

$$F(x, \nabla^2 u(x)) - F(x, 0) = -F(x, 0),$$

we have

$$a_{ij} u_{ij} = -F(x, 0),$$

where

$$a_{ij}(x) = \int_0^1 F_{ij}(x, t\nabla^2 u(x))dt.$$

It is easy to check that, for any $x \in B_R$ and any $\xi \in \mathbb{R}^n$,

$$\lambda|\xi|^2 \le a_{ij}(x)\xi_i\xi_j \le \Lambda|\xi|^2.$$

Now we apply Lemma 5.2.6 to the matrix (a_{ij}) and let $\gamma_1, \ldots, \gamma_N$ be the unit vectors as in Lemma 5.2.6, the choice of which depends only on n, λ, and Λ. There exist functions $\beta_1, \ldots, \beta_N$ in B_R such that

$$\sum_{k=1}^N \beta_k u_{\gamma_k\gamma_k} = -F(x, 0) \quad \text{in } B_R \tag{7}$$

and

$$\lambda_* \le \beta_k \le \Lambda_* \quad \text{in } B_R, \tag{8}$$

where λ_* and Λ_* are positive constants depending only on n, λ, and Λ. We point out that $\{\gamma_1, \ldots, \gamma_N\}$ includes e_i, for $i = 1, \ldots, n$, and $(e_i \pm e_j)/\sqrt{2}$, for $i \ne j$. Set

$$K = \sup\left\{\Lambda_*\varphi(x)|u_{\gamma\gamma}(x)| : x \in B_R, \gamma \in \mathbb{S}^{n-1}\right\}, \tag{9}$$

and assume that K is attained at $x_0 \in B_R$ and $\gamma_0 \in \mathbb{S}^{n-1}$. Then, take

$$b_k = \beta_k(x_0) \quad \text{for } k = 1, \ldots, N \tag{10}$$

and

$$w = \sum_{k=1}^N (b_k\varphi u_{\gamma_k\gamma_k} + K)^2 - NK^2. \tag{11}$$

Note that $\lambda_* \le b_k \le \Lambda_*$ by (8) and hence $b_k\varphi|u_{\gamma_k\gamma_k}| \le K$, for $k = 1, \ldots, N$. Then, (5) holds for each such b_k and $u_{\gamma_k\gamma_k}$. A simple addition yields

$$\begin{aligned} Lw \ge {} & \frac{1}{2}\lambda\lambda_*^2\varphi^2 \sum_{l=1}^n \sum_{k=1}^N u_{l\gamma_k\gamma_k}^2 - \varepsilon\Lambda_*^2\varphi^2 \sum_{i,j=1}^n \sum_{k=1}^N u_{ij\gamma_k}^2 \\ & - \frac{B_1}{\varphi^2} \sum_{k=1}^N (b_k\varphi u_{\gamma_k\gamma_k})^2 - \frac{B_1}{\varphi^2} NK^2 - \frac{\lambda N\Lambda_*^2}{R^2}|\nabla^2 u|^2 - N\Lambda_*^2 R^2 F_{20}^2. \end{aligned}$$

We note that

$$\sum_{i,j=1}^{n}\sum_{k=1}^{N} u_{ij\gamma_k}^2 \le c_1|\nabla^3 u|^2,$$

where c_1 is a positive constant depending only on n, λ, and Λ. We now claim

$$|\nabla^3 u|^2 \le \sum_{l=1}^{n}\sum_{k=1}^{N} u_{l\gamma_k\gamma_k}^2.$$

If γ_k is given by e_i, then

$$\sum_{l=1}^{n} u_{l\gamma_k\gamma_k}^2 = \sum_{l=1}^{n} u_{iil}^2.$$

Next, if γ_k and $\gamma_{k'}$ are given by $(e_i+e_j)/\sqrt{2}$ and $(e_i-e_j)/\sqrt{2}$, for $i\ne j$, then

$$\begin{aligned}
&\sum_{l=1}^{n} u_{l\gamma_k\gamma_k}^2 + \sum_{l=1}^{n} u_{l\gamma_{k'}\gamma_{k'}}^2\\
&\qquad = \frac{1}{4}\sum_{l=1}^{n}\left\{(u_{iil}+u_{jjl}+2u_{ijl})^2 + (u_{iil}+u_{jjl}-2u_{ijl})^2\right\}\\
&\qquad = \frac{1}{4}\sum_{l=1}^{n}\left\{2(u_{iil}+u_{jjl})^2 + 8u_{ijl}^2\right\} \ge 2\sum_{l=1}^{n} u_{ijl}^2.
\end{aligned}$$

This finishes the proof of the claim. By taking ε such that $c_1\Lambda_*^2\varepsilon = \lambda\lambda_*^2/2$, we obtain

$$(12)\quad Lw \ge -\frac{B_1}{\varphi^2}\sum_{k=1}^{N}(b_k\varphi u_{\gamma_k\gamma_k})^2 - \frac{B_1}{\varphi^2}NK^2 - \frac{\lambda N\Lambda_*^2}{R^2}|\nabla^2 u|^2 - N\Lambda_*^2R^2F_{20}^2.$$

Next, we note that

$$L(|\nabla u|^2) = 2F_{ij}u_{ki}u_{kj} - 2u_kF_{x_k}.$$

Hence,

$$L(|\nabla u|^2) \ge \lambda|\nabla^2 u|^2 - B_2|\nabla u|^2 - R^2F_{10}^2,$$

where

$$(13)\qquad B_2 = \frac{1}{R^2} + \frac{1}{\lambda}F_{11}^2.$$

Therefore, for a constant A to be determined, we have

$$\begin{aligned}
L\left(w + \frac{A}{R^2}|\nabla u|^2\right) &\ge \frac{\lambda}{R^2}(A - N\Lambda_*^2)|\nabla^2 u|^2 - \frac{B_1}{\varphi^2}\sum_{k=1}^{N}(b_k\varphi u_{\gamma_k\gamma_k})^2\\
&\quad - \frac{B_1}{\varphi^2}NK^2 - \frac{AB_2}{R^2}|\nabla u|^2 - (AF_{10}^2 + N\Lambda_*^2R^2F_{20}^2).
\end{aligned}$$

We also note that

$$|\nabla^2 u|^2 \geq c_2 \sum_{k=1}^{N} (b_k u_{\gamma_k\gamma_k})^2,$$

where c_2 is a positive constant depending only on n, λ, and Λ. Hence,

$$\begin{aligned} L\left(w + \frac{A}{R^2}|\nabla u|^2\right) \geq \frac{1}{R^2\varphi^2}\left(\lambda c_2(A - N\Lambda_*^2) - R^2 B_1\right) \sum_{k=1}^{N} (b_k \varphi u_{\gamma_k\gamma_k})^2 \\ - \frac{B_1}{\varphi^2} N K^2 - \frac{AB_2}{R^2}|\nabla u|^2 - (AF_{10}^2 + N\Lambda_*^2 R^2 F_{20}^2). \end{aligned}$$

By taking $A \geq 2N\Lambda_*^2$, we obtain

$$\begin{aligned} L\left(w + \frac{A}{R^2}|\nabla u|^2\right) \geq \frac{1}{R^2\varphi^2}(c_3 A - R^2 B_1) \sum_{k=1}^{N} (b_k \varphi u_{\gamma_k\gamma_k})^2 - \frac{B_1}{\varphi^2} N K^2 \\ - \frac{AB_2}{R^2}|\nabla u|^2 - (AF_{10}^2 + N\Lambda_*^2 R^2 F_{20}^2), \end{aligned} \tag{14}$$

where c_3 is a positive constant depending only on n, λ, and Λ. We now relate the summation in the right-hand side to w. By the definition of w in (11) and the definition of K in (9), we have

$$w = \sum_{k=1}^{N} b_k \varphi u_{\gamma_k\gamma_k}(2K + b_k \varphi u_{\gamma_k\gamma_k}) \leq 3K \sum_{k=1}^{N} b_k \varphi |u_{\gamma_k\gamma_k}|.$$

The Cauchy inequality implies, for any $\delta > 0$,

$$w \leq \frac{1}{\delta} \sum_{k=1}^{N} (b_k \varphi u_{\gamma_k\gamma_k})^2 + 3\delta N K^2,$$

and hence

$$\sum_{k=1}^{N} (b_k \varphi u_{\gamma_k\gamma_k})^2 \geq \delta w - 3\delta^2 N K^2.$$

A simple substitution in (14) yields

$$\begin{aligned} L\left(w + \frac{A}{R^2}|\nabla u|^2\right) \geq \frac{1}{R^2\varphi^2}(c_3 A - R^2 B_1)(\delta w - 3\delta^2 N K^2) - \frac{B_1}{\varphi^2} N K^2 \\ - \frac{AB_2}{R^2}|\nabla u|^2 - (AF_{10}^2 + N\Lambda_*^2 R^2 F_{20}^2). \end{aligned}$$

In the following, we will also take $A \geq 2R^2 B_1/c_3$.

Now, we assume that $w + A|\nabla u|^2/R^2$ attains its maximum at $x_* \in \bar{B}_R$. We consider the case $x_* \in B_R$. Then, at x_*,

$$\begin{aligned} \delta(c_3 A - R^2 B_1)(w - 3\delta N K^2) \leq R^2 B_1 N K^2 + AB_2|\nabla u|^2 \\ + AR^2 F_{10}^2 + N\Lambda_*^2 R^4 F_{20}^2. \end{aligned}$$

This implies, at x_*,

$$w + \frac{A}{R^2}|\nabla u|^2 \le 3\delta N K^2 + \frac{R^2 B_1 N}{\delta(c_3 A - R^2 B_1)} K^2 + D,$$

where

$$D = \left(\frac{A}{R^2} + \frac{AB_2}{\delta(c_3 A - R^2 B_1)}\right) \sup_{B_R} |\nabla u|^2 + \frac{AR^2 F_{10}^2 + N\Lambda_*^2 R^4 F_{20}^2}{\delta(c_3 A - R^2 B_1)}. \tag{15}$$

By the maximality of the expression in the left-hand side at x_*, we obtain

$$w + \frac{A}{R^2}|\nabla u|^2 \le 3\delta N K^2 + \frac{R^2 B_1 N}{\delta(c_3 A - R^2 B_1)} K^2 + D \quad \text{in } B_R,$$

and hence

$$w \le 3\delta N K^2 + \frac{R^2 B_1 N}{\delta(c_3 A - R^2 B_1)} K^2 + D \quad \text{in } B_R. \tag{16}$$

By (11), we write w as

$$w = \sum_{k=1}^{N} (b_k \varphi u_{\gamma_k \gamma_k})^2 + 2K \sum_{k=1}^{N} b_k \varphi u_{\gamma_k \gamma_k}.$$

Then, for any unit vector γ,

$$w \ge \lambda_*^2 (\varphi u_{\gamma\gamma})^2 + 2K \sum_{k=1}^{N} b_k \varphi u_{\gamma_k \gamma_k}.$$

We now evaluate this at x_0 with $\gamma = \gamma_0$, where K in (9) is attained. By (7) and the choice of b_k in (10), we have

$$w(x_0) \ge \frac{\lambda_*^2}{\Lambda_*^2} K^2 - 2KF(x_0, 0).$$

The Cauchy inequality implies

$$w(x_0) \ge \frac{\lambda_*^2}{2\Lambda_*^2} K^2 - \frac{2\Lambda_*^2}{\lambda_*^2} |F(x_0, 0)|^2.$$

A simple substitution in (16) yields

$$\frac{\lambda_*^2}{2\Lambda_*^2} K^2 \le 3\delta N K^2 + \frac{R^2 B_1 N}{\delta(c_3 A - R^2 B_1)} K^2 + \frac{2\Lambda_*^2}{\lambda_*^2} |F(\cdot, 0)|_{L^\infty(B_R)}^2 + D.$$

We first fix a $\delta > 0$ such that

$$3\delta N \le \frac{\lambda_*^2}{8\Lambda_*^2}$$

and then take $A \ge C_* R^2 B_1$ with

$$\frac{N}{\delta(c_3 C_* - 1)} \le \frac{\lambda_*^2}{8\Lambda_*^2}.$$

Therefore, by taking the square root, we obtain

$$K \le C\left\{\sqrt{D} + |F(\cdot,0)|_{L^\infty(B_R)}\right\}. \tag{17}$$

We now examine D given by (15). By the expression for B_1 in (6), we take

$$A = C_*\left(1 + R^4F_{21}^2 + R^2F_{22}^2\right),$$

for a sufficiently large constant C_*. Then, A satisfies all the earlier requirements. By the expression of D in (15) and the expression of B_2 in (13), we get

$$\begin{aligned} D \le C\bigg\{&\frac{1}{R^2}(1 + R^4F_{21}^2 + R^2F_{22}^2 + R^2F_{11}^2)|\nabla u|^2_{L^\infty(B_R)} \\ &+ R^2F_{10}^2 + \left(1 + R^4F_{21}^2 + R^2F_{22}^2\right)^{-1} R^4F_{20}^2\bigg\}. \end{aligned}$$

By a simple substitution in (17) and the definition of K in (9), we obtain, for any unit vector γ,

$$\begin{aligned} \varphi|u_{\gamma\gamma}| \le C\bigg\{&\frac{1}{R}(1 + RF_{11} + R^2F_{21} + RF_{22})|\nabla u|_{L^\infty(B_R)} + |F(\cdot,0)|_{L^\infty} \\ &+ RF_{10} + \left(1 + R^2F_{21} + RF_{22}\right)^{-1} R^2F_{20}\bigg\} \quad \text{in } B_R. \end{aligned}$$

This implies the desired result. □

To end this section, we make a remark on the assumptions on F in Theorems 5.2.7 and 5.2.8, as well as Theorem 5.4.3 to be proved later.

Remark 5.2.9. In applications in Section 5.6, $F(x,M)$ has a specific form. For $F = F(M) \in C^2(\mathcal{S})$, we assume, for any $M \in \mathcal{S}$,

$$|\nabla_M F(M)| \le \Lambda,$$

for a positive constant Λ. Let Ω be a domain in $\mathbb{R}^n$. For some functions $f \in C^2(\Omega)$ and $\eta \in C^2(\Omega)$ with $0 \le \eta \le 1$, we define, for any $x \in \Omega$ and $M \in \mathcal{S}$,

$$F(x,M) = \eta(x)F(M) + (1-\eta(x))\mathrm{Tr}(M) - f(x).$$

Then, for any $x \in \Omega$ and $M \in \mathcal{S}$,

$$\begin{aligned} |F(x,M)| &\le |f(x)| + |F(0)| + c\max\{\Lambda,1\}|M|, \\ |\nabla_x F(x,M)| &\le |\nabla f(x)| + |\nabla\eta(x)||F(0)| + c\max\{\Lambda,1\}|\nabla\eta(x)||M|, \\ |\nabla_x^2 F(x,M)| &\le |\nabla^2 f(x)| + |\nabla^2\eta(x)||F(0)| + c\max\{\Lambda,1\}|\nabla^2\eta(x)||M|, \end{aligned}$$

and

$$|\nabla_{xM}^2 F(x,M)| \le c\max\{\Lambda,1\}|\nabla\eta(x)|,$$

where c is a positive constant depending only on n.

Refer to [**23**] and [**97**] for more general F.

5.3. Global C^2-Estimates

In this section, we derive global C^2-estimates for solutions of Dirichlet problems for fully nonlinear uniformly elliptic differential equations of the concave type.

We first derive a global C^1-estimate.

Theorem 5.3.1. *Let Ω be a bounded domain in $\mathbb{R}^n$ with a C^2-boundary and F be a uniformly elliptic C^1-function in $\mathcal{S}$. Suppose that u is a $C^3(\bar{\Omega})$-solution of*

$$\begin{aligned} F(\nabla^2 u) &= f \quad \text{in } \Omega, \\ u &= \varphi \quad \text{on } \partial\Omega, \end{aligned}$$

for some $f \in C^1(\bar{\Omega})$ and $\varphi \in C^2(\bar{\Omega})$. Then,

$$|u|_{C^1(\bar{\Omega})} \le C \left\{ |\varphi|_{C^2(\bar{\Omega})} + |f|_{C^1(\bar{\Omega})} + |F(0)| \right\},$$

where C is a positive constant depending only on n, λ, Λ, and Ω.

Instead of the uniform ellipticity of F in $\mathcal{S}$, it suffices to assume, for any $x \in \Omega$ and $t \in [0,1]$,

$$\lambda|\xi|^2 \le F_{ij}\big(t\nabla^2 u(x)\big)\xi_i\xi_j \le \Lambda|\xi|^2,$$

for some positive constants $\lambda \le \Lambda$.

Proof. In the following, we always denote by C a constant depending only on n, λ, Λ, and Ω.

By Lemma 5.1.2, u satisfies a linear equation of the form

$$a_{ij}u_{ij} = f - F(0) \quad \text{in } \Omega,$$

where a_{ij} satisfies, for any $x \in \Omega$ and $\xi \in \mathbb{R}^n$,

$$\lambda|\xi|^2 \le a_{ij}(x)\xi_i\xi_j \le \Lambda|\xi|^2.$$

Hence,

$$\begin{aligned} a_{ij}u_{ij} &= f - F(0) \quad \text{in } \Omega, \\ u &= \varphi \quad \text{on } \partial\Omega. \end{aligned}$$

By Theorem 1.1.10, we have

$$\sup_{\Omega} |u| \le |\varphi|_{L^\infty(\partial\Omega)} + C\left\{|f|_{L^\infty(\Omega)} + |F(0)|\right\}. \tag{1}$$

By (1) and Theorem 1.1.14, we also have

$$\max_{\partial\Omega} |\nabla u| \le C\left\{|\varphi|_{C^2(\Omega)} + |f|_{L^\infty(\Omega)} + |F(0)|\right\}.$$

By Lemma 5.1.1(2), for any $k = 1, \ldots, n$, u_k satisfies

$$Lu_k = f_k \quad \text{in } \Omega.$$

Again by Theorem 1.1.10, we get

$$\sup_{\Omega} |u_k| \leq |u_k|_{L^\infty(\partial\Omega)} + C|f_k|_{L^\infty(\Omega)},$$

and hence

$$\max_{\bar{\Omega}} |\nabla u| \leq C\left\{|\varphi|_{C^2(\bar{\Omega})} + |f|_{C^1(\bar{\Omega})} + |F(0)|\right\}. \tag{2}$$

We obtain the desired result by combining (1) and (2). □

Next, we derive an estimate of the second derivatives on boundary.

Theorem 5.3.2. *Let Ω be a bounded domain in $\mathbb{R}^n$ with a C^3-boundary and F be a uniformly elliptic C^1-function in $\mathcal{S}$. Suppose that u is a $C^3(\bar{\Omega})$-solution of*

$$\begin{aligned} F(\nabla^2 u) &= f \quad \text{in } \Omega, \\ u &= \varphi \quad \text{on } \partial\Omega, \end{aligned} \tag{1}$$

for some $f \in C^2(\bar{\Omega})$ and $\varphi \in C^3(\bar{\Omega})$. Then,

$$|\nabla^2 u|_{L^\infty(\partial\Omega)} \leq C\left\{|u|_{C^1(\bar{\Omega})} + |\varphi|_{C^3(\bar{\Omega})} + |f|_{C^2(\bar{\Omega})} + |F(0)|\right\},$$

where C is a positive constant depending only on n, λ, Λ, and Ω.

It suffices to assume, for any $x \in \Omega$ and $t \in [0,1]$,

$$\lambda|\xi|^2 \leq F_{ij}\big(t\nabla^2 u(x)\big)\xi_i\xi_j \leq \Lambda|\xi|^2,$$

for some positive constants $\lambda \leq \Lambda$. We point out that the concavity of F is not assumed in Theorem 5.3.2.

Proof. Set

$$K = |u|_{C^1(\bar{\Omega})} + |\varphi|_{C^3(\bar{\Omega})} + |f|_{C^2(\bar{\Omega})} + |F(0)|.$$

We will prove

$$|\nabla^2 u|_{L^\infty(\partial\Omega)} \leq CK, \tag{2}$$

where C is a positive constant depending only on n, λ, Λ, and Ω. In the following, a universal constant is a positive constant depending only on n, λ, Λ, and Ω.

For a fixed point $x_0 \in \partial\Omega$, we flatten the boundary $\partial\Omega$ near x_0. In other words, we consider a neighborhood U of x_0 which contains a ball centered at x_0, with a universal radius, and C^3-diffeomorphisms

$$x = \chi(y), \quad y = \chi^{-1}(x) = \eta(x) \quad \text{for any } x \in U \text{ and } y \in A = \eta(U),$$

so that $\eta(x_0) = 0$ and

$$\begin{aligned} \eta(U \cap \Omega) &= B_1^+ = \{y \in \mathbb{R}^n : |y| < 1, y_n > 0\}, \\ \eta(U \cap \partial\Omega) &= \Sigma_1 = \{y \in \mathbb{R}^n : |y| < 1, y_n = 0\}. \end{aligned}$$

We also require that $|\eta|_{C^3(\bar{U})}$ and $|\chi|_{C^3(\bar{A})}$ are bounded by a universal constant. We consider the function

$$v(y) = u(x) - \varphi(x) \quad \text{for any } y \in \bar{B}_1^+.$$

Now we estimate $\nabla_y^2 v(0)$.

We first estimate $v_{pq}(0)$ for any $p, q = 1, \ldots, n-1$. Note that v vanishes on $\Sigma_1 \subset \{y_n = 0\}$. Hence, for any $p, q = 1, \ldots, n-1$,

$$v_{pq}(0) = 0. \tag{3}$$

Next, we estimate the mixed derivative $v_{pn}(0)$, for some $p = 1, \ldots, n-1$. The basic idea is to treat v_p as a solution of some linear elliptic differential equation and proceed to derive a boundary gradient estimate for v_p. In the following, we write $\eta = (\eta^1, \ldots, \eta^n)$. We first note that

$$u_i = \eta_i^k v_k + \varphi_i$$

and

$$u_{ij} = \eta_i^k \eta_j^l v_{kl} + \eta_{ij}^k v_k + \varphi_{ij}.$$

Then, v satisfies

$$F\left(\left[\eta_i^k \eta_j^l v_{kl} + \eta_{ij}^k v_k + \varphi_{ij}\right]_{ij}\right) = f \quad \text{in } B_1^+, \tag{4}$$

where η_i^k, η_{ij}^k, φ_{ij}, and f are evaluated at $\chi(y)$. By differentiating (4) with respect to y_p, we have

$$F_{ij}\eta_i^k \eta_j^l v_{klp} + F_{ij} v_{kl} \partial_p(\eta_i^k \eta_j^l) + F_{ij}\partial_p\left(\eta_{ij}^k v_k + \varphi_{ij}\right) = \partial_p f.$$

Set

$$L = a_{kl}\partial_{kl} = F_{ij}\eta_i^k \eta_j^l \partial_{kl}.$$

Then, L is uniformly elliptic in B_1^+; namely, for any $y \in B_1^+$ and $\xi \in \mathbb{R}^n$,

$$\lambda^*|\xi|^2 \leq a_{kl}(y)\xi_k\xi_l \leq \Lambda^*|\xi|^2,$$

for some universal constants λ^* and Λ^*. Here, we used the fact that $\nabla\eta$ is an invertible matrix and hence, for any $x \in U$ and any $\xi \in \mathbb{R}^n$,

$$c_1|\xi|^2 \leq |\nabla\eta(x) \cdot \xi|^2 \leq c_2|\xi|^2,$$

where c_1 and c_2 are positive constants. The equation for v_p has the form

$$Lv_p = \widetilde{f} \quad \text{in } B_1^+,$$

where $\widetilde{f}$ satisfies

$$|\widetilde{f}| \leq C\left\{|\nabla_x f| + |\nabla_x^3 \varphi| + |\nabla_y v| + |\nabla_y^2 v|\right\}.$$

Hence,

$$L(\pm v_p) \geq -C\left(|\nabla_y^2 v| + K\right) \quad \text{in } B_1^+.$$

Next, for any $q = 1, \ldots, n-1$,

$$L\big(v_q^2\big) = 2a_{kl}v_{qk}v_{ql} + 2v_qLv_q.$$

By the uniform ellipticity, we have

$$L\big(v_q^2\big) \geq 2\lambda^* \sum_{k=1}^{n} v_{qk}^2 - CK\left(|\nabla_y^2 v| + K\right).$$

By setting

$$w = \pm v_p + \frac{1}{K}\sum_{q=1}^{n-1} v_q^2,$$

it follows that

$$Lw \geq \frac{2\lambda^*}{K} \sum_{(k,l)\neq(n,n)} v_{kl}^2 - C\left(|\nabla_y^2 v| + K\right).$$

We note that v_{nn}^2 does not appear in the summation in the right-hand side. We rewrite (4) as

$$F - F(0) = f - F(0).$$

By using the mean-value theorem, we get

$$\widetilde{F}_{ij} \cdot \left(\eta_i^k \eta_j^l v_{kl} + \eta_{ij}^k v_k + \varphi_{ij}\right) = f - F(0),$$

where $(\widetilde{F}_{ij})$ satisfies, for any $y \in B_1^+$ and any $\xi \in \mathbb{R}^n$,

$$\lambda|\xi|^2 \leq \widetilde{F}_{ij}(y)\xi_i\xi_j \leq \Lambda|\xi|^2.$$

Hence,

$$|v_{nn}| \leq C\left\{\sum_{(k,l)\neq(n,n)} |v_{kl}| + |\nabla_y v| + |\nabla_x^2\varphi| + |f| + |F(0)|\right\},$$

or

$$|v_{nn}| \leq C\left\{\sum_{(k,l)\neq(n,n)} |v_{kl}| + K\right\}. \tag{5}$$

Therefore,

$$Lw \geq \frac{2\lambda^*}{K} \sum_{(k,l)\neq(n,n)} v_{kl}^2 - C\left\{\sum_{(k,l)\neq(n,n)} |v_{kl}| + K\right\}.$$

By the Cauchy inequality, we obtain

$$Lw \geq -CK \quad \text{in } B_1^+. \tag{6}$$

Moreover,

$$w = 0 \quad \text{on } \partial B_1^+ \cap \{y_n = 0\}, \tag{7}$$

and

$$w \leq CK \quad \text{on } \partial B_1^+ \cap \{y_n > 0\}. \tag{8}$$

Consider, for some positive constants a and b to be determined,

$$\overline{w}(y) = a(y_n - y_n^2 + b|y|^2).$$

Then,

$$L\overline{w} = 2a\Big(- a_{nn} + b\sum_{k=1}^{n} a_{kk}\Big) \leq 2a(-\lambda^* + nb\Lambda^*) \quad \text{in } B_1^+.$$

By taking

$$b = \frac{\lambda^*}{2n\Lambda^*},$$

we have

$$L\overline{w} \leq -a\lambda^*.$$

Hence,

$$L\overline{w} \leq -CK \quad \text{in } B_1^+ \tag{9}$$

if

$$a\lambda^* \geq CK.$$

Next, it is obvious that

$$\overline{w} \geq 0 \quad \text{on } \partial B_1^+ \cap \{y_n = 0\}. \tag{10}$$

On $\partial B_1^+ \cap \{y_n > 0\}$,

$$\overline{w} \geq ab = \frac{\lambda^*}{2n\Lambda^*}a.$$

Hence,

$$\overline{w} \geq CK \quad \text{on } \partial B_1 \cap \{y_n > 0\} \tag{11}$$

if

$$\frac{\lambda^*}{2n\Lambda^*}a \geq CK.$$

By choosing $a = C_*K$ for some positive constant C_* sufficiently large, we have (9) and (11). With (6)–(8) and (9)–(11), we obtain

$$L w \geq L\overline{w} \quad \text{in } B_1^+,$$
$$w \leq \overline{w} \quad \text{on } \partial B_1^+.$$

The maximum principle implies

$$w \leq \overline{w} \quad \text{in } B_1^+$$

and, in particular,

$$\pm v_p \le C_* K \left(y_n - y_n^2 + \frac{\lambda^*}{2n\Lambda^*}|y|^2 \right) \quad \text{in } B_1^+.$$

By letting $y' = 0$, dividing by y_n, and letting $y_n \to 0$, we get

$$|v_{pn}(0)| \le C_* K. \tag{12}$$

By substituting (3) and (12) in (5), we have

$$|v_{nn}(0)| \le CK,$$

and hence

$$|\nabla_y^2 v(0)| \le CK.$$

Transforming back to x-coordinates, we obtain

$$|\nabla^2 u(x_0)| \le CK.$$

This implies (2) since x_0 is an arbitrary point on $\partial\Omega$. □

Next, we derive a global estimate of the second derivatives.

Theorem 5.3.3. *Let Ω be a bounded domain in $\mathbb{R}^n$ with a C^2-boundary and F be a uniformly elliptic and concave C^2-function in $\mathcal{S}$. Suppose that u is a $C^2(\bar\Omega) \cap C^4(\Omega)$-solution of*

$$\begin{aligned} F(\nabla^2 u) &= f \quad \text{in } \Omega, \\ u &= \varphi \quad \text{on } \partial\Omega, \end{aligned} \tag{1}$$

for some $f \in C^2(\bar\Omega)$ and $\varphi \in C^2(\bar\Omega)$. Then,

$$|\nabla^2 u|_{L^\infty(\Omega)} \le C\left\{ |\nabla^2 u|_{L^\infty(\partial\Omega)} + |f|_{C^2(\bar\Omega)} + |F(0)| \right\},$$

where C is a positive constant depending only on n, λ, Λ, and Ω.

It suffices to assume that F is uniformly elliptic in the range of $\nabla^2 u$, i.e., for any $x \in \Omega$ and $\xi \in \mathbb{R}^n$,

$$\lambda|\xi|^2 \le F_{ij}(\nabla^2 u(x))\xi_i\xi_j \le \Lambda|\xi|^2,$$

for some positive constants λ and Λ, and that F is a concave function in the range of $\nabla^2 u$, i.e., for any $x \in \Omega$ and $(m_{ij}) \in \mathcal{S}$,

$$F_{ij,kl}(\nabla^2 u(x)) m_{ij} m_{kl} \le 0.$$

Proof. Set

$$L = F_{ij}\partial_{ij}.$$

Let $\gamma \in \mathbb{R}^n$ be an arbitrary unit vector. By the concavity of F and Lemma 5.1.1(3), we have

$$Lu_{\gamma\gamma} \ge f_{\gamma\gamma}.$$

We note that L is uniformly elliptic. By Theorem 1.1.9, we obtain

$$\begin{aligned}\sup_{\Omega} u_{\gamma\gamma} &\leq \sup_{\partial\Omega} u_{\gamma\gamma}^{+} + C \sup_{\Omega} f_{\gamma\gamma}^{-} \\ &\leq |\nabla^2 u|_{L^\infty(\partial\Omega)} + C|\nabla^2 f|_{L^\infty(\Omega)}.\end{aligned}$$

Then, Corollary 5.1.4 implies

$$\sup_{\Omega} |\nabla^2 u| \leq C\left\{|\nabla^2 u|_{L^\infty(\partial\Omega)} + |f|_{C^2(\bar{\Omega})} + |F(0)|\right\}.$$

This is the desired estimate. □

By combining Theorems 5.3.1, 5.3.2, and 5.3.3, we obtain the following global C^2-estimates.

Theorem 5.3.4. *Let Ω be a bounded domain in $\mathbb{R}^n$ with a C^3-boundary and F be a uniformly elliptic and concave C^2-function in $\mathcal{S}$. Suppose that u is a $C^3(\bar{\Omega}) \cap C^4(\Omega)$-solution of*

$$\begin{aligned} F(\nabla^2 u) &= f \quad \text{in } \Omega, \\ u &= \varphi \quad \text{on } \partial\Omega, \end{aligned}$$

for some $f \in C^2(\bar{\Omega})$ and $\varphi \in C^3(\bar{\Omega})$. Then,

$$|u|_{C^2(\bar{\Omega})} \leq C\left\{|\varphi|_{C^3(\bar{\Omega})} + |f|_{C^2(\bar{\Omega})} + |F(0)|\right\},$$

where C is a positive constant depending only on n, λ, Λ, and Ω.

5.4. Interior $C^{2,\alpha}$-Estimates

In this section, we derive the fundamental interior estimates of the Hölder semi-norms of the second derivatives of solutions to fully nonlinear uniformly elliptic differential equations of the concave type. This result is due to Evans [**51**] and Krylov [**96**]. Krylov-Safonov's weak Harnack inequality plays an essential role in our derivation.

Theorem 5.4.1. *Let B_R be a ball in $\mathbb{R}^n$ and F be a uniformly elliptic and concave C^2-function on $\mathcal{S}$. Suppose u is a $C^4(B_R)$-solution of*

$$F(\nabla^2 u) = f \quad \text{in } B_R, \tag{1}$$

for some $f \in C^2(B_R)$, with $|f|_{C^2(B_R)} < \infty$ and $|\nabla^2 u|_{C^2(B_R)} < \infty$. Then, for some constant $\alpha \in (0,1)$ depending only on n, λ, and Λ,

$$R^\alpha [\nabla^2 u]_{C^\alpha(B_{R/2})} \leq C\left\{|\nabla^2 u|_{L^\infty(B_R)} + R|\nabla f|_{L^\infty(B_R)} + R^2|\nabla^2 f|_{L^\infty(B_R)}\right\},$$

where C is a positive constant depending only on n, λ, and Λ.

As we will see in the proof, we need only assume that F is uniformly elliptic in the range of $\nabla^2 u$, i.e., for any $x \in B_R$ and $\xi \in \mathbb{R}^n$,

$$\lambda|\xi|^2 \leq F_{ij}\big(\nabla^2 u(x)\big)\xi_i\xi_j \leq \Lambda|\xi|^2,$$

for some positive constants λ and Λ, and that F is a concave function in the range of $\nabla^2 u$, i.e., for any $x \in B_R$ and $(m_{ij}) \in \mathcal{S}$,

$$F_{ij,kl}(\nabla^2 u(x))m_{ij}m_{kl} \leq 0$$

and, for any $x, y \in B_R$,

$$F(\nabla^2 u(x)) \leq F(\nabla^2 u(y)) + F_{ij}(\nabla^2 u(y))\big(u_{ij}(x) - u_{ij}(y)\big).$$

Proof. Let γ be an arbitrary unit vector in $\mathbb{R}^n$. Since F is concave, Lemma 5.1.1(3) implies

$$F_{ij}\partial_{ij}(u_{\gamma\gamma}) \geq f_{\gamma\gamma} \quad \text{in } B_R. \tag{2}$$

For any $r < R$, set

$$M(r) = \sup_{B_r} u_{\gamma\gamma}, \quad m(r) = \inf_{B_r} u_{\gamma\gamma}.$$

Then,

$$F_{ij}\partial_{ij}\big(M(r) - u_{\gamma\gamma}\big) \leq -f_{\gamma\gamma} \quad \text{in } B_r.$$

In the following, we set $\tau = 1/8\sqrt{n}$. Applying the weak Harnack inequality provided by Theorem 1.2.11 to the function $M(r) - u_{\gamma\gamma}$ in B_r, we obtain

$$r^{-\frac{n}{p}}\|M(r) - u_{\gamma\gamma}\|_{L^p(B_{\tau r})} \leq C\left\{M(r) - M(\tau r) + r^2|\nabla^2 f|_{L^\infty(B_R)}\right\},$$

and hence

$$\begin{aligned} &r^{-\frac{n}{p}}\|M(r) - u_{\gamma\gamma}\|_{L^p(B_{\tau r})} \\ &\qquad \leq C\left\{\big(M(r) - m(r)\big) - \big(M(\tau r) - m(\tau r)\big) + r^2|\nabla^2 f|_{L^\infty(B_R)}\right\}, \end{aligned} \tag{3}$$

where p and C are positive constants depending only on n, λ, and Λ. We point out that p may be less than one.

To conclude an estimate of the Hölder semi-norm of $u_{\gamma\gamma}$ from (3), we need an estimate of $u_{\gamma\gamma} - m(r)$, which we obtain by considering (1) as a functional relation among the second derivatives of u. By the concavity of F again, we have, for any $x, y \in B_r$,

$$F(\nabla^2 u(x)) \leq F(\nabla^2 u(y)) + F_{ij}(\nabla^2 u(y))\big(u_{ij}(x) - u_{ij}(y)\big),$$

and hence

$$F_{ij}(\nabla^2 u(y))\big(u_{ij}(y) - u_{ij}(x)\big) \leq f(y) - f(x). \tag{4}$$

Now we apply Lemma 5.2.6 to the matrix (F_{ij}) and let $\gamma_1, \ldots, \gamma_N$ be the unit vectors as in Lemma 5.2.6, the choice of which depends only on n, λ, and Λ. We write (4) as

$$\sum_{k=1}^{N} \beta_k \big(u_{\gamma_k\gamma_k}(y) - u_{\gamma_k\gamma_k}(x)\big) \leq f(y) - f(x), \tag{5}$$

where $\beta_k = \beta_k(y)$ satisfies

$$0 < \lambda^* \leq \beta_k \leq \Lambda^*,$$

for some positive constants λ^* and Λ^* depending only on n, λ, and Λ. Set, for $k = 1, \ldots, N$,

$$M_k(r) = \sup_{B_r} u_{\gamma_k\gamma_k}, \quad m_k(r) = \inf_{B_r} u_{\gamma_k\gamma_k},$$

and

$$\omega_k(r) = M_k(r) - m_k(r).$$

We also set

$$\omega(r) = \sum_{k=1}^{N} \omega_k(r).$$

Then, each of $u_{\gamma_k\gamma_k}$ satisfies (3), so that

$$\begin{aligned} r^{-\frac{n}{p}} &\|M_k(r) - u_{\gamma_k\gamma_k}\|_{L^p(B_{\tau r})} \\ &\leq C\left\{\omega_k(r) - \omega_k(\tau r) + r^2|\nabla^2 f|_{L^\infty(B_R)}\right\}. \end{aligned} \tag{6}$$

Fix an $l \in \{1, \ldots, N\}$. By a summation over $k \neq l$, we obtain

$$\begin{aligned} r^{-\frac{n}{p}} &\Big\|\sum_{k\neq l}(M_k(r) - u_{\gamma_k\gamma_k})\Big\|_{L^p(B_{\tau r})} \\ &\leq C\Big\{\sum_{k\neq l}(\omega_k(r) - \omega_k(\tau r)) + r^2|\nabla^2 f|_{L^\infty(B_R)}\Big\} \\ &\leq C\left\{\omega(r) - \omega(\tau r) + r^2|\nabla^2 f|_{L^\infty(B_R)}\right\}. \end{aligned}$$

By (5), we have, for any $x \in B_r$ and $y \in B_{\tau r}$,

$$\beta_l\big(u_{\gamma_l\gamma_l}(y) - u_{\gamma_l\gamma_l}(x)\big) \leq f(y) - f(x) + \sum_{k\neq l} \beta_k\big(u_{\gamma_k\gamma_k}(x) - u_{\gamma_k\gamma_k}(y)\big).$$

By taking the maximum over $x \in B_r$, we get

$$u_{\gamma_l\gamma_l}(y) - m_l(r) \leq \frac{1}{\lambda^*}\Big\{2r|\nabla f|_{L^\infty(B_r)} + \Lambda^* \sum_{k\neq l}\big(M_k(r) - u_{\gamma_k\gamma_k}(y)\big)\Big\}.$$

A simple integration in $B_{\tau r}$ implies

$$\begin{aligned} r^{-\frac{n}{p}} &\|u_{\gamma_l\gamma_l} - m_l(r)\|_{L^p(B_{\tau r})} \\ &\leq C\left\{\omega(r) - \omega(\tau r) + r|\nabla f|_{L^\infty(B_R)} + r^2|\nabla^2 f|_{L^\infty(B_R)}\right\}, \end{aligned} \tag{7}$$

where C is a positive constant depending only on n, λ, and Λ. By taking $k = l$ in (6) and adding this to (7), we get

$$M_l(r) - m_l(r) \leq C\left\{\omega(r) - \omega(\tau r) + r|\nabla f|_{L^\infty(B_R)} + r^2|\nabla^2 f|_{L^\infty(B_R)}\right\}.$$

Summing over $l = 1, \ldots, N$, we obtain

$$\omega(r) \leq C\left\{\omega(r) - \omega(\tau r) + r|\nabla f|_{L^\infty(B_R)} + r^2|\nabla^2 f|_{L^\infty(B_R)}\right\},$$

and hence

$$\omega(\tau r) \leq \gamma\omega(r) + r|\nabla f|_{L^\infty(B_R)} + r^2|\nabla^2 f|_{L^\infty(B_R)},$$

for some constant $\gamma \in (0,1)$. Lemma 1.2.13 implies, for any $r < R$,

$$\omega(r) \leq C\left(\frac{r}{R}\right)^\alpha \left\{\omega(R) + R|\nabla f|_{L^\infty(B_R)} + R^2|\nabla^2 f|_{L^\infty(B_R)}\right\},$$

where $C > 0$ and $\alpha \in (0,1)$ are constants depending only on n, λ, and Λ. Since $\{\gamma_1, \ldots, \gamma_N\}$ includes all e_i, for $i = 1, \ldots, n$, and $(e_i \pm e_j)/2$, for $i \neq j$, as shown in Lemma 5.2.6, we obtain, for any $r \leq R$,

$$\operatorname*{osc}_{B_r} \nabla^2 u \leq C\left(\frac{r}{R}\right)^\alpha \left\{\operatorname*{osc}_{B_R} \nabla^2 u + R|\nabla f|_{L^\infty(B_R)} + R^2|\nabla^2 f|_{L^\infty(B_R)}\right\}.$$

This implies the desired estimate. □

In Theorem 5.4.1, we are mainly concerned with the estimate of the Hölder semi-norms of the second derivatives and derive such an estimate under the assumption that f in the right-hand side of the equation is C^2. In fact, if f is assumed to be C^α only, then any C^2-solutions of (1) will be $C^{2,\alpha}$. A much stronger regularity result asserts that any viscosity solutions of (1) will be $C^{2,\alpha}$ under the assumption that f is C^α. The proof of this result is quite involved. We will not pursue along this line but instead refer the reader to [**20**] for details.

We now prove a Liouville type theorem.

Theorem 5.4.2. *Let F be a uniformly elliptic and concave C^2-function in $\mathcal{S}$ with $F(0) = 0$. Suppose that u is a $C^4(\mathbb{R}^n)$-solution of*

$$F(\nabla^2 u) = 0 \quad \text{in } \mathbb{R}^n.$$

If u has a quadratic growth in $\mathbb{R}^n$, then u is a quadratic polynomial.

Proof. First, by Theorems 5.2.4 and 5.2.5, $\nabla^2 u$ is bounded. Next, by Theorem 5.4.1, we have

$$R^\alpha[\nabla^2 u]_{C^\alpha(B_{R/2})} \leq C|\nabla^2 u|_{L^\infty(B_R)},$$

where $C > 0$ and $\alpha \in (0,1)$ are constants depending only on n, λ, and Λ. By letting $R \to \infty$, we conclude that $\nabla^2 u$ is constant in $\mathbb{R}^n$. □

To finish this section, we extend Theorem 5.4.1 to more general fully nonlinear elliptic equations.

Theorem 5.4.3. *Let B_R be a ball in $\mathbb{R}^n$ and F be a $C^2(B_R \times \mathcal{S})$-function, uniformly elliptic and concave in $M \in \mathcal{S}$. Suppose u is a $C^4(B_R)$-solution of*

$$F(x, \nabla^2 u) = 0 \quad \text{in } B_R. \tag{1}$$

Then, for some constant $\alpha \in (0,1)$ depending only on n, λ, and Λ,

$$R^\alpha [\nabla^2 u]_{C^\alpha(B_{R/2})} \leq C \left\{ |\nabla^2 u|_{L^\infty(B_R)} + c_1 R + c_2 R^2 \right\},$$

where C is a positive constant depending only on n, λ, and Λ and

$$c_1 = \sup_{y,z \in B_R} |\nabla_x F(y, \nabla^2 u(z))|,$$
$$c_2 = \sup_{B_R} |\nabla^2_{xM} F(x, \nabla^2 u)|^2 \sup_{B_R} |\nabla^2 u| + \sup_{B_R} |\nabla^2_x F(x, \nabla^2 u)|.$$

Proof. We set $K = \sup_{B_R} |\nabla^2 u|$. Let γ be an arbitrary unit vector in $\mathbb{R}^n$. By differentiating (1) with respect to x_γ twice, we have

$$F_{ij} u_{ij\gamma} + F_{x_\gamma} = 0$$

and

$$F_{ij} u_{ij\gamma\gamma} + F_{ij,kl} u_{ij\gamma} u_{kl\gamma} + 2F_{ij,x_\gamma} u_{ij\gamma} + F_{x_\gamma x_\gamma} = 0.$$

The concavity of F implies

$$F_{ij} u_{ij\gamma\gamma} + 2F_{ij,x_\gamma} u_{ij\gamma} + F_{x_\gamma x_\gamma} \geq 0 \quad \text{in } B_R.$$

Now we apply Lemma 5.2.6 to the matrix (F_{ij}) and let $\gamma_1, \ldots, \gamma_N$ be the unit vectors as in Lemma 5.2.6, the choice of which depends only on n, λ, and Λ. Set, for $k = 1, \ldots, N$,

$$h_k = 2 + \frac{1}{K} u_{\gamma_k \gamma_k}.$$

Then,

$$1 \leq h_k \leq 3,$$

and

$$F_{ij} \partial_{ij} h_k + \frac{1}{K} \left(2F_{ij,x_{\gamma_k}} u_{ij\gamma_k} + F_{x_{\gamma_k} x_{\gamma_k}} \right) \geq 0 \quad \text{in } B_R.$$

Hence,

$$F_{ij} \partial_{ij} h_k + \frac{1}{K} \left(2A_0 |\nabla^3 u| + B_0 \right) \geq 0 \quad \text{in } B_R, \tag{2}$$

where

$$A_0 = \sup_{B_R} |\nabla^2_{xM} F(x, \nabla^2 u(x))|, \quad B_0 = \sup_{B_R} |\nabla^2_x F(x, \nabla^2 u(x))|. \tag{3}$$

We note that A_0 and B_0 depend on K. Next, we set

$$v = \sum_{k=1}^N h_k^2.$$

Then,

$$v_i = 2\sum_{k=1}^N h_k \partial_i h_k,$$

$$v_{ij} = 2\sum_{k=1}^N h_k \partial_{ij} h_k + 2\sum_{k=1}^N \partial_i h_k \partial_j h_k.$$

By multiplying (2) by $2h_k$, a simple summation, and a substitution, we have

$$F_{ij} v_{ij} + \frac{c_*}{K}\left(2A_0|\nabla^3 u| + B_0\right) \geq 2F_{ij}\partial_i h_k \partial_j h_k \quad \text{in } B_R, \tag{4}$$

where c_* is a positive constant depending only on n, λ, and Λ. For some constant $\varepsilon \in (0,1)$ to be determined, we set, for $k = 1, \ldots, N$,

$$w_k = h_k + \varepsilon v.$$

Then, a simple addition of (2) and the ε-multiple of (4) yields

$$F_{ij}\partial_{ij} w_k + \frac{c_*}{K}\left(2A_0|\nabla^3 u| + B_0\right) \geq 2\varepsilon F_{ij}\partial_i h_k \partial_j h_k \quad \text{in } B_R. \tag{5}$$

By the uniform ellipticity, we have

$$F_{ij}\partial_i h_k \partial_j h_k \geq \lambda \sum_{k=1}^N |\nabla h_k|^2.$$

We note that $\gamma_1, \ldots, \gamma_N$ include e_i, for $i = 1, \ldots, n$, and $(e_i \pm e_j)/\sqrt{2}$, for $i \neq j$. As in the proof of Theorem 5.2.8, we have

$$K^2 \sum_{k=1}^N |\nabla h_k|^2 \geq \sum_{i,j,l=1}^n (\partial_{ijl} u)^2 = |\nabla^3 u|^2.$$

Now (5) has the form

$$F_{ij}\partial_{ij} w_k + c_* A_0 \left(\sum_{k=1}^N |\nabla h_k|^2\right)^{\frac{1}{2}} + \frac{c_* B_0}{K} \geq 2\varepsilon\lambda \sum_{k=1}^N |\nabla h_k|^2 \quad \text{in } B_R.$$

The Cauchy inequality yields

$$F_{ij}\partial_{ij} w_k + \frac{c_*^2 A_0^2}{\varepsilon\lambda} + \frac{c_* B_0}{K} \geq 0 \quad \text{in } B_R.$$

We write this as

$$F_{ij}\partial_{ij}w_k \geq -\frac{c_2}{K} \quad \text{in } B_R,$$

where, by renaming c_*,

$$c_2 = c_* \left(A_0^2 K + B_0\right). \tag{6}$$

In the following, we set $\tau = 1/8\sqrt{n}$. For any $r < R$, applying the weak Harnack inequality provided by Theorem 1.2.11 to the function $\sup_{B_r} w_k - w_k$ in B_r, we obtain

$$r^{-\frac{n}{p}} \| \sup_{B_r} w_k - w_k \|_{L^p(B_{\tau r})} \leq C \left\{ \sup_{B_r} w_k - \sup_{B_{\tau r}} w_k + \frac{c_2}{K} r^2 \right\}.$$

We now express the above estimate in terms of h_k by employing the inequality

$$\sup a + \inf b \leq \sup(a+b) \leq \sup a + \sup b.$$

First, we note that

$$\begin{aligned} \sup_{B_r} w_k - w_k &= \sup_{B_r}(h_k + \varepsilon v) - (h_k + \varepsilon v) \\ &\geq \sup_{B_r} h_k - h_k - \varepsilon \operatorname*{osc}_{B_r} v. \end{aligned}$$

Similarly,

$$\sup_{B_r} w_k - \sup_{B_{\tau r}} w_k \leq \sup_{B_r} h_k - \sup_{B_{\tau r}} h_k + \varepsilon \operatorname*{osc}_{B_r} v.$$

Next, by the definition of v and the bound of h_k, we have

$$\operatorname*{osc}_{B_r} v \leq 6 \sum_{k=1}^{N} \operatorname*{osc}_{B_r} h_k.$$

In the following, we set, for $k = 1, \ldots, N$,

$$M_k(r) = \sup_{B_r} h_k, \quad m_k(r) = \inf_{B_r} h_k,$$

and

$$\omega_k(r) = M_k(r) - m_k(r).$$

We also set

$$\omega(r) = \sum_{k=1}^{N} \omega_k(r).$$

Therefore, we obtain

$$r^{-\frac{n}{p}} \|M_k(r) - h_k\|_{L^p(B_{\tau r})} \leq C \left\{ M_k(r) - M_k(\tau r) + \varepsilon\omega(r) + \frac{c_2}{K} r^2 \right\},$$

and hence

$$r^{-\frac{n}{p}} \|M_k(r) - h_k\|_{L^p(B_{\tau r})} \leq C \left\{ \omega_k(r) - \omega_k(\tau r) + \varepsilon\omega(r) + \frac{c_2}{K} r^2 \right\}. \tag{7}$$

Now we fix an $l \in \{1, \dots, N\}$. By summing over $k \neq l$, we get

$$r^{-\frac{n}{p}} \|\sum_{k \neq l} (M_k(r) - h_k)\|_{L^p(B_{\tau r})}$$
$$\leq C\Big\{\sum_{k \neq l} (\omega_k(r) - \omega_k(\tau r)) + \varepsilon\omega(r) + \frac{c_2}{K} r^2\Big\}.$$

Hence,

$$(8) \qquad r^{-\frac{n}{p}} \|\sum_{k \neq l} (M_k(r) - h_k)\|_{L^p(B_{\tau r})} \leq C\left\{(1+\varepsilon)\omega(r) - \omega(\tau r) + \frac{c_2}{K} r^2\right\}.$$

Next, we proceed similarly as in the proof of Theorem 5.4.1. By the concavity of F, we have, for any $x, y \in B_r$,

$$\begin{aligned} F_{ij}(y, \nabla^2 u(y))\big(u_{ij}(y) - u_{ij}(x)\big) &\leq F(y, \nabla^2 u(y)) - F(y, \nabla^2 u(x)) \\ &= F(x, \nabla^2 u(x)) - F(y, \nabla^2 u(x)). \end{aligned}$$

Applying Lemma 5.2.6 to the matrix (F_{ij}), we write this inequality as

$$\sum_{k=1}^{N} \beta_k \big(u_{\gamma_k \gamma_k}(y) - u_{\gamma_k \gamma_k}(x)\big) \leq F(x, \nabla^2 u(x)) - F(y, \nabla^2 u(x)),$$

where $\beta_k = \beta_k(y)$ satisfies

$$0 < \lambda^* \leq \beta_k \leq \Lambda^*,$$

for some positive constants λ^* and Λ^* depending only on n, λ, and Λ. By the definition of h_k, we have, for any $x \in B_r$ and $y \in B_{\tau r}$,

$$\begin{aligned} \beta_l \big(h_l(y) - h_l(x)\big) \leq \frac{1}{K} &\left(F(x, \nabla^2 u(x)) - F(y, \nabla^2 u(x))\right) \\ &+ \sum_{k \neq l} \beta_k \big(h_k(x) - h_k(y)\big). \end{aligned}$$

By taking the maximum over $x \in B_r$, we get

$$h_l(y) - m_l(r) \leq \frac{1}{\lambda^*}\Big\{\frac{c_1}{K} r + \Lambda^* \sum_{k \neq l} \big(M_k(r) - h_k(y)\big)\Big\},$$

where

$$(9) \qquad c_1 = \sup_{y, z \in B_R} |\nabla_x F\big(y, \nabla^2 u(z)\big)|.$$

With (8), this implies

$$(10)\quad r^{-\frac{n}{p}}\|h_l - m_l(r)\|_{L^p(B_{\tau r})} \leq C\left\{(1+\varepsilon)\omega(r) - \omega(\tau r) + \frac{c_1}{K}r + \frac{c_2}{K}r^2\right\},$$

where C is a positive constant depending only on n, λ, and Λ. By taking $k = l$ in (7) and adding this to (10), we get

$$M_l(r) - m_l(r) \leq C\left\{(1+\varepsilon)\omega(r) - \omega(\tau r) + \frac{c_1}{K}r + \frac{c_2}{K}r^2\right\}.$$

Summing over $l = 1, \ldots, N$, we obtain

$$\omega(r) \leq C\left\{(1+\varepsilon)\omega(r) - \omega(\tau r) + \frac{c_1}{K}r + \frac{c_2}{K}r^2\right\},$$

and hence

$$\omega(\tau r) \leq \left(1 - \frac{1}{C} + \varepsilon\right)\omega(r) + C\left(\frac{c_1}{K}r + \frac{c_2}{K}r^2\right).$$

It is worth mentioning again that C is a positive constant depending only on n, λ, and Λ, independent of ε. By taking $\varepsilon = 1/(2C)$, we have

$$\omega(\tau r) \leq \gamma\omega(r) + C\left(\frac{c_1}{K}r + \frac{c_2}{K}r^2\right),$$

for some constant $\gamma \in (0,1)$, depending only on n, λ, and Λ. Lemma 1.2.13 implies, for any $r < R$,

$$\omega(r) \leq C\left(\frac{r}{R}\right)^{\alpha}\left\{\omega(R) + \frac{c_1}{K}r + \frac{c_2}{K}r^2\right\},$$

where $C > 0$ and $\alpha \in (0,1)$ are constants depending only on n, λ, and Λ. We note that $\{\gamma_1, \ldots, \gamma_N\}$ includes all e_i, for $i = 1, \ldots, n$, and $(e_i \pm e_j)/2$, for $i \neq j$, as shown in Lemma 5.2.6. By the definition of h_k, we obtain

$$\operatorname*{osc}_{B_r} \nabla^2 u \leq C\left(\frac{r}{R}\right)^{\alpha}\left\{\operatorname*{osc}_{B_R} \nabla^2 u + (c_1 R + c_2 R^2)\right\}.$$

By (3), (6), and (9), we note that c_1 and c_2 here are those in the statement of Theorem 5.4.3. Then, we have the desired estimate for $\nabla^2 u$. □

5.5. Global $C^{2,\alpha}$-Estimates

In this section, we derive global estimates of the Hölder semi-norms of the second derivatives of solutions to Dirichlet problems for fully nonlinear uniformly elliptic differential equations of the concave type.

Theorem 5.5.1. *Let Ω be a bounded domain in $\mathbb{R}^n$ with a C^3-boundary $\partial\Omega$ and F be a uniformly elliptic and concave C^2-function in $\mathcal{S}$. Suppose that u is a $C^3(\bar{\Omega})\cap C^4(\Omega)$-solution of*

$$(1)\qquad \begin{aligned} F(\nabla^2 u) &= f \quad \text{in } \Omega,\\ u &= \varphi \quad \text{on } \partial\Omega, \end{aligned}$$

for some $f\in C^2(\bar{\Omega})$ and $\varphi\in C^3(\bar{\Omega})$. Then, for some constant $\alpha\in(0,1)$ depending only on n, λ, and Λ,

$$[\nabla^2 u]_{C^\alpha(\bar{\Omega})}\le C\left\{|u|_{C^2(\bar{\Omega})}+|f|_{C^2(\bar{\Omega})}+|\varphi|_{C^3(\bar{\Omega})}\right\},$$

where C is a positive constant depending only on n, λ, Λ, and Ω.

As we will see in the proof, we need only assume that F is uniformly elliptic in the range of $\nabla^2 u$, i.e., for any $x\in\bar{\Omega}$ and $\xi\in\mathbb{R}^n$,

$$\lambda|\xi|^2\le F_{ij}\big(\nabla^2 u(x)\big)\xi_i\xi_j\le\Lambda|\xi|^2,$$

for some positive constants λ and Λ, and that F is a concave function in the range of $\nabla^2 u$, i.e., for any $x\in\bar{\Omega}$ and $(m_{ij})\in\mathcal{S}$,

$$F_{ij,kl}(\nabla^2 u(x))m_{ij}m_{kl}\le 0$$

and, for any $x,y\in\bar{\Omega}$,

$$F(\nabla^2 u(y))+F_{ij}(\nabla^2 u(y))\big(u_{ij}(x)-u_{ij}(y)\big)\ge F(\nabla^2 u(x)).$$

Proof. Set

$$K=|u|_{C^2(\bar{\Omega})}+|f|_{C^2(\bar{\Omega})}+|\varphi|_{C^3(\bar{\Omega})}.$$

We will prove

$$[\nabla^2 u]_{C^\alpha(\bar{\Omega})}\le CK,$$

where C is a positive constant depending only on n, λ, Λ, and Ω. We divide the proof into several steps. In the following, a universal constant is a positive constant depending only on n, λ, Λ, and Ω.

Step 1. We estimate the C^β-norm of $\nabla^2 u|_{\partial\Omega}$, for a constant $\beta\in(0,1)$, depending only on n, λ, and Λ, and prove, for any $x_0,x_1\in\partial\Omega$,

$$(2)\qquad |\nabla^2 u(x_1)-\nabla^2 u(x_0)|\le C_1K|x_1-x_0|^\beta.$$

For a fixed point $x_0\in\partial\Omega$, we flatten the boundary $\partial\Omega$ near x_0. In other words, we consider a neighborhood U of x_0 which contains a ball centered at x_0, with a universal radius, and C^3-diffeomorphisms

$$x=\chi(y),\quad y=\chi^{-1}(x)=\eta(x)\quad\text{for any } x\in U \text{ and } y\in A=\eta(U),$$

so that $\eta(x_0) = 0$ and

$$\eta(U \cap \Omega) = B_4^+ = \{y \in \mathbb{R}^n : |y| < 4, y_n > 0\},$$
$$\eta(U \cap \partial\Omega) = \Sigma_4 = \{y \in \mathbb{R}^n : |y| < 4, y_n = 0\}.$$

We also require that $|\eta|_{C^3(\bar{U})}$ and $|\chi|_{C^3(\bar{A})}$ are bounded by a universal constant. We consider the function

$$v(y) = u(x) - \varphi(x) \quad \text{for any } y \in B_4^+ \cup \Sigma_4.$$

In the following, we write $\eta = (\eta^1, \ldots, \eta^n)$. Then, v vanishes on Σ_4 and satisfies the equation

$$F\left(\left[\eta_i^k \eta_j^l v_{kl} + \eta_{ij}^k v_k + \varphi_{ij}\right]_{ij}\right) = f \quad \text{in } B_4^+, \tag{3}$$

where η_i^k, η_{ij}^k, φ_{ij}, and f are evaluated at $\chi(y)$. We prove (2) by showing, for any $y \in \Sigma_1$ and any $p, q = 1, \ldots, n$,

$$|v_{pq}(y) - v_{pq}(0)| \le CK|y|^\beta, \tag{4}$$

where

$$\Sigma_1 = \{y \in \mathbb{R}^n : |y| < 1, y_n = 0\}.$$

Since $v = 0$ on Σ_1, then $v_{pq} = 0$ on Σ_1 for any $p, q = 1, \ldots, n-1$. Hence, (4) holds trivially for $p, q = 1, \ldots, n-1$.

Next, we prove (4) for any $p = 1, \ldots, n-1$ and $q = n$. We differentiate (3) with respect to y_p and get

$$F_{ij}\eta_i^k \eta_j^l v_{klp} + F_{ij} v_{kl} \partial_p(\eta_i^k \eta_j^l) + F_{ij}\partial_p\left(\eta_{ij}^k v_k + \varphi_{ij}\right) = \partial_p f.$$

Set

$$\widetilde{L} = a_{kl}\partial_{kl} = F_{ij}\eta_i^k \eta_j^l \partial_{kl}.$$

Then, $\widetilde{L}$ is uniformly elliptic in B_4^+; namely, for any $y \in B_4^+$ and $\xi \in \mathbb{R}^n$,

$$\lambda^*|\xi|^2 \le a_{kl}(y)\xi_k\xi_l \le \Lambda^*|\xi|^2,$$

for some positive constants λ^* and Λ^*, depending only on n, λ, and Λ. The equation for v_p has the form

$$\widetilde{L}v_p = \widetilde{f} \quad \text{in } B_4^+,$$

where $\widetilde{f}$ satisfies

$$|\widetilde{f}|_{L^\infty(B_4^+)} \le CK,$$

for a universal constant C. We therefore have, for any $p = 1, \ldots, n-1$,

$$\widetilde{L}v_p = \widetilde{f} \quad \text{in } B_4^+,$$
$$v_p = 0 \quad \text{on } \Sigma_4^+.$$

Note that

$$|v_p|_{L^\infty(B_4^+)} + |\nabla v_p|_{L^\infty(B_4^+)} + |\widetilde{f}|_{L^\infty(B_4^+)} \le CK.$$

Theorem 1.2.16 implies, for some constant $\beta \in (0,1)$, depending only on n, λ, and Λ,

$$|\partial_n v_p|_{C^\beta(\Sigma_1)} \leq CK,$$

or, for any $y, y' \in \Sigma_1$,

$$|v_{np}(y) - v_{np}(y')| \leq CK|y - y'|^\beta,$$

where C is a universal constant. This implies (4) for $p = 1, \ldots, n-1$ and $q = n$.

Last, to prove (4) for $p = q = n$, we write (3) as

$$F\left(\left[\eta_i^k(\chi(y))\eta_j^l(\chi(y))v_{kl}(y) + \eta_{ij}^k(\chi(y))v_k(y) + \varphi_{ij}(\chi(y))\right]_{ij}\right) = f(\chi(y)),$$

for any $y \in B_4^+$. Now take the difference of the above expression and a similar one evaluated at $y = 0$. We obtain, for any $y \in B_4^+$,

$$\begin{aligned}\widetilde{F}_{ij} \cdot \big[&\eta_i^k(\chi(y))\eta_j^l(\chi(y))v_{kl}(y) + \eta_{ij}^k(\chi(y))v_k(y) + \varphi_{ij}(\chi(y)) \\ &- \eta_i^k(\chi(0))\eta_j^l(\chi(0))v_{kl}(0) - \eta_{ij}^k(\chi(0))v_k(0) - \varphi_{ij}(\chi(0))\big] \\ = f&(\chi(y)) - f(\chi(0)),\end{aligned}$$

where $\widetilde{F}_{ij}$ satisfies, for any $y \in B_4^+$ and any $\xi \in \mathbb{R}^n$,

$$\lambda|\xi|^2 \leq \widetilde{F}_{ij}(y)\xi_i\xi_j \leq \Lambda|\xi|^2.$$

A simple calculation yields

$$|v_{nn}(y) - v_{nn}(0)| \leq C\left\{\sup_{(k,l)\neq(n,n)} |v_{kl}(y) - v_{kl}(0)| + K|y|\right\}.$$

This ends the proof of (4).

Note that we did not use the concavity in this step.

Step 2. We prove, for any $x \in \bar{\Omega}$ and any $x_0 \in \partial\Omega$,

$$(5) \qquad |\nabla^2 u(x) - \nabla^2 u(x_0)| \leq C_2 K|x - x_0|^{\frac{\beta}{1+\beta}},$$

where β is as in (2).

Set

$$L = F_{ij}\partial_{ij}.$$

Let $\gamma \in \mathbb{R}^n$ be a unit vector. Since F is concave, Lemma 5.1.1(3) implies

$$Lu_{\gamma\gamma} \geq f_{\gamma\gamma} \geq -K.$$

Note that the linear operator L is uniformly elliptic, with ellipticity constants λ and Λ. Fix a point $x_0 \in \partial\Omega$. By (2), we have, for any $x \in \partial\Omega$,

$$u_{\gamma\gamma}(x) - u_{\gamma\gamma}(x_0) \leq CK|x - x_0|^\beta.$$

It is obvious that, for any $x \in \Omega$,

$$u_{\gamma\gamma}(x) - u_{\gamma\gamma}(x_0) \le CK.$$

Theorem 1.1.15 implies, for any $x \in \bar{\Omega}$,

$$u_{\gamma\gamma}(x) - u_{\gamma\gamma}(x_0) \le CK|x - x_0|^{\frac{\beta}{1+\beta}}. \tag{6}$$

We point out that Theorem 1.1.15 is formulated for solutions of uniformly elliptic linear equations. However, its proof clearly yields a one-sided estimate for subsolutions. Note that (6) holds for arbitrary unit vector $\gamma \in \mathbb{R}^n$. By Corollary 5.1.6, we obtain

$$\begin{aligned}|\nabla^2 u(x) - \nabla^2 u(x_0)| &\le C\Big\{\sup_{|\gamma|=1}\,(u_{\gamma\gamma}(x) - u_{\gamma\gamma}(x_0))^+ + |f(x) - f(x_0)|\Big\}\\ &\le C\left\{K|x - x_0|^{\frac{\beta}{1+\beta}} + |x - x_0||\nabla f|_{L^\infty(\Omega)}\right\}.\end{aligned}$$

This yields (5).

Step 3. Take α to be the minimum of α in Theorem 5.4.1 and $\beta/(1+\beta)$ in Step 2. We prove, for any $x, y \in \bar{\Omega}$,

$$|\nabla^2 u(x) - \nabla^2 u(y)| \le CK|x - y|^\alpha. \tag{7}$$

We first recall Theorem 5.4.1, the interior estimates of the Hölder seminorms. For any $B_R(x_*) \subset \Omega$ and any $x, y \in B_{R/2}(x_*)$, we have

$$\begin{aligned}R^\alpha \frac{|\nabla^2 u(x) - \nabla^2 u(y)|}{|x - y|^\alpha} &\le C\left\{|\nabla^2 u|_{L^\infty(B_R(x_*))}\right.\\ &\qquad \left.+ R|\nabla f|_{L^\infty(\Omega)} + R^2|\nabla^2 f|_{L^\infty(\Omega)}\right\}.\end{aligned} \tag{8}$$

For any $x, y \in \Omega$, set $d_x = \operatorname{dist}(x, \partial\Omega)$ and $d_y = \operatorname{dist}(y, \partial\Omega)$. Suppose $d_y \le d_x$. Take $x_0, y_0 \in \partial\Omega$ such that $|x - x_0| = d_x$ and $|y - y_0| = d_y$. First, we assume $|x - y| < d_x/2$. Then, $y \in B_{d_x/2}(x) \subset B_{d_x}(x) \subset \Omega$. Consider

$$w = u - u(x_0) - \nabla u(x_0)(x - x_0) - \frac{1}{2}(x - x_0)^t \nabla^2 u(x_0)(x - x_0).$$

Then, w satisfies

$$G(\nabla^2 w) = f \quad \text{in } \Omega,$$

where

$$G(M) = F\big(M + \nabla^2 u(x_0)\big).$$

Obviously, G is uniformly elliptic and concave with the same ellipticity constants as F. We apply (8) to w in $B_{d_x}(x)$ and get

$$\begin{aligned}d_x^\alpha \frac{|\nabla^2 u(x) - \nabla^2 u(y)|}{|x - y|^\alpha} &\le C\Big\{|\nabla^2 u - \nabla^2 u(x_0)|_{L^\infty(B_{d_x}(x))}\\ &\qquad + d_x|\nabla f|_{L^\infty(\Omega)} + d_x^2|\nabla^2 f|_{L^\infty(\Omega)}\Big\}.\end{aligned}$$

By (5), we obtain

$$|\nabla^2 u - \nabla^2 u(x_0)|_{L^\infty(B_{d_x}(x))} \le C_2 K d_x^\alpha,$$

and hence

$$\begin{aligned}|\nabla^2 u(x) - \nabla^2 u(y)| &\le C|x-y|^\alpha \left\{C_2 K + |\nabla f|_{L^\infty(\Omega)} + |\nabla^2 f|_{L^\infty(\Omega)}\right\}\\ &\le CK|x-y|^\alpha.\end{aligned}$$

Next, we assume $d_y \le d_x \le 2|x-y|$. Then,

$$|x_0 - y_0| \le d_x + |x-y| + d_y \le 5|x-y|.$$

By (2) and (5), we have

$$\begin{aligned}|\nabla^2 u(x) - \nabla^2 u(y)| \le\ & |\nabla^2 u(x) - \nabla^2 u(x_0)| + |\nabla^2 u(x_0) - \nabla^2 u(y_0)|\\ &+ |\nabla^2 u(y_0) - \nabla^2 u(y)|\\ \le\ & CK\left\{d_x^\alpha + |x_0 - y_0|^\alpha + d_y^\alpha\right\}\\ \le\ & CK|x-y|^\alpha.\end{aligned}$$

This finishes the proof of (7). □

5.6. Dirichlet Problems

In this section, we employ the method of continuity to solve Dirichlet problems for fully nonlinear uniformly elliptic differential equations of the concave type given by

$$\begin{aligned}F(\nabla^2 u) &= f \quad \text{in } \Omega,\\ u &= \varphi \quad \text{on } \partial\Omega.\end{aligned}$$

The global $C^{2,\alpha}$-estimates we derived earlier play an essential role.

The method of continuity consists of three steps:

Step 1. Prove that the given Dirichlet problem, labeled by $(*)$, can be embedded in a family of Dirichlet problems $\{(*)_t\}$ indexed by a parameter t in a bounded closed interval, say $[0,1]$, with $(*)_1$ corresponding to the given Dirichlet problem and $(*)_0$ corresponding to a Dirichlet problem which can be solved.

Step 2. Prove that the set of $t \in [0,1]$ for which the Dirichlet problem $(*)_t$ can be solved is open; that is, if $(*)_{t_0}$ can be solved, then there exists a small neighborhood of t_0, say $|t - t_0| < \varepsilon(t_0)$, such that $(*)_t$ can be solved for all t in this neighborhood.

Step 3. Prove that the set of $t \in [0,1]$ for which $(*)_t$ can be solved is closed.

Steps 1–3 imply that the set of t for which $(*)_t$ can be solved is the whole segment $0 \le t \le 1$ and hence contains $t = 1$. Therefore, the given Dirichlet problem, corresponding to $(*)_1$, is solved.

Step 1 is usually fulfilled by connecting the given Dirichlet problem with a simple Dirichlet problem, for example, the Dirichlet problem of the Laplace operator. Step 2, referred to as the "openness", is based on the implicit function theorem or iterations. It reduces the solvability of nonlinear equations to that of linear equations. Step 3, referred to as the "closedness", is based on a priori estimates for solutions independent of the parameter t.

Now we solve the Dirichlet problem for fully nonlinear uniformly elliptic differential equations by the method of continuity.

Theorem 5.6.1. *Let $\alpha \in (0,1)$ be a constant, Ω be a bounded domain in $\mathbb{R}^n$ with a $C^{4,\alpha}$-boundary, and F be a uniformly elliptic and concave C^3-function in $\mathcal{S}$. Then for any $f \in C^{2,\alpha}(\bar{\Omega})$ and $\varphi \in C^{4,\alpha}(\bar{\Omega})$, there exists a unique solution $u \in C^{4,\alpha}(\bar{\Omega})$ of*

$$(1)\qquad \begin{aligned} F(\nabla^2 u) &= f \quad \text{in } \Omega, \\ u &= \varphi \quad \text{on } \partial\Omega. \end{aligned}$$

Proof. For simplicity, we assume the given α is the same α in Theorem 5.5.1. We need only prove the existence of a $C^{2,\alpha}(\bar{\Omega})$-solution of (1). Then, the uniqueness follows from Corollary 5.1.8 and the higher regularity $u \in C^{4,\alpha}(\bar{\Omega})$ from Proposition 5.1.10. With $u \in C^4(\bar{\Omega})$, we can apply Theorem 5.3.4 and Theorem 5.5.1 to conclude

$$(2)\qquad |u|_{C^{2,\alpha}(\bar{\Omega})} \le C\Big\{|\varphi|_{C^3(\bar{\Omega})} + |f|_{C^2(\bar{\Omega})} + |F(0)|\Big\},$$

where C is a positive constant depending only on n, λ, Λ, α, and Ω.

For each $t \in [0,1]$, consider a family of problems

$$(3)\qquad \begin{aligned} tF(\nabla^2 u) + (1-t)\Delta u &= f \quad \text{in } \Omega, \\ u &= \varphi \quad \text{on } \partial\Omega. \end{aligned}$$

For $t = 0$, (3) corresponds to the Dirichlet problem for the Laplace operator. In general, (3) is the Dirichlet problem for the uniformly elliptic and concave C^2-operator $tF(\nabla^2 u)+(1-t)\Delta u$. In fact, the uniform ellipticity holds since, for any $M \in \mathcal{S}$ and $\xi \in \mathbb{R}^n$,

$$\min\{1,\lambda\}|\xi|^2 \le \big(tF_{ij}(M) + (1-t)\delta_{ij}\big)\xi_i\xi_j \le \max\{1,\Lambda\}|\xi|^2.$$

Moreover, any $C^{2,\alpha}$-solutions of (3) are $C^4(\bar{\Omega})$ and satisfy the estimate (2).

Letting $v = u - \varphi$, (3) is equivalent to

$$(4)\qquad \begin{aligned} tF(\nabla^2 v + \nabla^2\varphi) + (1-t)(\Delta v + \Delta\varphi) &= f \quad \text{in } \Omega, \\ v &= 0 \quad \text{on } \partial\Omega. \end{aligned}$$

Hence, any solutions of (4) satisfy the estimate (2).

Next, we set

$$\mathcal{X} = \{v \in C^{2,\alpha}(\bar{\Omega}) : \ v = 0 \text{ on } \partial\Omega\}$$

and

$$\mathcal{F}(v,t) = tF(\nabla^2 v + \nabla^2\varphi) + (1-t)(\Delta v + \Delta\varphi) - f.$$

Solving (4) is equivalent to finding a function $v \in \mathcal{X}$ such that $\mathcal{F}(v,t) = 0$ in Ω.

Set

$$I = \{t \in [0,1] : \text{there exists a } v \in \mathcal{X} \text{ such that } \mathcal{F}(v,t) = 0\}.$$

By the Schauder theory for the Laplace operator, we know $0 \in I$. To prove $1 \in I$, we prove that I is both open and closed in $[0,1]$. We first prove that I is open. We note that $\mathcal{F} : \mathcal{X} \times [0,1] \to C^\alpha(\bar{\Omega})$ is of class C^1 and its Frèchet derivative with respect to $v \in \mathcal{X}$ is given by

$$\mathcal{F}_v(v,t)w = \big(tF_{ij}(\nabla^2 v + \nabla^2\varphi) + (1-t)\delta_{ij}\big)w_{ij}.$$

This is a uniformly elliptic linear operator with C^α-coefficients. By the Schauder theory, $\mathcal{F}_v(v,t)$ is an invertible operator from $\mathcal{X}$ to $C^\alpha(\bar{\Omega})$. Suppose $t_0 \in I$; i.e., $\mathcal{F}(v^{t_0}, t_0) = 0$ for some $v^{t_0} \in \mathcal{X}$. By the implicit function theorem, for any t close to t_0, there is a unique $v^t \in \mathcal{X}$, close to v^{t_0} in the $C^{2,\alpha}(\bar{\Omega})$-norm, satisfying $\mathcal{F}(v^t, t) = 0$. Hence, $t \in I$ for all such t, and therefore I is open. For the closedness, we note that any solution $v \in \mathcal{X}$ of $\mathcal{F}(v,t) = 0$ in Ω satisfies a uniform $C^{2,\alpha}(\bar{\Omega})$-estimate, independent of t; i.e.,

$$|v^t|_{C^{2,\alpha}(\bar{\Omega})} \leq K, \text{ independent of } t.$$

Hence, the closedness of I follows from the compactness in $C^2(\bar{\Omega})$ of bounded sets in $C^{2,\alpha}(\bar{\Omega})$, a consequence of the Arzela-Ascoli theorem. Therefore, I is the whole unit interval. The function v^1 is then our desired solution of (4) corresponding to $t = 1$. □

In Theorem 5.6.1, we proved the existence of $C^{4,\alpha}$-solutions in $\bar{\Omega}$ if boundary values are $C^{4,\alpha}$ on $\partial\Omega$. By an approximation, we can conclude the existence of solutions under weaker conditions. In particular, when boundary values are only continuous, we have solutions which are continuous up to the boundary.

Theorem 5.6.2. *Let $\alpha \in (0,1)$ be a constant, Ω be a bounded domain in $\mathbb{R}^n$ with a $C^{4,\alpha}$-boundary, and F be a uniformly elliptic and concave C^2-function in $\mathcal{S}$. Then for any $f \in C(\bar{\Omega}) \cap C^2(\Omega)$ and $\varphi \in C(\bar{\Omega})$, there exists a unique solution $u \in C(\bar{\Omega}) \cap C^2(\Omega)$ of*

$$\begin{aligned} F(\nabla^2 u) &= f \quad \text{in } \Omega, \\ u &= \varphi \quad \text{on } \partial\Omega. \end{aligned} \tag{1}$$

Theorem 5.6.2 can be proved as a corollary of Theorem 5.6.1 by an approximation. In the following, we provide another approximation which does not require Theorem 5.6.1. In the proof below, we do not need the boundary estimates of $\nabla^2 u$ in Theorem 5.3.3 or global Hölder estimates of $\nabla^2 u$ in Theorem 5.5.1. We still need the global L^∞-estimates in Theorem 5.3.1. We can relax requirements on $\partial\Omega$ and assume that Ω satisfies a uniform exterior sphere condition.

Proof. For simplicity, we assume that the given α is the same α in Theorem 5.4.3. Let $\{\eta_m\}$ be a sequence of $C_0^2(\Omega)$-functions satisfying $0 \le \eta_m \le 1$ and $\eta_m(x) = 1$ for any $x \in \Omega$ with $\mathrm{dist}(x, \partial\Omega) \ge 1/m$. For each $m \ge 1$, consider, for any $x \in \bar{\Omega}$ and $M \in \mathcal{S}$,

$$F_m(x, M) = \eta_m(x) F(M) + (1 - \eta_m(x)) \operatorname{Tr}(M).$$

Then, $F_m(x, \cdot)$ defines a uniformly elliptic and concave C^2-operator. In fact, the uniform ellipticity holds since, for any $x \in \bar{\Omega}$, $M \in \mathcal{S}$, and $\xi \in \mathbb{R}^n$,

$$\begin{aligned} \min\{1, \lambda\} |\xi|^2 &\le \big(\eta_m(x) \partial_{m_{ij}} F_m(M) + (1 - \eta_m(x)) \delta_{ij}\big) \xi_i \xi_j \\ &\le \max\{1, \Lambda\} |\xi|^2. \end{aligned}$$

We point out that F_m depends also on $x \in \Omega$. Next, we take $f_m \in C^{4,\alpha}(\bar{\Omega})$ and $\varphi_m \in C^{2,\alpha}(\bar{\Omega})$ such that

$$\begin{aligned} f_m &\to f \quad \text{uniformly in } \bar{\Omega} \text{ as } m \to \infty, \\ \varphi_m &\to \varphi \quad \text{uniformly in } \bar{\Omega} \text{ as } m \to \infty, \end{aligned}$$

and, for any $\Omega' \Subset \Omega$,

$$f_m \to f \quad \text{in } C^2(\bar{\Omega}') \text{ as } m \to \infty.$$

Now, we consider the Dirichlet problem

$$\begin{aligned} F_m(x, \nabla^2 u) &= f_m \quad \text{in } \Omega, \\ u &= \varphi_m \quad \text{on } \partial\Omega. \end{aligned} \tag{2}$$

We prove the existence of a $C^{2,\alpha}$-solution of (2) by the method of continuity.

We fix an m and proceed to derive a $C^{2,\alpha}$-estimate for any $C^{2,\alpha}$-solutions u_m of (2). We first proceed as in the proof of Theorem 5.3.1. By the definition of F_m, we write the equation in (2) as

$$\big(\eta_m a_{ij} + (1 - \eta_m) \delta_{ij}\big) \partial_{ij} u_m = f_m - F(0) \eta_m \quad \text{in } \Omega,$$

where a_{ij} satisfies, for any $x \in \Omega$ and $\xi \in \mathbb{R}^n$,

$$\lambda |\xi|^2 \le a_{ij}(x) \xi_i \xi_j \le \Lambda |\xi|^2.$$

By Theorem 1.1.10, we have

$$|u_m|_{L^\infty(\Omega)} \leq |\varphi_m|_{L^\infty(\partial\Omega)} + C\left\{|f_m|_{L^\infty(\Omega)} + |F(0)|\right\}, \tag{3}$$

where C is a positive constant depending only on n, λ, Λ, and Ω. It is obvious that the right-hand side of (3) can be made independent of m by the uniform convergence of f_m and φ_m. Moreover, by Theorem 1.1.12, we have, for any $x \in \Omega$ and $x_0 \in \partial\Omega$,

$$|u_m(x) - u_m(x_0)| \leq \omega(|x - x_0|), \tag{4}$$

where ω is a nondecreasing continuous function on $(0, D)$ with $D = \mathrm{diam}(\Omega)$ and $\lim_{r\to 0}\rho(r) = 0$, depending only on n, λ, Λ, $|F(0)|$, Ω, the L^∞-norm of f_m in Ω, the L^∞-norm of φ_m on $\partial\Omega$, and the modulus of continuity of φ_m on $\partial\Omega$. Again, ω can be made independent of m. We point out that (3) and (4) are the only global estimates related to fully nonlinear equations, which are needed in this proof.

For the fixed m above, we consider $\Omega_1 \Subset \Omega_2 \Subset \Omega_3 \Subset \Omega$ such that $\mathrm{supp}\,\eta_m \subset \Omega_1$. Then in $\Omega \setminus \bar{\Omega}_1$, F_m reduces to

$$\Delta u_m = f_m \quad \text{in } \Omega \setminus \bar{\Omega}_1.$$

By the boundary Schauder estimates, we have

$$\begin{aligned} |u_m|_{C^{2,\alpha}(\bar{\Omega}\setminus\Omega_2)} &\leq C\left\{|u_m|_{L^\infty(\Omega\setminus\Omega_1)} + |\varphi_m|_{C^{2,\alpha}(\bar{\Omega}\setminus\Omega_1)} + |f_m|_{C^\alpha(\bar{\Omega}\setminus\Omega_1)}\right\} \\ &\leq C\left\{|u_m|_{L^\infty(\Omega)} + |\varphi_m|_{C^{2,\alpha}(\bar{\Omega})} + |f_m|_{C^\alpha(\bar{\Omega})}\right\}, \end{aligned} \tag{5}$$

where C is a positive constant depending only on α, λ, Λ, $\Omega \setminus \bar{\Omega}_1$, and $\mathrm{dist}(\Omega_1, \partial\Omega_2)$. We note that this constant C depends on m. Next, we consider

$$F_m(x, \nabla^2 u) = f_m \quad \text{in } \Omega.$$

By using Theorems 5.2.7, 5.2.8, and 5.4.3 successively, we obtain

$$|u_m|_{C^{2,\alpha}(\Omega_3)} \leq C\left\{|u_m|_{L^\infty(\Omega)} + |f_m|_{C^2(\bar{\Omega})}\right\}, \tag{6}$$

where C is a positive constant depending only on n, λ, Λ, Ω_3, and $\mathrm{dist}(\Omega_3, \partial\Omega)$. Here, we refer to Remark 5.2.9 for estimates on derivatives of F_m. By combining (3), (5), and (6), we obtain

$$|u_m|_{C^{2,\alpha}(\bar{\Omega})} \leq C_*\left\{|\varphi_m|_{C^{2,\alpha}(\bar{\Omega})} + |f_m|_{C^2(\bar{\Omega})} + |F(0)|\right\},$$

where C_* is a positive constant depending only on n, λ, Λ, η_m, and Ω. In general, the constant C_* depends on m. Now, we can employ the method of continuity as in the proof of Theorem 5.6.1 to conclude the existence of a $C^{2,\alpha}(\bar{\Omega})$-solution u_m of (2). We note that (4) is not used in this step.

Next, we discuss the limit of u_m as $m \to \infty$. For any fixed subdomain $\Omega' \Subset \Omega$, we take Ω'' such that $\Omega' \Subset \Omega'' \Subset \Omega$ and $\eta_m = 1$ in Ω'' for sufficiently large m. Then, for such m, we have

$$F(\nabla^2 u_m) = f \quad \text{in } \Omega''.$$

By using Theorems 5.2.4, 5.2.5, and 5.4.1 successively, we obtain

$$|u_m|_{C^{2,\alpha}(\Omega')} \leq C\left\{|u_m|_{L^\infty(\Omega)} + |f_m|_{C^2(\Omega'')} + |F(0)|\right\}, \tag{7}$$

where C is a positive constant depending only on n, λ, Λ, Ω', and Ω'', independent of m. Note that the right-hand side of (7) can be made independent of m. By (3), (4), (7), and the Arzela-Ascoli theorem, there exists a subsequence of $\{u_m\}$ convergent in C^2 in any subset $\Omega' \Subset \Omega$ to a function $u \in C^{2,\alpha}(\Omega)$, which also satisfies

$$|u(x) - \varphi(x_0)| \leq \omega(|x - x_0|),$$

where ω is the function as in (4). Hence, u can be extended to a $C(\bar{\Omega})$-function and satisfies (1), by passing the limit. □

Chapter 6

Monge-Ampère Equations

In this chapter, we discuss Monge-Ampère equations. We will derive a priori estimates for solutions and their derivatives up to the second order and solve the corresponding Dirichlet problem by the method of continuity.

In Section 6.1, we discuss basic properties of Monge-Ampère equations. We prove a comparison principle and the higher regularity for their solutions.

In Section 6.2, we discuss the Dirichlet problem for Monge-Ampère equations in uniformly convex domains. We first derive a global C^2-estimate and then solve the Dirichlet problem by the method of continuity. We establish the estimate of the second-order derivatives on the boundary by a careful construction of barrier functions.

In Section 6.3, we discuss interior estimates of second derivatives. Different from uniformly elliptic equations, boundary conditions are needed even for interior estimates. For simplicity, we discuss only the constant boundary value.

In Section 6.4, we discuss the global convex solutions of Monge-Ampère equations and prove a type of the Bernstein theorem. Geometric properties of convex sets and the structure of Monge-Ampère equations play a fundamental role.

6.1. Basic Properties

Suppose Ω is a domain in $\mathbb{R}^n$. The Monge-Ampère operator M is defined by

$$M(u) = \det(\nabla^2 u),$$

for any $u \in C^2(\Omega)$. Obviously, $M(u) \geq 0$ if u is convex. Now, we define an important notion in this chapter.

Definition 6.1.1. Let Ω be a domain in $\mathbb{R}^n$ and u be a function in Ω. Then, u is *uniformly convex* in Ω if $u \in C^2(\Omega)$ and $\nabla^2 u$ is positive definite in Ω. Similarly, u is *uniformly convex* in $\bar{\Omega}$ if $u \in C^2(\bar{\Omega})$ and $\nabla^2 u$ is positive definite in $\bar{\Omega}$.

We point out that the uniform convexity defined in Definition 6.1.1 is referred to as the strict convexity in [**27**]. In this book, strictly convex functions have a different meaning and are defined in Definition 8.1.2.

For some uniformly convex function u, it is convenient to introduce

$$F(\nabla^2 u) = \log\det(\nabla^2 u).$$

We claim

$$\begin{aligned} F_{ij} &= u^{ij}, \\ F_{ij,kl} &= -\ u^{il}u^{kj}, \end{aligned}$$

where (u^{ij}) is the inverse of the Hessian matrix $H = (u_{ij})$. To check this, we denote by $A = (A^{ij})$ the cofactor matrix of H; i.e., $A = (\det H)H^{-1}$. For a fixed $i = 1, \ldots, n$, we expand the determinant according to the ith-row,

$$\det(\nabla^2 u) = A^{i1}u_{i1} + \cdots + A^{in}u_{in}.$$

Then,

$$F_{ij} = \frac{1}{\det(\nabla^2 u)} \cdot A^{ij} = u^{ij}.$$

Next, for fixed $i, j = 1, \ldots, n$, we have

$$u^{ik}u_{jk} = \delta^i_j.$$

Differentiating with respect to u_{pq}, we get

$$(u^{ik})_{u_{pq}}u_{jk} + u^{ik}(u_{jk})_{u_{pq}} = 0.$$

Multiplying by u^{jl} and summing over j, we have

$$(u^{il})_{u_{pq}} = (u^{ik})_{u_{pq}}u_{jk}u^{jl} = -u^{ik}u^{jl}(u_{jk})_{u_{pq}} = -u^{iq}u^{pl},$$

or

$$\partial_{u_{kl}}u^{ij} = -u^{il}u^{kj}.$$

Hence,

$$F_{ij,kl} = \partial_{u_{kl}}u^{ij} = -u^{il}u^{kj}.$$

We now verify that F is a concave function of its argument, the positive definite matrices $\nabla^2 u = (u_{ij})$. This means, for any symmetric matrix $M = (m_{ij})$,

$$F_{ij,kl}m_{ij}m_{kl} \leq 0.$$

To check this, we diagonalize the matrix (u_{ij}). Then, (u^{ij}) is a diagonal matrix diag $(\lambda^1, \ldots, \lambda^n)$, with $\lambda^i > 0$, $i = 1, \ldots, n$. Hence,

$$F_{ij,kl} m_{ij} m_{kl} = -u^{il} u^{kj} m_{ij} m_{kl} = -\lambda^i \lambda^j m_{ij}^2 \leq 0.$$

We now recall a simple result on positive definite matrices, which will be needed later. If $\nabla^2 u = (u_{ij})$ is a positive definite matrix, then

$$|u_{ij}| \leq \frac{1}{2}(u_{ii} + u_{jj}).$$

This can be checked easily as follows. Since $\nabla^2 u$ is positive definite, any 2×2 diagonal minor has a positive determinant. Then,

$$u_{ij}^2 \leq u_{ii} u_{jj}.$$

The Cauchy inequality implies the desired result.

We proceed to consider the *Monge-Ampère equation*

$$\det(u_{ij}) = f,$$

for a uniformly convex C^4-function u. We write it as

$$\log \det(u_{ij}) = \log f.$$

Suppose γ is an arbitrary unit vector in $\mathbb{R}^n$. Differentiating the equation above with respect to γ, we obtain

$$u^{ij} u_{ij\gamma} = (\log f)_\gamma.$$

This leads to the linear differential operator

$$L = u^{ij} \partial_{ij}.$$

Since u is uniformly convex, L is elliptic. Then,

$$Lu_\gamma = (\log f)_\gamma.$$

Differentiating again with respect to γ, we obtain

$$u^{ij} u_{ij\gamma\gamma} - u^{il} u^{jk} u_{ij\gamma} u_{kl\gamma} = (\log f)_{\gamma\gamma},$$

or

$$Lu_{\gamma\gamma} - u^{il} u^{jk} u_{ij\gamma} u_{kl\gamma} = (\log f)_{\gamma\gamma}.$$

The second term in the left-hand side is nonpositive. Then,

$$Lu_{\gamma\gamma} \geq (\log f)_{\gamma\gamma}.$$

Difficulties in studying Monge-Ampère equations arise from the lack of the uniform ellipticity. Although elliptic for uniformly convex functions, the Monge-Ampère operator $\det(u_{ij})$ defined on the convex function u has ellipticity constants determined by the convexity of u. These ellipticity constants are controlled only after the C^2-norms of u are derived.

Even without the uniform ellipticity, we still have the following comparison principle.

Lemma 6.1.2. *Let Ω be a bounded domain in $\mathbb{R}^n$. Suppose that u, v are convex $C(\bar{\Omega}) \cap C^2(\Omega)$-functions and satisfy $\det(u_{ij}) \geq \det(v_{ij})$ in Ω and $u \leq v$ on $\partial\Omega$. Then, $u \leq v$ in Ω.*

The uniqueness of solutions of the Dirichlet problem follows easily.

Proof. First, we assume u is uniformly convex in $\bar{\Omega}$. Then,

$$\begin{aligned}\det(u_{ij}) - \det(v_{ij}) &= \int_0^1 \frac{d}{dt} \det\big((tu + (1-t)v)_{ij}\big)dt \\ &= \int_0^1 a^{ij}(t)dt(u-v)_{ij},\end{aligned}$$

where $(a^{ij}(t))$ is the cofactor matrix of $(tu_{ij}+(1-t)v_{ij})$. Hence, $L(u-v) \geq 0$ in Ω for some linear uniformly elliptic operator L since u is uniformly convex in $\bar{\Omega}$. We have the desired result by the maximum principle.

If u is only convex, we consider, for some $\varepsilon > 0$,

$$u_\varepsilon = u + \varepsilon(|x|^2 - \max_{\partial\Omega} |x|^2).$$

Then, u_ε is uniformly convex in $\bar{\Omega}$. Also, $\det(\partial_{ij}u_\varepsilon) \geq \det(v_{ij})$ in Ω and $u_\varepsilon \leq v$ on $\partial\Omega$. By what we just proved, we have $u_\varepsilon \leq v$ in Ω. Hence, we obtain the desired result by letting $\varepsilon \to 0$. □

The higher regularity of solutions can also be established once they are known to be $C^{2,\alpha}$. The following results are simple corollaries of Proposition 5.1.9 and Proposition 5.1.10.

Proposition 6.1.3. *Let $\alpha \in (0,1)$ be a constant and Ω be a domain in $\mathbb{R}^n$. Suppose that u is a uniformly convex $C^{2,\alpha}(\Omega)$-solution of*

$$\det(u_{ij}) = f \quad \text{in } \Omega,$$

for some $f \in C^\alpha(\Omega)$, with $f > 0$ in Ω. For any integer $m \geq 1$, if $f \in C^{m,\alpha}(\Omega)$, then $u \in C^{m+2,\alpha}(\Omega)$. In particular, if $f \in C^\infty(\Omega)$, then $u \in C^\infty(\Omega)$.

Proposition 6.1.4. *Let $\alpha \in (0,1)$ be a constant and Ω be a bounded domain in $\mathbb{R}^n$ with a $C^{2,\alpha}$-boundary. Suppose that u is a uniformly convex $C^{2,\alpha}(\bar{\Omega})$-solution of*

$$\begin{aligned}\det(u_{ij}) &= f \quad \text{in } \Omega, \\ u &= \varphi \quad \text{on } \partial\Omega,\end{aligned}$$

for some $f \in C^\alpha(\bar{\Omega})$ and $\varphi \in C^{2,\alpha}(\bar{\Omega})$, with $f > 0$ in $\bar{\Omega}$. For any integer $m \geq 1$, if $\partial\Omega \in C^{m+2,\alpha}$, $f \in C^{m,\alpha}(\bar{\Omega})$, and $\varphi \in C^{m+2,\alpha}(\bar{\Omega})$, then $u \in C^{m+2,\alpha}(\bar{\Omega})$. In particular, if $\partial\Omega \in C^\infty$, $f \in C^\infty(\bar{\Omega})$, and $\varphi \in C^\infty(\bar{\Omega})$, then $u \in C^\infty(\bar{\Omega})$.

6.2. Global C^2-Estimates

In this section, we discuss the Dirichlet problem for the Monge-Ampère equation in uniformly convex domains. We will first derive a global C^2-estimate and then employ the method of continuity to find uniformly convex solutions.

Let Ω be a bounded domain in $\mathbb{R}^n$ with a C^k-boundary, for some positive integer k. Then, $\Omega \subset \mathbb{R}^n$ is called *uniformly convex* if there exists a uniformly convex $C^k(\bar{\Omega})$-function σ such that $\sigma < 0$ in Ω and $\sigma = 0$, $\nabla\sigma \neq 0$ on $\partial\Omega$. Such a σ is called a *defining function* of Ω.

Let Ω be a bounded uniformly convex domain in $\mathbb{R}^n$, f be a positive continuous function in $\bar{\Omega}$, and φ be a continuous function on $\partial\Omega$. We consider the Dirichlet problem

$$\begin{aligned} \det(u_{ij}) &= f \quad \text{in } \Omega, \\ u &= \varphi \quad \text{on } \partial\Omega. \end{aligned}$$

We will derive a global C^2-estimate through a series of theorems, under extra assumptions on f and φ. We first derive a C^1-estimate.

Theorem 6.2.1. *Let Ω be a bounded domain in $\mathbb{R}^n$ with a C^2-boundary. Suppose u and $\underline{u}$ are convex $C^2(\bar{\Omega})$-functions and satisfy*

$$\begin{aligned} \det(u_{ij}) &\leq \det(\underline{u}_{ij}) \quad \text{in } \Omega, \\ u &= \underline{u} \quad \text{on } \partial\Omega. \end{aligned}$$

Then,

$$|u|_{C^1(\bar{\Omega})} \leq K,$$

where K is a positive constant depending only on Ω and the C^2-norm of $\underline{u}$ in $\bar{\Omega}$.

Proof. By the convexity of u, we have

$$u \leq \max_{\partial\Omega} \underline{u}.$$

Furthermore, by applying Lemma 6.1.2 to u and $\underline{u}$, we obtain

$$\underline{u} \leq u.$$

Hence,

$$|u| \leq K_0 \quad \text{in } \bar{\Omega},$$

where K_0 is a positive constant depending only on $\max_{\partial\Omega} \underline{u}$ and $\inf_{\Omega} \underline{u}$.

Since u is convex, $|\nabla u|$ takes its maximum on the boundary. The tangential derivatives of u on $\partial\Omega$ are the same as those of $\underline{u}$ on $\partial\Omega$ and hence are known, so it suffices to estimate the exterior normal derivative of u on $\partial\Omega$. Note that the convex function u is subharmonic. Let h be the harmonic function in Ω which equals $\underline{u}$ on $\partial\Omega$. The maximum principle implies $u \le h$ in Ω. Therefore,

$$\underline{u} \le u \le h \quad \text{in } \Omega.$$

Since these three functions have common values on $\partial\Omega$, then

$$\frac{\partial h}{\partial \nu} \le \frac{\partial u}{\partial \nu} \le \frac{\partial \underline{u}}{\partial \nu} \quad \text{on } \partial\Omega,$$

where ν is the exterior unit normal to $\partial\Omega$. Therefore, we obtain

$$|\nabla u| \le K_1 \quad \text{on } \partial\Omega \text{ and hence in } \bar{\Omega},$$

where K_1 is a positive constant depending only on the $L^\infty(\partial\Omega)$-norms of $\nabla\underline{u}$ and ∇h, the latter of which depends on Ω and the C^2-norm of $\underline{u}$ by Theorem 1.1.14. □

Next, we derive the fundamental boundary estimate of the second derivatives due to Caffarelli, Nirenberg, and Spruck [**27**]. We present two proofs for the second normal derivatives, the first proof by Guan [**63**] and the second the original proof by Caffarelli, Nirenberg, and Spruck.

Theorem 6.2.2. *Let Ω be a bounded uniformly convex domain in $\mathbb{R}^n$ with a C^4-boundary. Suppose that u is a uniformly convex $C^4(\bar{\Omega})$-solution of*

$$\begin{aligned} \det(u_{ij}) &= f \quad \text{in } \Omega, \\ u &= \varphi \quad \text{on } \partial\Omega, \end{aligned}$$

for some $f \in C^2(\bar{\Omega})$ and $\varphi \in C^4(\bar{\Omega})$, with $f > 0$ in $\bar{\Omega}$. Then,

$$|\nabla^2 u|_{L^\infty(\partial\Omega)} \le K,$$

where K is a positive constant depending only on Ω, the C^1-norm of u in $\bar{\Omega}$, the C^2-norm of f in $\bar{\Omega}$, $\max_{\bar{\Omega}} f^{-1}$, and the C^4-norm of φ in $\bar{\Omega}$.

The proof is based on the construction of suitable barrier functions.

Proof. In the following, a universal constant is a positive constant depending only on Ω, the C^1-norm of u in $\bar{\Omega}$, the C^2-norm of f in $\bar{\Omega}$, $\max_{\bar{\Omega}} f^{-1}$, and the C^4-norm of φ in $\bar{\Omega}$.

We write the equation in the form

$$\log\det(u_{ij}) = \log f$$

and set

$$L = u^{ij}\partial_{ij},$$

where (u^{ij}) is the inverse of the matrix (u_{ij}). Then, L is elliptic since u is uniformly convex. By $\det(u^{ij}) = f^{-1}$ and the inequality for arithmetic and geometric means, we get

$$\frac{1}{n}\sum_{i=1}^{n} u^{ii} \geq f^{-\frac{1}{n}}. \tag{1}$$

For each $k = 1, \ldots, n$, u_k satisfies

$$Lu_k = u^{ij}u_{ijk} = (\log f)_k. \tag{2}$$

Note that

$$\begin{aligned} L(x_l u_k) &= u^{ij}\partial_{ij}(x_l u_k) = u^{ij}\partial_i(\delta_{jl}u_k + x_l u_{jk}) \\ &= u^{ij}(\delta_{jl}u_{ik} + \delta_{il}u_{jk} + x_l u_{ijk}) = u^{il}u_{ik} + u^{lj}u_{jk} + x_l u^{ij}u_{ijk} \\ &= 2\delta_k^l + x_l(\log f)_k. \end{aligned}$$

Hence,

$$L(x_l u_k - x_k u_l) = (x_l\partial_k - x_k\partial_l)\log f. \tag{3}$$

This simply reflects the fact that the operator $x_l\partial_k - x_k\partial_l$ is an angular derivative (on $|x| =$ constant) and the expression $\det(u_{ij})$ is invariant under the rotation of coordinates.

Consider any boundary point; without loss of generality, we take it to be the origin and we take the x_n-axis to be the interior normal. Let σ be a uniformly convex C^4-defining function of Ω. Then, $\sigma_j(0) = 0$ for $j < n$. Without loss of generality, we assume $\sigma_n(0) = -1$. Writing the Taylor expansion of σ up to the second order, we obtain

$$\sigma = -x_n + \sum_{i,j=1}^{n} a_{ij}x_i x_j + O(|x|^3).$$

Since σ is uniformly convex in $\bar{\Omega}$, (a_{ij}) is positive definite and the eigenvalues of the symmetric matrix (a_{ij}) have uniform upper and lower bounds, independent of the points on $\partial\Omega$. Then, near the origin, $\partial\Omega$ is represented by

$$x_n = \rho(x') = \frac{1}{2}b_{\alpha\beta}x_\alpha x_\beta + O(|x'|^3), \tag{4}$$

where $x' = (x_1, \ldots, x_{n-1})$ and $(b_{\alpha\beta})$ is an $(n-1)\times(n-1)$ positive definite matrix. In the summation, Greek letters α, β, etc., range from 1 to $n-1$.

Step 1. We first estimate $u_{\alpha\beta}(0)$ for $\alpha, \beta = 1, \ldots, n-1$. On $\partial\Omega$, we have

$$u - \varphi = 0,$$

or, for small x',

$$(u - \varphi)(x', \rho(x')) = 0.$$

Recall that φ is defined in $\bar{\Omega}$. By differentiating with respect to x_α and then x_β, we get

$$(\partial_\alpha + \rho_\alpha \partial_n)(u - \varphi) = 0 \quad \text{on } \partial\Omega$$

and

$$(\partial_\beta + \rho_\beta \partial_n)(\partial_\alpha + \rho_\alpha \partial_n)(u - \varphi) = 0 \quad \text{on } \partial\Omega.$$

Note that $\rho_\alpha(0) = 0$ and $\rho_{\alpha\beta}(0) = b_{\alpha\beta}$. Hence, at 0, we have, for any $\alpha, \beta = 1, \ldots, n-1$,

$$(u - \varphi)_{\alpha\beta}(0) + b_{\alpha\beta}(u - \varphi)_n(0) = 0.$$

We obtain, for any $\alpha, \beta = 1, \ldots, n-1$,

$$|u_{\alpha\beta}(0)| \leq K_1, \tag{5}$$

for a universal constant K_1.

Step 2. Next, we estimate the mixed derivative $u_{\alpha n}(0)$, for $\alpha = 1, \ldots, n-1$. Consider the vector field (directional derivative)

$$T = \partial_\alpha + \sum_{\beta < n} b_{\alpha\beta}(x_\beta \partial_n - x_n \partial_\beta).$$

In view of (2) and (3), we have

$$L(Tu) = T(\log f).$$

This implies

$$L\big(T(u - \varphi)\big) = L(Tu) - L(T\varphi) = T(\log f) - u^{ij}\partial_{ij}(T\varphi).$$

Since (u^{ij}) is positive definite, we get

$$|L(T(u - \varphi))| \leq C\Big(1 + \sum_{i=1}^n u^{ii}\Big) \quad \text{in } \Omega. \tag{6}$$

Recall that $\partial_\alpha + \rho_\alpha \partial_n$ is a tangential operator along $\partial\Omega$. On $\partial\Omega$ close to the origin, we have, for any $\alpha = 1, \ldots, n-1$,

$$(\partial_\alpha + \rho_\alpha \partial_n)(u - \varphi) = 0,$$

and hence,

$$|(u - \varphi)_\alpha + (u - \varphi)_n b_{\alpha\beta} x_\beta| \leq C|x|^2.$$

This implies

$$|T(u - \varphi)| \leq C|x|^2 \quad \text{on } \partial\Omega \cap B_\varepsilon, \tag{7}$$

for some constant $\varepsilon > 0$ sufficiently small.

Now we claim

(i) $\pm L\big(T(u - \varphi)\big) \geq L(Ax_n - B|x|^2)$ in $\Omega \cap B_\varepsilon$;

(ii) $\pm T(u - \varphi) \leq Ax_n - B|x|^2$ on $\partial(\Omega \cap B_\varepsilon)$, for suitably chosen universal constants A and B. We first note that

$$L(Ax_n - B|x|^2) = -2B\sum_{i=1}^{n} u^{ii}.$$

To prove (i), we need, in view of (6),

$$(2B - C)\sum_{i=1}^{n} u^{ii} \geq C.$$

This can be achieved by (1) and choosing B sufficiently large. To prove (ii), we consider first $\partial\Omega \cap B_\varepsilon$. By the uniform convexity of $\partial\Omega$ or (4) specifically, we have, for some constant $c_0 > 0$,

$$x_n \geq c_0|x|^2 \quad \text{on } \partial\Omega \cap B_\varepsilon.$$

To prove (ii) on $\partial\Omega \cap B_\varepsilon$, we need, in view of (7),

$$(B + C)|x|^2 \leq Ax_n \quad \text{on } \partial\Omega \cap B_\varepsilon.$$

This can be achieved by taking A sufficiently large. Next, we note that $x_n \geq c_0'$ on $\Omega \cap \partial B_\varepsilon$ for some constant $c_0' > 0$, by the uniform convexity again. We choose A large further so that (ii) holds on $\Omega\cap\partial B_\varepsilon$. This finishes the proof of (i) and (ii). By the maximum principle, we obtain

$$|T(u - \varphi)| \leq Ax_n - B|x|^2 \quad \text{in } \Omega \cap B_\varepsilon.$$

By taking $x' = 0$, dividing by x_n, and then letting $x_n \to 0$, we get

$$|\partial_n T(u - \varphi)| \leq A \quad \text{at } 0,$$

or

$$\Big|(u - \varphi)_{\alpha n}(0) - \sum_{\beta<n} b_{\alpha\beta}(u - \varphi)_\beta(0)\Big| \leq A.$$

Thus, we obtain

$$|u_{\alpha n}(0)| \leq K_2, \tag{8}$$

for a universal constant K_2.

Step 3. Next, we consider $u_{nn}(0)$. By the equation $\det(u_{ij}) = f$ at 0, we have

$$\sum_{i=1}^{n} A^{ni}(0)u_{in}(0) = \det(u_{ij})(0) = f(0),$$

where (A^{ij}) is the cofactor matrix of (u_{ij}). By the estimates already established, we note that the $n - 1$ terms in the sum corresponding to $i = 1, \ldots, n - 1$ are bounded so that

$$A^{nn}(0)u_{nn}(0) \leq C.$$

If we have a lower bound for $A^{nn}(0)$, then

$$u_{nn}(0) \leq K_3,$$

for a universal constant K_3. Note that $u_{nn}(0) > 0$. In view of (5) and (8), we then have the desired estimate.

To prove a lower bound of $A^{nn}(0)$, we proceed to establish the following estimate: for any vector $\xi' = (\xi_1, \ldots, \xi_{n-1}) \in \mathbb{R}^{n-1}$,

$$\sum_{\alpha,\beta=1}^{n-1} u_{\alpha\beta}(0)\xi_\alpha\xi_\beta \geq c_0|\xi'|^2, \tag{9}$$

where c_0 is a universal constant.

Let σ be a uniformly convex $C^4(\bar{\Omega})$-defining function of Ω; i.e.,

$$\Omega = \{\sigma < 0\}, \quad \partial\Omega = \{\sigma = 0\}, \quad \nabla\sigma|_{\partial\Omega} \neq 0.$$

For convenience, we assume $|\nabla\sigma| = 1$ on $\partial\Omega$. For any $p \in \partial\Omega$, the tangent space of $\partial\Omega$ at p is given by

$$T_p\partial\Omega = \{\xi \in \mathbb{R}^n : \nabla\sigma(p) \cdot \xi = 0\}.$$

Consider

$$m_0 = \min_{p\in\partial\Omega} \min_{\xi\in T_p\partial\Omega\cap\mathbb{S}^{n-1}} u_{ij}(p)\xi_i\xi_j.$$

We assume that m_0 is attained at $p = 0$ and $\xi = (1, 0, \ldots, 0)$. We choose coordinates $x = (x_1, \ldots, x_n)$ as before so that the x_n-axis is the interior normal to $\partial\Omega$ at 0. We will prove

$$m_0 = u_{11}(0) \geq c_0, \tag{10}$$

for some universal constant c_0.

Without loss of generality, we assume that φ is uniformly convex in $\bar{\Omega}$; otherwise, we replace φ by $\varphi + t\sigma$ for sufficiently large constant t. There exists a function $h \in C^3(\bar{\Omega})$ such that, near $\partial\Omega$,

$$u - \varphi = h\sigma.$$

A simple differentiation yields

$$(u - \varphi)_i = h_i\sigma + h\sigma_i$$

and

$$(u - \varphi)_{ij} = h_{ij}\sigma + h_i\sigma_j + h_j\sigma_i + h\sigma_{ij}.$$

Hence,

$$(u - \varphi)_i\sigma_i = \sigma h_i\sigma_i + h|\nabla\sigma|^2,$$

and, for any $\xi \in \mathbb{R}^n$,

$$(u - \varphi)_{ij}\xi_i\xi_j = h_{ij}\xi_i\xi_j\sigma + h_i\sigma_j\xi_i\xi_j + h_j\sigma_i\xi_i\xi_j + h\sigma_{ij}\xi_i\xi_j.$$

First, by restricting to $\partial\Omega$, we have

$$(u-\varphi)_i\sigma_i = h|\nabla\sigma|^2 \quad \text{on } \partial\Omega.$$

Next, we fix a point $p \in \partial\Omega$ and take $\xi \in T_p\partial\Omega$. Then,

$$(u-\varphi)_{ij}\xi_i\xi_j = h\sigma_{ij}\xi_i\xi_j \quad \text{at } p,$$

and hence,

$$(u-\varphi)_{ij}\xi_i\xi_j = \frac{1}{|\nabla\sigma|^2}(u-\varphi)_k\sigma_k\sigma_{ij}\xi_i\xi_j \quad \text{at } p.$$

Moreover, at $p=0$, we get, for any $\alpha, \beta = 1, \ldots, n-1$,

$$(u-\varphi)_{\alpha\beta}(0) = -(u-\varphi)_n(0)\sigma_{\alpha\beta}(0),$$

where we used $\nabla\sigma(0) = (0, \ldots, 0, -1)$. In particular,

$$u_{11}(0) = \varphi_{11}(0) - (u-\varphi)_n(0)\sigma_{11}(0).$$

If $u_{11}(0) \geq \varphi_{11}(0)/2$, then (10) holds by the uniform convexity of φ. Next, we assume $u_{11}(0) \leq \varphi_{11}(0)/2$. Then,

$$(u-\varphi)_n(0)\sigma_{11}(0) \geq \frac{1}{2}\varphi_{11}(0),$$

and hence,

$$\sigma_{11}(0) \geq \frac{1}{2C}\varphi_{11}(0). \tag{11}$$

Let $\varepsilon > 0$ be a constant small enough so that

$$\sigma_1^2 + \sigma_n^2 > 0 \quad \text{in } \bar{\Omega} \cap \bar{B}_\varepsilon.$$

Define $\xi = (\xi_1, \ldots, \xi_n)$ by

$$\begin{aligned} \xi_1 &= -\sigma_n(\sigma_1^2 + \sigma_n^2)^{-\frac{1}{2}}, \\ \xi_i &= 0 \quad \text{for } i = 2, \ldots, n-1, \\ \xi_n &= \sigma_1(\sigma_1^2 + \sigma_n^2)^{-\frac{1}{2}}. \end{aligned}$$

Then, ξ is a unit tangent vector to $\partial\Omega$ in B_ε and $\xi(0) = (1, 0, \ldots, 0)$. Set

$$w = \varphi_{ij}\xi_i\xi_j + \frac{1}{|\nabla\sigma|^2}(u-\varphi)_k\sigma_k\sigma_{ij}\xi_i\xi_j - u_{11}(0) \quad \text{in } \Omega \cap B_\varepsilon.$$

Then,

$$w = u_{ij}\xi_i\xi_j - u_{11}(0) \quad \text{on } \partial\Omega \cap B_\varepsilon,$$

and hence, $w(0) = 0$ and

$$w \geq 0 \quad \text{on } \partial\Omega \cap B_\varepsilon.$$

Next, we calculate Lw. It is easy to check that

$$L\Big(\varphi_{ij}\xi_i\xi_j - \frac{1}{|\nabla\sigma|^2}\varphi_k\sigma_k\sigma_{ij}\xi_i\xi_j\Big) \leq C\sum_{i=1}^n u^{ii}.$$

For each $k = 1, \ldots, n$, we set

$$v = \frac{1}{|\nabla\sigma|^2}\sigma_k\sigma_{ij}\xi_i\xi_j.$$

Then,

$$\begin{aligned} L(u_k v) &= u_k Lv + vLu_k + 2u^{pq}u_{pk}v_q \\ &= u_k Lv + v(\log f)_k + 2v_k \le C\Big(1 + \sum_{i=1}^n u^{ii}\Big). \end{aligned}$$

Therefore,

$$\begin{aligned} Lw &\le C\Big(1 + \sum_{i=1}^n u^{ii}\Big) \quad \text{in } \Omega \cap B_\varepsilon, \\ w &\ge 0 \quad \text{on } \partial\Omega \cap B_\varepsilon. \end{aligned}$$

Arguing as in Step 2, we obtain

$$-w \le Ax_n - B|x|^2 \quad \text{in } \Omega \cap B_\varepsilon,$$

for appropriate universal constants A and B. Hence,

$$-w_n(0) \le A.$$

This implies, with $\sigma_n(0) = -1$,

$$(u - \varphi)_{nn}(0)\sigma_{11}(0) \le C,$$

and hence, by (11),

$$u_{nn}(0) \le C. \tag{12}$$

In view of (5), (8), and (12), we have an a priori upper bound for all eigenvalues of the Hessian matrix $(u_{ij}(0))$. Since $\det(u_{ij}(0)) = f(0) > 0$, the eigenvalues of $(u_{ij}(0))$ also admit a positive lower bound; i.e.,

$$\min_{\xi \in \mathbb{S}^{n-1}} u_{ij}(0)\xi_i\xi_j \ge c_0.$$

Hence,

$$m_0 = \min_{\xi \in T_0\partial\Omega\cap\mathbb{S}^{n-1}} u_{ij}(0)\xi_i\xi_j \ge \min_{\xi\in\mathbb{S}^{n-1}} u_{ij}(0)\xi_i\xi_j \ge c_0.$$

This establishes (10). □

We note that (9) is proved by a global argument. In fact, we can prove (9) at each point of the boundary. We now provide such a proof.

Alternative proof of Theorem 6.2.2. We provide an alternative proof of (9). Without loss of generality, we assume $\xi' = (1, 0, \ldots, 0) \in \mathbb{R}^{n-1}$ and prove

$$u_{11}(0) \ge c_0. \tag{9'}$$

Furthermore, we suppose, for $\alpha = 1, \ldots, n-1$,

$$u(0) = 0, \quad u_\alpha(0) = 0. \tag{13}$$

To prove (9′), we make use of a carefully constructed barrier function. Recall that we have (4) on $\partial\Omega$.

Set $\widetilde{u} = u - \lambda x_n$ and choose λ such that

$$\partial_{11}\widetilde{u}(x', \rho(x')) = 0 \quad \text{at } 0.$$

By

$$\partial_{11}\widetilde{u} = \widetilde{u}_{11} + 2\widetilde{u}_{1n}\rho_1 + \widetilde{u}_{nn}\rho_1^2 + \widetilde{u}_n\rho_{11},$$

we have

$$u_{11}(0) + \widetilde{u}_n(0)\rho_{11}(0) = u_{11}(0) + (u_n(0) - \lambda)\rho_{11}(0) = 0,$$

and hence

$$u_{11}(0) = -\widetilde{u}_n(0)\rho_{11}(0). \tag{14}$$

In the following, we will estimate $\widetilde{u}_n(0)$ from above by a negative constant. Note that $\widetilde{u}$ also satisfies $\det(\widetilde{u}_{ij}) = f$. We claim, for some small constant $\varepsilon > 0$,

$$\widetilde{u} \le \sum_{j=2}^{n} a_j x_1 x_j + C\sum_{j=2}^{n} x_j^2 \quad \text{on } \partial\Omega \cap B_\varepsilon, \tag{15}$$

where $a_2, \ldots, a_n$ are constants with a universal bound and C is a universal constant. To prove (15), we consider the Taylor expansion of $\widetilde{u}(x', \rho(x'))$, given by

$$\widetilde{u}(x', \rho(x')) = u(x', \rho(x')) - \lambda\rho(x') = \varphi(x', \rho(x')) - \lambda\rho(x').$$

In view of (4) and (13), there are no linear terms. For quadratic terms $x_\alpha x_\beta$, there is no x_1^2 term by the definition of $\widetilde{u}$. Hence, the quadratic part of $\widetilde{u}$ can be written as

$$\sum_{1<\alpha<n} a_{1\alpha}x_1x_\alpha + \sum_{1<\alpha,\beta<n} a_{\alpha\beta}x_\alpha x_\beta,$$

which is a part of the right-hand side in (15). Now we consider the cubic terms and higher-order terms. By (4), we have

$$x_1^2 = \frac{2x_n}{b_{11}} - \sum_{(\alpha,\beta)\neq(1,1)} \frac{b_{\alpha\beta}}{b_{11}} x_\alpha x_\beta + O(|x'|^3) \quad \text{on } \partial\Omega \cap B_\varepsilon,$$

and hence

$$x_1^3 = \frac{2}{b_{11}} x_1 x_n - \sum_{(\alpha,\beta)\neq(1,1)} \frac{b_{\alpha\beta}}{b_{11}} x_1 x_\alpha x_\beta + O(|x'|^4) \quad \text{on } \partial\Omega \cap B_\varepsilon.$$

Note that the x_1x_n term goes to the first sum in the right-hand side of (15). The rest of the cubic terms in $\widetilde{u}(x', \rho(x'))$ have the forms $x_1^2x_\alpha$, $x_1x_\alpha x_\beta$, and $x_\alpha x_\beta x_\gamma$, for $1 < \alpha, \beta, \gamma < n$. For any $1 < \alpha, \beta < n$, we have

$$x_1^2 x_\alpha \le \frac{1}{2}(x_\alpha^2 + x_1^4)$$

and, for $|x_1| \le 1$,

$$x_1 x_\alpha x_\beta \le \frac{1}{2}(x_\alpha^2 + x_\beta^2).$$

For the fourth-order term, we note that, for $i \ge 2$,

$$x_1^4 + |x_1^3 x_i| \le \sum_{1<\alpha<n} x_\alpha^2 + x_n^2.$$

Therefore, (15) is proved for any $x \in \partial\Omega$ close to the origin.

For some positive constants δ_0, δ_1, and μ to be determined, we set

$$\begin{aligned} h &= -\delta_0 x_n + \delta_1|x|^2 + \frac{1}{2\mu}\sum_{j=2}^n (a_j x_1 + \mu x_j)^2 \\ &= -\delta_0 x_n + \delta_1|x|^2 + \frac{1}{2\mu}\sum_{j=2}^n a_j^2 x_1^2 + \sum_{j=2}^n a_j x_1 x_j + \frac{\mu}{2}\sum_{j=2}^n x_j^2, \end{aligned}$$

where a_j are given in (15). We claim, by choosing δ_0, δ_1, and μ appropriately,

$$\begin{aligned} \det(\widetilde{u}_{ij}) &\ge \det(h_{ij}) \quad \text{in } \Omega \cap B_\varepsilon, \\ \widetilde{u} &\le h \quad \text{on } \partial(\Omega \cap B_\varepsilon). \end{aligned} \tag{16}$$

First, we note that $x_n \ge \varepsilon_0$ on $\partial B_\varepsilon \cap \Omega$, for some positive constant ε_0, and

$$h \ge -\delta_0 x_n + \frac{1}{2\mu}(a_n x_1 + \mu x_n)^2 \ge \frac{\mu}{2}x_n^2 + (a_n x_1 - \delta_0)x_n.$$

For any $\delta_0, \delta_1 \in (0, 1)$, by choosing μ sufficiently large, we have

$$\widetilde{u} \le h \quad \text{on } \partial B_\varepsilon \cap \Omega.$$

By choosing also $\mu \ge 2C$ for the constant C in (15), we have by (15)

$$\widetilde{u} \le \frac{1}{2\mu}\sum_{j=2}^n (a_j x_1 + \mu x_j)^2 \quad \text{on } \partial\Omega \cap B_\varepsilon.$$

Next, we calculate $\det(h_{ij})$. A straightforward calculation yields

$$\nabla^2 h = \begin{pmatrix} 2\delta_1 + \frac{1}{\mu}\sum_{j=2}^n a_j^2 & a_2 & \cdots & a_n \\ a_2 & 2\delta_1 + \mu & & \\ \vdots & & \ddots & \\ a_n & & & 2\delta_1 + \mu \end{pmatrix}.$$

The eigenvalues of (h_{ij}) are given by

$$2\delta_1,\ 2\delta_1 + \mu + \frac{1}{\mu}\sum_{j=2}^{n} a_j^2,\ 2\delta_1 + \mu, \ldots, 2\delta_1 + \mu.$$

This implies that h is uniformly convex in $\bar{\Omega}$ and

$$\det(h_{ij}) = 2\delta_1(2\delta_1 + \mu)^{n-2}\left(2\delta_1 + \mu + \frac{1}{\mu}\sum_{j=2}^{n} a_j^2\right).$$

Therefore, by choosing δ_1 small, we have

$$\det(h_{ij}) \le f \quad \text{in } \Omega.$$

Last, on $\partial\Omega \cap B_\varepsilon$, we require

$$-\delta_0 x_n + \delta_1 |x|^2 \ge 0,$$

or

$$x_n \le \frac{\delta_1}{\delta_0}|x|^2.$$

This can be achieved by taking δ_0 relatively small compared with δ_1. Then,

$$\widetilde{u} \le h \quad \text{on } \partial\Omega \cap B_\varepsilon.$$

This finishes the proof of (16). By Lemma 6.1.2, we have

$$\widetilde{u} \le h \quad \text{in } \Omega \cap B_\varepsilon.$$

Since $\widetilde{u}(0) = h(0) = 0$, we get

$$\widetilde{u}_n(0) \le h_n(0) = -\delta_0.$$

By (14), we obtain

$$u_{11}(0) = -\widetilde{u}_n(0)\rho_{11}(0) \ge \delta_0\rho_{11}(0).$$

Hence, (9′) is proved. □

Next, we derive a global estimate of the second derivatives.

Theorem 6.2.3. *Let Ω be a bounded uniformly convex domain in $\mathbb{R}^n$ with a C^2-boundary. Suppose that u is a uniformly convex $C^2(\bar{\Omega})$-solution of*

$$\det(u_{ij}) = f \quad \text{in } \Omega,$$
$$u = \varphi \quad \text{on } \partial\Omega,$$

for some $f \in C^2(\bar{\Omega})$ and $\varphi \in C^2(\bar{\Omega})$, with $f > 0$ in $\bar{\Omega}$. Then,

$$|\nabla^2 u|_{L^\infty(\Omega)} \le K,$$

where K is a positive constant depending only on Ω, the C^1-norm of u in $\bar{\Omega}$, the L^∞-norm of $\nabla^2 u$ on $\partial\Omega$, $\max_{\bar{\Omega}} f^{-1}$, *and the C^2-norm of f in $\bar{\Omega}$.*

Proof. We write the equation in the form

$$\log \det (u_{ij}) = \log f.$$

Consider any unit vector γ in $\mathbb{R}^n$. By differentiating the equation above twice with respect to x_γ, we have

$$Lu_{\gamma\gamma} \geq (\log f)_{\gamma\gamma} \geq -nC,$$

where C is a positive constant depending only on $\max_{\bar\Omega} f^{-1}$ and the C^2-norm of f in $\bar\Omega$. Since $Lu = n$, we get

$$L(u_{\gamma\gamma} + Cu) \geq 0.$$

By the maximum principle,

$$\sup_{\Omega} \big(u_{\gamma\gamma} + Cu\big) \leq \max_{\partial\Omega} \big(u_{\gamma\gamma} + Cu\big),$$

and hence

$$u_{\gamma\gamma} \leq K \quad \text{in } \Omega.$$

Since (u_{ij}) is positive definite, we have $u_{ii} > 0$ and then,

$$|u_{ij}| \leq K \quad \text{in } \Omega.$$

This finishes the proof. □

Now we can prove the main estimate in this section.

Theorem 6.2.4. *Let Ω be a bounded uniformly convex domain in $\mathbb{R}^n$ with a C^4-boundary. Suppose that u is a uniformly convex $C^4(\bar\Omega)$-solution of*

$$\begin{aligned} \det(u_{ij}) &= f \quad \text{in } \Omega, \\ u &= \varphi \quad \text{on } \partial\Omega, \end{aligned}$$

for some $f \in C^2(\bar\Omega)$ and $\varphi \in C^4(\bar\Omega)$, with $f > 0$ in $\bar\Omega$. Then,

$$|u|_{C^2(\bar\Omega)} \leq K,$$

where K is a positive constant depending only on Ω, the C^2-norm of f in $\bar\Omega$, $\max_{\bar\Omega} f^{-1}$, and the C^4-norm of φ in $\bar\Omega$.

Proof. Let $\underline{u}$ be a uniformly convex $C^4(\bar\Omega)$-function satisfying

$$\begin{aligned} \det(\underline{u}_{ij}) &\geq f \quad \text{in } \Omega, \\ \underline{u} &= \varphi \quad \text{on } \partial\Omega. \end{aligned}$$

To construct $\underline{u}$, we let σ be a uniformly convex $C^4(\bar\Omega)$-defining function of Ω; i.e., $\sigma < 0$ in Ω, and $\sigma = 0$ and $\nabla\sigma \neq 0$ on $\partial\Omega$. Then, we take $\underline{u} = \varphi + t\sigma$ for t sufficiently large. We have the desired estimate by combining Theorems 6.2.1, 6.2.2, and 6.2.3. □

Remark 6.2.5. Since the eigenvalues of $\nabla^2 u$ are bounded from above by K and their product is equal to f, we obtain a positive lower bound for each eigenvalue. Therefore, the linearized operator L is uniformly elliptic. Let T be an arbitrary constant directional derivative $T = \sum c_j\partial_j$ with $\sum c_j^2 = 1$. We have

$$\begin{aligned}L(T^2u) &= u^{ik}u^{jl}Tu_{ij}Tu_{kl} + T^2(\log f)\\ &\geq C_0\sum_{i,j}|Tu_{ij}|^2 - C,\end{aligned}$$

where C_0 and C are positive constants under control. In particular, we have a positive lower bound for C_0.

We now employ the method of continuity to solve the Dirichlet problem in a uniformly convex domain $\Omega \subset \mathbb{R}^n$.

Theorem 6.2.6. *Let $\alpha \in (0,1)$ be a constant and Ω be a bounded uniformly convex domain in $\mathbb{R}^n$ with a $C^{4,\alpha}$-boundary. Then for any $f \in C^{2,\alpha}(\bar{\Omega})$ with $f > 0$ in $\bar{\Omega}$ and any $\varphi \in C^{4,\alpha}(\bar{\Omega})$, there exists a unique uniformly convex solution $u \in C^{4,\alpha}(\bar{\Omega})$ of*

$$\begin{aligned}\det(u_{ij}) &= f \quad \text{in } \Omega,\\ u &= \varphi \quad \text{on } \partial\Omega.\end{aligned} \tag{1}$$

The proof is similar to that of Theorem 5.6.1.

Proof. Let $u^0 \in C^{4,\alpha}(\bar{\Omega})$ be a uniformly convex function satisfying

$$\begin{aligned}\det(u^0_{ij}) &\geq f \quad \text{in } \Omega,\\ u^0 &= \varphi \quad \text{on } \partial\Omega.\end{aligned}$$

Set $f^0 = \det(u^0_{ij})$. Then, $f^0 \geq f$ in Ω.

For each $t \in [0,1]$, we intend to find a uniformly convex solution $u^t \in C^{2,\alpha}(\bar{\Omega})$ of

$$\begin{aligned}\det(u^t_{ij}) &= tf + (1-t)f^0 \quad \text{in } \Omega,\\ u^t &= \varphi \quad \text{on } \partial\Omega.\end{aligned} \tag{$(1)_t$}$$

We set

$$I = \{t \in [0,1] : (1)_t \text{ has a uniformly convex solution } u^t \in C^{2,\alpha}(\bar{\Omega})\}.$$

Obviously $0 \in I$ since $(1)_0$ has a solution u^0.

Now we prove that I is open. Set

$$\mathcal{X} = \{w \in C^{2,\alpha}(\bar{\Omega}) : w = 0 \text{ on } \partial\Omega\}$$

and

$$G(w,t) = \det\big((w+\varphi)_{ij}\big) - tf - (1-t)f^0.$$

Then, $G : \mathcal{X} \to C^\alpha(\bar\Omega)$ is a C^1-map and its Frèchet derivative with respect to $w \in \mathcal{X}$ is given by

$$G_w(w,t)v = A^{ij}\partial_{ij}v,$$

where (A^{ij}) is the cofactor matrix of $((w+\varphi)_{ij})$. If $w+\varphi$ is a uniformly convex $C^{2,\alpha}(\bar\Omega)$-function, $G_w(w,t)$ is a uniformly elliptic linear operator with $C^\alpha(\bar\Omega)$-coefficients. By the Schauder theory, $G_w(w,t) : \mathcal{X} \to C^\alpha(\bar\Omega)$ is an invertible operator. Suppose $t_0 \in I$ and $u^{t_0} \in C^{2,\alpha}(\bar\Omega)$ is the uniformly convex solution of $(1)_{t_0}$. By writing $u^{t_0} = w^{t_0} + \varphi$ for some $w^{t_0} \in \mathcal{X}$, we have $G(w^{t_0}, t_0) = 0$. By the implicit function theorem, for any t close to t_0, there is a unique $w^t \in \mathcal{X}$, close to w^{t_0} in the $C^{2,\alpha}$-norm and satisfying $G(w^t, t) = 0$. Obviously, $w^t + \varphi$ is uniformly convex for t close to t_0. Then, $u^t = w^t + \varphi$ is the desired solution of $(1)_t$. Hence, $t \in I$ for all such t, and therefore I is open.

Next, we claim

$$|u^t|_{C^{2,\alpha}(\bar\Omega)} \le K, \quad \text{independent of } t. \tag{2}$$

Then, it follows that I is also closed, by the Arzela-Ascoli theorem, and therefore, I is the whole unit interval. The function u^1 is then our desired solution of (1).

To prove (2), we first note that, by Theorem 6.2.4,

$$|u^t|_{C^2(\bar\Omega)} \le K_2,$$

where K_2 is a positive constant depending only on Ω, the C^2-norms of f and f^0 in $\bar\Omega$, $\max_{\bar\Omega} f^{-1}$, and the C^4-norm of φ in $\bar\Omega$, independent of t. With this C^2-estimate, the ellipticity constants of the equation in $(1)_t$ are independent of t. Then, by Theorem 5.5.1, we have

$$|\nabla^2 u^t|_{C^\alpha(\bar\Omega)} \le K_{2,\alpha},$$

where $K_{2,\alpha}$ is a positive constant depending only on Ω, the C^2-norms of f and f^0 in $\bar\Omega$, $\max_{\bar\Omega} f^{-1}$, and the C^4-norm of φ in $\bar\Omega$, independent of t. This ends the proof of (2). We note that the concavity is needed in order to apply Theorem 5.5.1. □

6.3. Interior C^2-Estimates

As we pointed out in Section 6.1, the regularity of solutions of Monge-Ampère equations is based on estimates of the second derivatives. Once an interior sup-norm estimate of second derivatives is established, it follows the interior Hölder estimate of second derivatives. Then, the Schauder theory can be applied to the equation satisfied by first derivatives, and the higher regularity of solutions can be derived. In Section 6.2, global C^2-estimates of solutions are derived by the maximum principle with appropriate regularity assumptions on the boundary of domains and boundary values. In this section, we prove an interior estimate of second derivatives due to Pogorelov [**120**]. Different from uniformly elliptic equations, boundary conditions are needed here even for interior estimates. For simplicity, we discuss a special class of boundary values: the constant boundary values. In contrast to results in Section 6.2, there is no regularity assumption on the boundary of the domain.

Theorem 6.3.1. *Let Ω be a bounded convex domain in $\mathbb{R}^n$. Suppose that u is a convex $C(\bar{\Omega}) \cap C^4(\Omega)$-solution of*

$$\det(u_{ij}) = f \quad \text{in } \Omega,$$
$$u = 0 \quad \text{on } \partial\Omega,$$

for some $f \in C(\bar{\Omega}) \cap C^2(\Omega)$, with $f > 0$ in $\bar{\Omega}$. Then,

$$(-u)|\nabla^2 u| \le C \quad \text{in } \Omega,$$

where C is a positive constant depending only on $|u|_{C^1(\Omega)}$, $|f|_{C^2(\Omega)}$, and $\max_{\bar{\Omega}} f^{-1}$.

Proof. We prove, for any fixed unit vector γ,

$$-uu_{\gamma\gamma} \le C \quad \text{in } \Omega.$$

We note that $u < 0$ in Ω by the maximum principle and $u_{\gamma\gamma} \ge 0$ by the convexity. Now, we set

$$M = \sup_{\Omega} |\nabla u|^2.$$

Take some functions $\eta \in C(\bar{\Omega}) \cap C^2(\Omega)$ and $\psi \in C^2([0, M])$ such that $\eta > 0$ in Ω and $\psi, \psi' > 0$ in $[0, M]$. For some unit vector $\gamma \in \mathbb{R}^n$, we set

$$w = \eta\psi(|\nabla u|^2)u_{\gamma\gamma}.$$

By differentiating $\log w$, we have

$$(\log w)_i = \frac{\eta_i}{\eta} + \frac{2\psi'}{\psi} u_k u_{ki} + \frac{u_{\gamma\gamma i}}{u_{\gamma\gamma}}$$

and

$$\begin{aligned}(\log w)_{ij} = {} & \frac{\eta_{ij}}{\eta} - \frac{\eta_i\eta_j}{\eta^2} + 4\left(\frac{\psi''}{\psi} - \frac{\psi'^2}{\psi^2}\right) u_k u_l u_{ki} u_{lj} \\ & + \frac{2\psi'}{\psi}(u_k u_{kij} + u_{ki}u_{kj}) + \frac{u_{\gamma\gamma ij}}{u_{\gamma\gamma}} - \frac{u_{\gamma\gamma i}u_{\gamma\gamma j}}{u_{\gamma\gamma}^2}.\end{aligned}$$

We point out that γ is fixed and there is no summation for γ. In the following, we take

$$\eta = -u.$$

Then,

$$\frac{\eta_{ij}}{\eta} - \frac{\eta_i\eta_j}{\eta^2} = \frac{u_{ij}}{u} - \frac{u_iu_j}{u^2}.$$

Note that (u_{ij}) is positive definite in Ω. Hence,

$$\begin{aligned} u^{ij}(\log w)_{ij} &= \frac{n}{u} - \frac{1}{u^2}u^{ij}u_iu_j + 4\left(\frac{\psi''}{\psi} - \frac{\psi'^2}{\psi^2}\right)u_{ij}u_iu_j \\ &\quad + \frac{2\psi'}{\psi}(u_ku^{ij}u_{kij} + \Delta u) + \frac{1}{u_{\gamma\gamma}}u^{ij}u_{\gamma\gamma ij} - \frac{1}{u_{\gamma\gamma}^2}u^{ij}u_{\gamma\gamma i}u_{\gamma\gamma j}, \end{aligned}$$

where we used $u^{ij}u_{ij} = n$ and $u^{ij}u_{ki}u_{kj} = \delta_k^j u_{kj} = \Delta u$. Recall that

$$u^{ij}u_{kij} = (\log f)_k$$

and

$$u^{ij}u_{\gamma\gamma ij} = u^{ik}u^{jl}u_{ij\gamma}u_{kl\gamma} + (\log f)_{\gamma\gamma}.$$

By a simple substitution and rearrangement, we have

$$\begin{aligned} u^{ij}(\log w)_{ij} &= \frac{n}{u} + \frac{2\psi'}{\psi}\nabla u\cdot\nabla\log f + \frac{1}{u_{\gamma\gamma}}(\log f)_{\gamma\gamma} \\ &\quad - \frac{1}{u^2}u^{ij}u_iu_j + 4\left(\frac{\psi''}{\psi} - \frac{\psi'^2}{\psi^2}\right)u_{ij}u_iu_j + \frac{2\psi'}{\psi}\Delta u \\ &\quad + \frac{1}{u_{\gamma\gamma}}u^{ik}u^{jl}u_{ij\gamma}u_{kl\gamma} - \frac{1}{u_{\gamma\gamma}^2}u^{ij}u_{\gamma\gamma i}u_{\gamma\gamma j}. \end{aligned}$$

Now, we view $w = w(x,\gamma)$ as a function of $x \in \bar\Omega$ and $\gamma \in \mathbb{S}^{n-1} \subset \mathbb{R}^n$. Since $w(\cdot,\gamma) = 0$ on $\partial\Omega$ for any $\gamma \in \mathbb{S}^{n-1}$, we assume w attains its (positive) maximum at some $p_0 \in \Omega$ and some $\gamma \in \mathbb{S}^{n-1}$. In the following, we fix such a γ. Then, at p_0,

$$(\log w)_i = 0, \quad u^{ij}(\log w)_{ij} \le 0.$$

To simplify our calculation, we assume $\nabla^2 u$ is diagonal at p_0 by a rotation. For the fixed point p_0, $u_{\gamma\gamma}$ at p_0 is the maximum of all pure second derivatives of u at p_0. Then, γ has to be one of the coordinate directions. By abusing notations, we write $\gamma \in \{1,\ldots,n\}$. Then, at p_0, we have

$$\begin{aligned} u^{ij}(\log w)_{ij} &= \frac{n}{u} + \frac{2\psi'}{\psi}\nabla u\cdot\nabla\log f + \frac{1}{u_{\gamma\gamma}}(\log f)_{\gamma\gamma} \\ &\quad - \frac{1}{u^2}u^{ii}u_i^2 + 4\left(\frac{\psi''}{\psi} - \frac{\psi'^2}{\psi^2}\right)u_{ii}u_i^2 + \frac{2\psi'}{\psi}\Delta u \\ &\quad + \frac{1}{u_{\gamma\gamma}}u^{ii}u^{jj}u_{ij\gamma}^2 - \frac{1}{u_{\gamma\gamma}^2}u^{ii}u_{\gamma\gamma i}^2. \end{aligned} \tag{1}$$

In the following, all calculations are performed at p_0.

Set

$$I = \frac{1}{u_{\gamma\gamma}} u^{ii} u^{jj} u_{ij\gamma}^2 - \frac{1}{u_{\gamma\gamma}^2} u^{ii} u_{\gamma\gamma i}^2.$$

We point out that the summation is only for i and j, not for γ. In the first summation in I, we write $i = \gamma$ and $i \neq \gamma$. The term corresponding to $i = \gamma$ cancels the second term in I. This implies

$$I = \frac{1}{u_{\gamma\gamma}} \sum_{i \neq \gamma} u^{ii} u^{jj} u_{ij\gamma}^2.$$

Then, we only keep $j = \gamma$. Hence,

$$I \geq \sum_{i \neq \gamma} u^{ii} \frac{u_{\gamma\gamma i}^2}{u_{\gamma\gamma}^2}.$$

By $(\log w)_i = 0$ at p_0, we have

$$\frac{u_{\gamma\gamma i}}{u_{\gamma\gamma}} = -\left(\frac{u_i}{u} + \frac{2\psi'}{\psi} u_k u_{ki} \right) = -\left(\frac{u_i}{u} + \frac{2\psi'}{\psi} u_i u_{ii} \right).$$

Then,

$$I \geq \sum_{i \neq \gamma} \left(\frac{1}{u^2} u^{ii} u_i^2 + \frac{4\psi'}{\psi} \frac{u_i^2}{u} + \frac{4\psi'^2}{\psi^2} u_i^2 u_{ii} \right).$$

A simple substitution in (1) yields

$$\begin{aligned} u^{ij} (\log w)_{ij} \geq \frac{n}{u} &+ \frac{2\psi'}{\psi} \left(\nabla u \cdot \nabla \log f + \sum_{i \neq \gamma} \frac{2u_i^2}{u} \right) \\ &+ \frac{1}{u_{\gamma\gamma}} \left((\log f)_{\gamma\gamma} - \frac{u_\gamma^2}{u^2} \right) \\ &+ \frac{4\psi''}{\psi} u_{ii} u_i^2 - \frac{4\psi'^2}{\psi^2} u_{\gamma\gamma} u_\gamma^2 + \frac{2\psi'}{\psi} \Delta u. \end{aligned}$$

We note that $u < 0$ and $u_{ii} \geq 0$. If all ψ, ψ', and ψ'' are positive, by keeping only $u_{\gamma\gamma}$ for the second derivatives, we obtain

$$\begin{aligned} u^{ij} (\log w)_{ij} \geq \frac{n}{u} &+ \frac{2\psi'}{\psi} \left(\nabla u \cdot \nabla \log f + \frac{2|\nabla u|^2}{u} \right) \\ &+ \frac{1}{u_{\gamma\gamma}} \left((\log f)_{\gamma\gamma} - \frac{|\nabla u|^2}{u^2} \right) \\ &+ 2u_{\gamma\gamma} \left(\frac{\psi'}{\psi} + 2\left(\frac{\psi''}{\psi} - \frac{\psi'^2}{\psi^2} \right) u_\gamma^2 \right). \end{aligned} \tag{2}$$

Now we take, for some constant $a > 0$,

$$\psi(t) = \left(1 - \frac{t}{2M}\right)^{-a}.$$

Then, $1 \leq \psi \leq 2^a$ in $[0, M]$. Moreover,

$$\psi'(t) = \frac{a}{2M}\left(1 - \frac{t}{2M}\right)^{-a-1},$$

and

$$\psi''(t) = \frac{a(a+1)}{4M^2}\left(1 - \frac{t}{2M}\right)^{-a-2}.$$

Hence,

$$\frac{a}{2M} \leq \frac{\psi'}{\psi} \leq \frac{a}{M} \quad \text{on } [0, M], \tag{3}$$

and

$$\frac{\psi''}{\psi} - \frac{\psi'^2}{\psi^2} = \frac{a}{4M^2}\left(1 - \frac{t}{2M}\right)^{-2} \geq \frac{a}{4M^2} \geq 0 \quad \text{on } [0, M].$$

Then,

$$2\left(\frac{\psi'}{\psi} + 2\left(\frac{\psi''}{\psi} - \frac{\psi'^2}{\psi^2}\right)u_\gamma^2\right) \geq 2\frac{\psi'}{\psi} \geq \frac{a}{M}.$$

In the following, we take $a = 1$ and obtain

$$2\left(\frac{\psi'}{\psi} + 2\left(\frac{\psi''}{\psi} - \frac{\psi'^2}{\psi^2}\right)u_\gamma^2\right) \geq \frac{1}{M}. \tag{4}$$

By substituting (4) in (2), we obtain, at p_0,

$$\frac{u_{\gamma\gamma}}{M} + \frac{n}{u} + \frac{2\psi'}{\psi}\left(\nabla u \cdot \nabla \log f + \frac{2|\nabla u|^2}{u}\right) + \frac{1}{u_{\gamma\gamma}}\left((\log f)_{\gamma\gamma} - \frac{|\nabla u|^2}{u^2}\right) \leq 0.$$

Multiplying by $Mu^2 u_{\gamma\gamma}$ and using (3), we get, at p_0,

$$\begin{aligned} u^2 u_{\gamma\gamma}^2 - |u| u_{\gamma\gamma}\big(nM + 2|u\nabla u \cdot \nabla \log f| + 4|\nabla u|^2\big) \\ -M\big(|u^2 \nabla^2 \log f| + |\nabla u|^2\big) \leq 0. \end{aligned}$$

Set

$$F_1 = \sup_\Omega |u \nabla \log f|, \quad F_2 = \sup_\Omega |u^2 \nabla^2 \log f|.$$

Then, at p_0,

$$u^2 u_{\gamma\gamma}^2 - A|u| u_{\gamma\gamma} - B \leq 0,$$

where

$$\begin{aligned} A &= (n+4)M + 2\sqrt{M} F_1, \\ B &= M^2 + M F_2. \end{aligned}$$

Therefore, at p_0,

$$|u|u_{\gamma\gamma} \leq \frac{1}{2}\big(A + \sqrt{A^2 + 4B}\big) \leq A + \sqrt{B},$$

and hence, with the expressions of A and B,

$$\begin{aligned}|u|u_{\gamma\gamma} &\leq (n+4)M + 2\sqrt{M}F_1 + M + \sqrt{MF_2}\\ &= (n+5)M + \sqrt{M}(2F_1 + \sqrt{F_2}).\end{aligned}$$

Then, we obtain

$$w(p_0) \leq (2n+10)M + \sqrt{M}(4F_1 + 2\sqrt{F_2}),$$

or

$$\sup_{\Omega} w \leq (2n+10)M + \sqrt{M}(4F_1 + 2\sqrt{F_2}).$$

This yields the desired result. □

It is clear from the proof that the estimate of $\nabla^2 u$ has the form

$$\begin{aligned}\sup_{\Omega}(-u|\nabla^2 u|) \leq C_0\Bigg\{ &\sup_{\Omega}|\nabla u|^2 + \sup_{\Omega}|\nabla u|\\ &\cdot\left(\sup_{\Omega}|u\nabla \log f| + \sup_{\Omega}|u^2\nabla^2 \log f|^{\frac{1}{2}}\right)\Bigg\},\end{aligned}$$

where C_0 is a positive constant depending only on n. In particular, if f is a constant, then,

$$\sup_{\Omega}(-u|\nabla^2 u|) \leq C_0 \sup_{\Omega}|\nabla u|^2.$$

6.4. The Bernstein Problem

In this section, we discuss global convex solutions of the Monge-Ampère equations and prove a type of Bernstein theorem. This result was proved for smooth solutions by Jörgens [**93**] for $n = 2$, Calabi [**31**] for $n \leq 5$, and Pogorelov [**121**] for general n and by Cheng and Yau [**37**] by a different augment. Such a result was extended to viscosity solutions by Caffarelli. In this section, we present the proof by Caffarelli in the context of smooth solutions, following Caffarelli and Li [**24**].

An ellipsoid centered at the origin can be written as

$$E(A) = \{x \in \mathbb{R}^n : \langle Ax, x\rangle < 1\},$$

for some positive definite matrix A. Here, $\langle \cdot, \cdot \rangle$ is the standard inner product in $\mathbb{R}^n$. The volume of $E(A)$ is given by

$$|E(A)| = \frac{|B_1|}{\sqrt{\det(A)}}.$$

Throughout this section, we always write, for any subset $S \subset \mathbb{R}^n$ and any $\tau > 0$,

$$\tau S = \{\tau x : x \in S\}.$$

We first prove a normalization lemma due to John [**91**], which plays an important role in this section.

Lemma 6.4.1. *Let Ω be a bounded convex domain in $\mathbb{R}^n$, with $0 \in \Omega$ as the center of mass of Ω. Then, there exists an ellipsoid E with its center at the origin such that*

$$n^{-3/2} E \subset \Omega \subset E.$$

Moreover, E can be taken as the ellipsoid of minimum volume containing Ω with its center at the origin.

Proof. *Step 1.* Let $\mathcal{E}$ be the collection of all ellipsoids E with their centers at the origin and $\Omega \subset E$, and set

$$\tau = \inf_{E \in \mathcal{E}} |E|.$$

We claim that there exists an ellipsoid $E_0 \in \mathcal{E}$ such that $\tau = |E_0|$.

To prove the claim, we let $\mathcal{A}$ be the collection of positive definite matrices A such that $E(A) \in \mathcal{E}$. We fix an $R > 0$ such that $\overline{B}_R \subset \Omega$. Take any $A \in \mathcal{A}$. Then, $\overline{B}_R \subset E(A)$. For any unit vector $\xi \in \mathbb{R}^n$, we have $x = R\xi \in \overline{B}_R \subset E(A)$ and hence

$$\langle A\xi, \xi \rangle < \frac{1}{R^2}.$$

This implies, for any $A = (a_{ij}) \in \mathcal{A}$,

$$|a_{ij}| \leq \frac{1}{R^2}. \tag{1}$$

By the definition of τ, we have $\tau \geq |\Omega| > 0$ and there exists a sequence $A_m = (a_{ij}^m) \in \mathcal{A}$ such that $|E(A_m)| \to \tau$. By (1), there exists a convergent subsequence $a_{ij}^m \to a_{ij}^0$ as $m \to \infty$. The matrix $A_0 = (a_{ij}^0)$ is symmetric and positive semi-definite. Since $\tau > 0$, we have $\det A_0 > 0$ and then A_0 is positive definite. Hence, $E(A_0)$ is the desired ellipsoid.

Step 2. Let E be the ellipsoid constructed in Step 1; namely, E is the ellipsoid of the minimum volume among all ellipsoids with their centers at the origin and $\Omega \subset E$. By an affine transform, we assume that E is the unit ball B_1. Rotating the coordinates, we assume

$$\sigma \equiv \operatorname{dist}(0, \partial\Omega) = \operatorname{dist}(0, P),$$

for some point P on the positive x_n-axis. By the convexity of Ω, the hyperplane $\{x_n = \sigma\}$ is a supporting hyperplane to Ω at P. Let $Q \in \overline{\Omega}$ be a point with the largest distance to the hyperplane $\{x_n = \sigma\}$. Then, $Q \in \partial\Omega$ and a supporting hyperplane to Ω at Q is given by $\{x_n = -\mu\}$ for some $\mu > 0$.

Next, consider the subset $S = \Omega \cap \{x_n = 0\}$, and let $\mathcal{C}$ be the cone with its vertex Q passing through S and contained in $\{-\mu \leq x_n \leq \sigma\}$.

For any $t \in [-\mu, \sigma]$, let S_t be the slice of $\mathcal{C}$ through $(0, \ldots, 0, t)$ and perpendicular to the x_n-axis. The slice S_t is obtained by dilating S with respect to the point Q; i.e.,

$$S_t = \frac{t+\mu}{\mu} S.$$

Then,

$$\text{area}(S_t) = \left(\frac{t+\mu}{\mu}\right)^{n-1} \text{area}(S).$$

Let $\widehat{x} = (\widehat{x}_1, \ldots, \widehat{x}_n)$ be the center of mass of $\mathcal{C}$; i.e.,

$$\widehat{x} = \frac{1}{|\mathcal{C}|} \int_{\mathcal{C}} x \, dx.$$

By integrating on slices, we have

$$\widehat{x}_n = \frac{1}{|\mathcal{C}|} \text{area}(S) \int_{-\mu}^{\sigma} t \left(\frac{t+\mu}{\mu}\right)^{n-1} dt.$$

Since Ω has its center of mass at 0, $\mathcal{C} \cap \Omega$ has its center of mass located above S. Since $\mathcal{C} \cap \Omega \subset \mathcal{C}$, the cone $\mathcal{C}$ also has its center of mass above S; i.e., $\widehat{x}_n > 0$. Hence,

$$\int_{-\mu}^{\sigma} t \left(\frac{t+\mu}{\mu}\right)^{n-1} dt > 0.$$

An evaluation of this integral yields

$$\frac{\sigma}{\mu} \geq \frac{1}{n}. \tag{2}$$

Consider

$$\widehat{B} = B_1 \cap \{-\mu < x_n < \sigma\}.$$

It is obvious that $\Omega \subset \widehat{B}$ since $\Omega \subset B_1$. We claim that if $\mu < 1/\sqrt{n}$, then there exist constants a and b with $\mu < b < 1 < a$ such that the ellipsoid E_0 given by

$$E_0 = \left\{(x', x_n) : \frac{|x'|^2}{a^2} + \frac{x_n^2}{b^2} < 1\right\}$$

satisfies

$$\widehat{B} \subset E_0 \quad \text{and} \quad |E_0| < |B_1|. \tag{3}$$

This contradicts the fact that B_1 is the ellipsoid of the minimum volume and therefore we must have $\mu \geq 1/\sqrt{n}$. Hence, (2) yields $\sigma \geq n^{-3/2}$, and the theorem is proved.

We now prove (3). Since $\mu \geq \sigma$, we have, for any $x \in B_1$,

$$\frac{|x'|^2}{a^2} + \frac{x_n^2}{b^2} \leq \frac{1 - x_n^2}{a^2} + \frac{x_n^2}{b^2} = \frac{1}{a^2} + \left(\frac{1}{b^2} - \frac{1}{a^2}\right) x_n^2$$
$$\leq \frac{1}{a^2} + \left(\frac{1}{b^2} - \frac{1}{a^2}\right) \mu^2 = \frac{1-\mu^2}{a^2} + \frac{\mu^2}{b^2}.$$

Hence, $\widehat{B} \subset E_0$ if

$$\frac{1-\mu^2}{a^2} + \frac{\mu^2}{b^2} < 1,$$

which is equivalent to

$$a^2 > \frac{b^2(1-\mu^2)}{b^2 - \mu^2}.$$

Also, $|E_0| = a^{n-1}b|B_1|$, and hence $|E_0| < |B_1|$ is equivalent to $a^{n-1}b < 1$. We will choose a and b such that

$$\mu < b < 1 < a$$

and

(4) $$\frac{b^2(1-\mu^2)}{b^2-\mu^2} < a^2 < \left(\frac{1}{b}\right)^{\frac{2}{n-1}}.$$

We note that

(5) $$\frac{b^2(1-\mu^2)}{b^2-\mu^2} < \left(\frac{1}{b}\right)^{\frac{2}{n-1}}$$

if and only if

$$b^2 - \mu^2 - b^{\frac{2n}{n-1}}(1-\mu^2) > 0.$$

Consider the function

$$f(s) = s - \mu^2 - s^{\frac{n}{n-1}}(1-\mu^2).$$

We have $f(1) = 0$ and

$$f'(1) = 1 - \frac{n}{n-1}(1-\mu^2).$$

The assumption $\mu < 1/\sqrt{n}$ implies $f'(1) < 0$. Hence, $f(s) > 0$ for $s < 1$ and s close to 1. By taking $s = b^2$ with $\mu < b < 1$, we obtain (5). With such a choice of b, the expression in the left-hand side in (5) is greater than 1. Then, we choose $a > 1$ satisfying (4). This proves (3). □

In the rest of this section, we always write

$$\tau_n = n^{-3/2}.$$

A convex domain Ω is *normalized* if $0 \in \Omega$ is the center of mass of Ω and

$$B_{\tau_n} \subset \Omega \subset B_1.$$

Lemma 6.4.1 asserts that any bounded convex domain can be transformed to a normalized one by an affine transform.

For any function $u \in C(\bar{\Omega})$, we set

$$S_t = S_t(u) = \{x \in \Omega : u(x) < t\}.$$

This is the *level set* of u. If Ω is a convex domain and u is a convex function in Ω, then S_t is convex for any $\min_\Omega u < t \le \max_\Omega u$.

We first estimate slopes of convex functions in terms of distances to the boundary.

Lemma 6.4.2. *Let Ω be a convex domain and u be a C^1-convex function in Ω with $u \le 0$ on $\partial\Omega$. Then, for any $x \in \Omega$,*

$$|\nabla u(x)| \le \frac{-u(x)}{\text{dist}(x, \partial\Omega)}.$$

Proof. We fix a point $x \in \Omega$ and assume $\nabla u(x) \neq 0$. The convexity of u implies, for any $y \in \Omega$,

$$u(y) \ge u(x) + \nabla u(x) \cdot (y - x).$$

For any $0 < r < \text{dist}(x, \partial\Omega)$, take

$$y = x + r\frac{\nabla u(x)}{|\nabla u(x)|} \in \Omega.$$

Then,

$$0 \ge u(y) \ge u(x) + r|\nabla u(x)|.$$

We have the desired result by letting $r \to \text{dist}(x, \partial\Omega)$. □

Next, we prove several results concerning level sets of convex solutions of Monge-Ampère equations.

Lemma 6.4.3. *Let Ω be a normalized convex domain in $\mathbb{R}^n$. Suppose that u is a nonnegative convex $C(\bar{\Omega}) \cap C^4(\Omega)$-solution of*

$$\begin{aligned} \det \nabla^2 u &= 1 \quad \textit{in } \Omega, \\ u &= h \quad \textit{on } \partial\Omega, \\ u(x_0) &= 0 \quad \textit{for some } x_0 \in \Omega, \end{aligned}$$

for some positive constant h. Then,

(i) *there exist two positive constants c_1 and c_2, depending only on n, such that*

$$c_1 \le h \le c_2;$$

(ii) *there exists a constant $\tau_0 \in (0,1)$, depending only on n, such that $B_{\tau_0}(x_0) \subset \Omega$ and*

$$u < \frac{1}{2}h \quad \textit{in } B_{\tau_0}(x_0);$$

(iii) *there exists a constant* $\alpha \in (0,1)$, *depending only on* n, *such that*

$$C_1 I \le \nabla^2 u \le C_2 I \quad \text{in } S_{h/2}$$

and

$$[\nabla^2 u]_{C^\alpha(B_{\tau_0}(x_0))} \le C,$$

where C_1, C_2, *and* C *are positive constants depending only on* n.

Lemma 6.4.3(ii) asserts $B_{\tau_0}(x_0) \subset S_{h/2}$.

Proof. Note that u assumes its minimum at x_0. Since u is convex in Ω, $h = \max_{\bar\Omega} u$, $u < h$ in Ω, and hence $\Omega = S_h$.

(i) Consider, for any $x \in \mathbb{R}^n$,

$$v(x) = \frac{1}{2}(|x|^2 - 1) + h.$$

Then, v is convex in $\mathbb{R}^n$ and $\det \nabla^2 v = 1$. Note that $v = h$ on ∂B_1 and $v \le h$ on $\partial\Omega$, and hence $v \le u$ on $\partial\Omega$. Applying the comparison principle, provided by Lemma 6.1.2, to u and v in Ω, we obtain

$$v \le u \quad \text{in } \Omega.$$

By evaluating at x_0, we have

$$\frac{1}{2}(|x_0|^2 - 1) + h = v(x_0) \le u(x_0) = 0,$$

and hence

$$h \le \frac{1}{2}.$$

For the lower bound, we consider

$$w(x) = \frac{1}{2}(|x|^2 - \tau_n^2) + h.$$

Then, w is convex in $\mathbb{R}^n$ and $\det \nabla^2 w = 1$. Note that $w = h$ on ∂B_{τ_n} and $w \ge h$ on $\partial\Omega$, and hence $w \ge u$ on $\partial\Omega$. Applying the comparison principle to u and w in Ω, we obtain

$$w \ge u \quad \text{in } \Omega.$$

By evaluating at 0, we obtain

$$-\frac{1}{2}\tau_n^2 + h = w(0) \ge u(0) \ge 0,$$

and hence

$$h \ge \frac{1}{2}\tau_n^2.$$

(ii) By applying Lemma 1.2.3 to $h - u$ in S_h, we have, for any $x_* \in S_h$,

$$\big(h - u(x_*)\big)^n \le C(\operatorname{diam}(S_h))^{n-1} \operatorname{dist}(x_*, \partial S_h)|S_h|.$$

By taking $x_* \in \overline{S}_{h/2}$ with $\operatorname{dist}(x_*, \partial S_h) = \operatorname{dist}(S_{h/2}, \partial S_h)$ and $u(x_*) \le h/2$, we obtain

$$\left(\frac{h}{2}\right)^n \le C \operatorname{dist}(S_{h/2}, \partial S_h),$$

and hence

$$\operatorname{dist}(S_{h/2}, \partial S_h) \ge c. \tag{1}$$

A similar argument yields

$$\operatorname{dist}(S_{h/4}, \partial S_{h/2}) \ge c. \tag{2}$$

By applying Lemma 6.4.2 to u in S_h and (1), we have

$$|\nabla u| \le C \quad \text{in } S_{h/2}. \tag{3}$$

Since $x_0 \in S_{h/4}$, we have $B_c(x_0) \subset S_{h/2}$ by (2). For any $x \in \Omega$, the convexity of u implies

$$u(x) + \nabla u(x) \cdot (x_0 - x) \le u(x_0) = 0.$$

Hence, if $x \in B_{\tau_0}(x_0) \subset B_c(x_0) \subset S_{h/2}$, we have by (3)

$$u(x) \le \nabla u(x) \cdot (x - x_0) \le C\tau_0 < \frac{1}{2}h,$$

by taking τ_0 sufficiently small such that $C\tau_0 < h/2$ and $\tau_0 < c$.

(iii) Proceeding as in the proof of (3), we obtain

$$|\nabla u| \le C \quad \text{in } S_{3h/4}. \tag{4}$$

Next, consider

$$v = u - \frac{3}{4}h.$$

Then,

$$\begin{aligned} \det \nabla^2 v &= 1 \quad \text{in } S_{3h/4}, \\ v &= 0 \quad \text{on } \partial S_{3h/4}. \end{aligned}$$

Take any unit vector $\gamma \in \mathbb{R}^n$. Since $\nabla v = \nabla u$, (4) also holds for v. By applying Theorem 6.3.1 to v in $S_{3h/4}$, we have

$$\sup_{S_{3h/4}} (-v v_{\gamma\gamma}) \le C.$$

If $x \in S_{h/2}$, then

$$v(x) = u(x) - \frac{3}{4}h < \frac{1}{2}h - \frac{3}{4}h = -\frac{1}{4}h,$$

and hence

$$-v \ge \frac{1}{4}h \quad \text{in } S_{h/2}.$$

Therefore,

$$u_{\gamma\gamma} \le C \quad \text{in } S_{h/2}.$$

Let $\lambda_1, \ldots, \lambda_n$ be the eigenvalues of $\nabla^2 u$. Then, for $j = 1, \ldots, n$,

$$\lambda_j \leq C \quad \text{in } S_{h/2}. \tag{5}$$

Since $\det \nabla^2 u = 1$, we have

$$\lambda_1 \ldots \lambda_n = 1,$$

and hence

$$\lambda_j = (\lambda_1 \cdots \lambda_{j-1}\lambda_{j+1} \cdots \lambda_n)^{-1} \geq C^{-(n-1)} \quad \text{in } S_{h/2}. \tag{6}$$

By combining (5) and (6), we have the desired estimate of eigenvalues of $\nabla^2 u$.

By taking the log function of the equation, we write the equation as

$$\log(\det \nabla^2 u) = \log 1 = 0.$$

By what we just proved, this is uniformly elliptic in $S_{h/2}$ and, in particular, in $B_{\tau_0}(x_0)$. By Theorem 5.4.1, we obtain

$$[\nabla^2 u]_{C^\alpha(B_{\tau_0/2}(x_0))} \leq C,$$

where $C > 0$ and $\alpha \in (0, 1)$ are constants depending only on n. This is the desired result if we rename τ_0. □

We now prove an important result on the interior Hölder semi-norms of the second derivatives.

Theorem 6.4.4. *Let Ω be a bounded convex domain in $\mathbb{R}^n$. Suppose that u is a nonnegative convex $C(\bar{\Omega}) \cap C^4(\Omega)$-solution of*

$$\det \nabla^2 u = 1 \quad \text{in } \Omega,$$
$$u = 1 \quad \text{on } \partial\Omega.$$

Assume $0 \in \Omega$ such that

$$u(0) = 0 \quad \text{and} \quad \nabla^2 u(0) = I.$$

Then, there exist constants $\mu > 0$ and $\alpha \in (0, 1)$, depending only on n, such that $B_\mu \subset \Omega$ and

$$[\nabla^2 u]_{C^\alpha(B_\mu)} \leq C,$$

where C is a positive constant depending only on n.

Proof. Let $x_0 = (x_{0,1}, \ldots, x_{0,n}) \in \mathbb{R}^n$ be the center of mass of Ω and E be the ellipsoid of minimum volume containing Ω with its center at x_0. By taking an appropriate rotation O and considering $u(O^{-1}\cdot)$ instead, we assume that the axes of the ellipsoid E are parallel to the coordinate axes and that E is given by

$$E = \left\{ x : \sum_{i=1}^{n} \frac{(x_i - x_{0,i})^2}{a_i^2} < 1 \right\},$$

for some positive constants $a_1, \ldots, a_n$. Consider the matrix

$$A = \operatorname{diag}\left(\frac{1}{a_1}, \ldots, \frac{1}{a_n}\right)$$

and the affine transform

$$y = Tx = A(x - x_0).$$

Then, $T(E) = B_1$ and hence, by Lemma 6.4.1,

$$B_{\tau_n} \subset T(\Omega) \subset B_1.$$

Set

$$v(y) = hu(x) \quad \text{for any } y \in T(\Omega),$$

where h is chosen such that $h^n(\det A^{-1})^2 = 1$. Then

$$\det \nabla_y^2 v = h^n(\det A^{-1})^2 \det \nabla_x^2 u = 1.$$

The function v is convex and satisfies

$$\begin{aligned} \det \nabla^2 v &= 1 \quad \text{in } T(\Omega), \\ v &= h \quad \text{on } \partial(T(\Omega)), \\ v(y_0) &= 0, \end{aligned}$$

where $y_0 = -Ax_0 \in T(\Omega)$. We now apply Lemma 6.4.3 to v. By Lemma 6.4.3(i), we have

$$c_1 \leq h \leq c_2, \tag{1}$$

where c_1 and c_2 are positive constants depending only on n. By Lemma 6.4.3(ii), there exists a positive constant τ_0, depending only on n, such that

$$B_{\tau_0}(y_0) \subset \{y \in T(\Omega) : v(y) < h/2\}.$$

Then by Lemma 6.4.3(iii), we obtain

$$C_1 I \leq \nabla^2 v \leq C_2 I \quad \text{in } B_{\tau_0}(y_0), \tag{2}$$

where C_1 and C_2 are positive constants depending only on n, and

$$[\nabla_y^2 v]_{C^\alpha(B_{\tau_0}(y_0))} \leq C, \tag{3}$$

where $C > 0$ and $\alpha \in (0, 1)$ are constants depending only on n.

With $\nabla^2 u(0) = I$, we get

$$\nabla_y^2 v(y_0) = h(A^{-1})^T \nabla_x^2 u(0) A^{-1} = h(A^{-1})^T A^{-1}.$$

Then, (2) implies

$$C_1 I \leq h(A^{-1})^T A^{-1} \leq C_2 I.$$

By (1) and renaming C_1 and C_2, we have, for $i = 1, \ldots, n$,

$$C_1 \le a_i \le C_2, \tag{4}$$

and hence

$$T(B_{c_0\tau_0}) \subset B_{\tau_0}(y_0),$$

for some positive constant c_0 depending only on n. Note that

$$v_{y_iy_j}(y) = ha_ia_ju_{x_ix_j}(x).$$

By combining (1) and (4), we obtain

$$[\nabla_x^2 u]_{C^\alpha(B_{c_0\tau_0})} \le C.$$

This implies the desired result. □

We now prove the main result in this section.

Theorem 6.4.5. *Suppose that u is a convex $C^4(\mathbb{R}^n)$-solution of*

$$\det \nabla^2 u = 1 \quad \text{in } \mathbb{R}^n.$$

Then, u is a quadratic polynomial.

Proof. Without loss of generality, we assume

$$u(0) = 0, \quad \nabla u(0) = 0, \quad \nabla^2 u(0) = I. \tag{1}$$

In fact, by subtracting an appropriate linear function from u, we may assume $u(0) = 0$ and $\nabla u(0) = 0$. Next, by a rotation, we may assume $\nabla^2 u(0)$ is diagonal and hence is given by $\text{diag}(a_1, \ldots, a_n)$, for some positive constants $a_1, \ldots, a_n$, with $a_1 \cdots a_n = 1$. Last, by an appropriate scaling, we may assume $\nabla^2 u(0) = I$. Meanwhile, we preserve $\det \nabla^2 u = 1$ in $\mathbb{R}^n$.

By the convexity, u assumes its minimum at 0 and hence $u \ge 0$. For an arbitrarily fixed constant $R > 0$, we set

$$\Omega = \{x \in \mathbb{R}^n : u(x) < R^2\}.$$

Set $x = Ry$ and

$$v(y) = \frac{1}{R^2} u(x).$$

Moreover, set

$$\widetilde{\Omega} = \{y \in \mathbb{R}^n : x = Ry \in \Omega\}.$$

Then,

$$\begin{aligned} \det \nabla^2 v &= 1 \quad \text{in } \widetilde{\Omega}, \\ v &= 1 \quad \text{on } \partial\widetilde{\Omega}, \end{aligned}$$

and

$$v(0) = \min_{\widetilde{\Omega}} v = 0 \quad \text{and} \quad \nabla^2 v(0) = I.$$

By Theorem 6.4.4, $B_\mu \subset \widetilde{\Omega}$ and

$$[\nabla^2 v]_{C^\alpha(B_\mu)} \le C,$$

where $\mu > 0$, $\alpha \in (0,1)$, and $C > 0$ are constants depending only on n. Back to u, we have $B_{\mu R} \subset \Omega$ and

$$R^\alpha [\nabla^2 u]_{C^\alpha(B_{\mu R})} \le C.$$

Letting $R \to \infty$, we conclude that $\nabla^2 u$ is constant in $\mathbb{R}^n$. □

Chapter 7

Complex Monge-Ampère Equations

In this chapter, we extend results in the previous chapter to the complex case and discuss complex Monge-Ampère equations.

In Section 7.1, we discuss basic properties of complex Monge-Ampère equations. We prove a comparison principle and the higher regularity for their solutions.

In Section 7.2, we discuss the Dirichlet problem for complex Monge-Ampère equations in strongly pseudo-convex domains. We first derive a global C^2-estimate and then solve the Dirichlet problem by the method of continuity.

7.1. Basic Properties

In this section, we discuss basic properties of complex Monge-Ampère equations. This section is similar to Section 6.1.

Let $z_1, \ldots, z_n$ be complex coordinates in $\mathbb{C}^n$, with $x_j = \mathrm{Re}(z_j)$ and $y_j = \mathrm{Im}(z_j)$. If u is a real-valued C^2-function in a domain in $\mathbb{C}^n$, we define u_j and $u_{\bar{j}}$ by

$$u_j = u_{z_j} = \frac{1}{2}(u_{x_j} - \sqrt{-1}u_{y_j}),$$
$$u_{\bar{j}} = u_{\bar{z}_j} = \frac{1}{2}(u_{x_j} + \sqrt{-1}u_{y_j}).$$

We also define

$$|\nabla_{\mathbb{C}} u|^2 = \sum_{j=1}^{n} |u_j|^2.$$

Then,

$$|\nabla_{\mathbb{C}} u|^2 = \sum_{j=1}^{n} u_j u_{\bar{j}} = \frac{1}{4}\sum_{j=1}^{n}(u_{x_j}^2 + u_{y_j}^2).$$

We also note that

$$u_{j\bar{k}} = u_{z_j\bar{z}_k} = \frac{1}{4}(u_{x_jx_k} + u_{y_jy_k}) + \frac{\sqrt{-1}}{4}(u_{x_jy_k} - u_{x_ky_j})$$

and in particular

$$\sum_{j=1}^{n} u_{j\bar{j}} = \frac{1}{4}\sum_{j=1}^{n}(u_{x_jx_j} + u_{y_jy_j}).$$

The $n \times n$ matrix $(u_{j\bar{k}})$ is called the *complex Hessian matrix* of u, denoted by $\nabla_{\mathbb{C}}^2 u$.

For a real-valued function u, $u_{j\bar{k}} = \overline{u_{k\bar{j}}}$. In other words, the complex Hessian matrix is a Hermitian matrix. In this chapter, we discuss only real-valued functions.

Definition 7.1.1. A real-valued C^2-function u is *pluri-subharmonic* if $(u_{i\bar{j}})$ is positive semi-definite, and u is *strictly pluri-subharmonic* if $(u_{i\bar{j}})$ is positive definite.

Now we recall some terminology for complex matrices. For convenience, we use $a_{i\bar{j}}$ to denote the (i, j)-entry in an $n \times n$ matrix $(a_{i\bar{j}})$. A matrix $(a_{i\bar{j}})$ is Hermitian if $a_{i\bar{j}} = \overline{a_{j\bar{i}}}$. For a Hermitian matrix, the quadratic form

$$a_{i\bar{j}} t_i \bar{t}_j \quad \text{for } t = (t_1, \ldots, t_n) \in \mathbb{C}^n$$

is real. If we write

$$a_{i\bar{j}} = b_{ij} + \sqrt{-1} c_{ij},$$

then (b_{ij}) is a real symmetric matrix and (c_{ij}) is a real skew symmetric matrix. With $t_i = \xi_i + \sqrt{-1}\eta_i$ for $\xi_i, \eta_i \in \mathbb{R}$, we have

$$a_{i\bar{j}} t_i \bar{t}_j = b_{ij}(\xi_i\xi_j + \eta_i\eta_j) + c_{ij}(\xi_i\eta_j - \eta_i\xi_j).$$

A Hermitian matrix $(a_{i\bar{j}})$ is positive definite if, for any $t \in \mathbb{C}^n \setminus \{0\}$,

$$a_{i\bar{j}} t_i \bar{t}_j > 0.$$

For a positive definite matrix $(a_{i\bar{j}})$, we have obviously $a_{i\bar{i}} > 0$. Moreover, any diagonal 2×2 minors have positive determinants. Hence,

$$|a_{i\bar{j}}|^2 \le a_{i\bar{i}} a_{j\bar{j}},$$

and then

$$|a_{i\bar{j}}| \leq \frac{1}{2}(a_{i\bar{i}} + a_{j\bar{j}}).$$

Suppose $(a_{i\bar{j}})$ is a Hermitian matrix. Define

$$L = a_{i\bar{j}}\partial_{i\bar{j}}.$$

This is a linear operator in $\mathbb{R}^{2n}$ with real coefficients. If $(a_{i\bar{j}})$ is positive definite, then L is elliptic. To check this, recall that

$$\partial_{i\bar{j}} = \frac{1}{4}\left((\partial_{x_ix_j} + \partial_{y_iy_j}) + \sqrt{-1}(\partial_{x_iy_j} - \partial_{y_ix_j})\right).$$

With $a_{i\bar{j}} = b_{ij} + \sqrt{-1}c_{ij}$, we have

$$L = a_{i\bar{j}}\partial_{i\bar{j}} = \frac{1}{4}\left(b_{ij}(\partial_{x_ix_j} + \partial_{y_iy_j}) - c_{ij}(\partial_{x_iy_j} - \partial_{y_ix_j})\right).$$

For any $\xi \in \mathbb{R}^n$ (corresponding to x) and $\eta \in \mathbb{R}^n$ (corresponding to y), we get

$$b_{ij}(\xi_i\xi_j + \eta_i\eta_j) - c_{ij}(\xi_i\eta_j - \xi_j\eta_i) = a_{i\bar{j}}(\xi_i - \sqrt{-1}\eta_i)(\xi_j + \sqrt{-1}\eta_j).$$

It is positive for nonzero $(\xi, \eta) \in \mathbb{R}^{2n}$.

For some strictly pluri-subharmonic function u, it is convenient to introduce

$$F(\nabla^2_{\mathbb{C}}u) = \log\det(u_{i\bar{j}}).$$

We claim

$$F_{i\bar{j}} \equiv \partial_{u_{i\bar{j}}}F = u^{i\bar{j}},$$

$$F_{i\bar{j},k\bar{l}} \equiv \partial_{u_{i\bar{j}}u_{k\bar{l}}}F = -u^{i\bar{l}}u^{k\bar{j}},$$

where $(u^{i\bar{j}})$ denotes the inverse of the complex Hessian matrix $H = (u_{i\bar{j}})$. To check this, we denote by $A = (A^{i\bar{j}})$ the cofactor matrix of H; i.e., $A = (\det H)H^{-1}$. Note that, for each fixed $i = 1, \ldots, n$,

$$\det(u_{i\bar{j}}) = A^{i\bar{1}}u_{i\bar{1}} + \cdots + A^{i\bar{n}}u_{i\bar{n}}.$$

Then,

$$F_{i\bar{j}} = \frac{1}{\det(u_{i\bar{j}})}A^{i\bar{j}} = u^{i\bar{j}}.$$

Next, for each fixed $i, j = 1, \ldots, n$, we have

$$u^{i\bar{k}}u_{j\bar{k}} = \delta^i_j.$$

Differentiating with respect to $u_{p\bar{q}}$, we get

$$(u^{i\bar{k}})_{u_{p\bar{q}}}u_{j\bar{k}} + u^{i\bar{k}}(u_{j\bar{k}})_{u_{p\bar{q}}} = 0.$$

Multiplying by $u^{j\bar{l}}$ and summing over j, we obtain

$$(u^{i\bar{l}})_{u_{p\bar{q}}} = (u^{i\bar{k}})_{u_{p\bar{q}}} u_{j\bar{k}} u^{j\bar{l}} = -u^{i\bar{k}} u^{j\bar{l}} (u_{j\bar{k}})_{u_{p\bar{q}}} = -u^{i\bar{q}} u^{p\bar{l}},$$

or

$$\partial_{u_{k\bar{l}}} u^{i\bar{j}} = -u^{i\bar{l}} u^{k\bar{j}}.$$

Hence,

$$F_{i\bar{j},k\bar{l}} = \partial_{u_{k\bar{l}}} u^{i\bar{j}} = -u^{i\bar{l}} u^{k\bar{j}}.$$

Therefore, for any Hermitian matrix $M = (m_{i\bar{j}})$, we obtain

$$F_{i\bar{j},k\bar{l}} m_{i\bar{j}} m_{k\bar{l}} \leq 0.$$

Now, we consider the *complex Monge-Ampère equation* for a strictly pluri-subharmonic function u:

$$\det(u_{i\bar{j}}) = f.$$

As in the real case, we write it as

$$\log\det(u_{i\bar{j}}) = \log f.$$

For complex constants c_j and d_j, $j = 1, \ldots, n$, set

$$\gamma = \sum_{j=1}^{n} (c_j \partial_j + d_j \partial_{\bar{j}}).$$

This is a first-order linear operator with constant coefficients. A simple differentiation yields

$$u^{i\bar{j}} u_{i\bar{j}\gamma} = (\log f)_\gamma.$$

This leads to the linear differential operator

$$L = u^{i\bar{j}} \partial_{i\bar{j}}.$$

Since u is strictly pluri-subharmonic, L is elliptic. We get

$$Lu_\gamma = (\log f)_\gamma.$$

Differentiating with respect to $\bar{\gamma}$, we have

$$Lu_{\gamma\bar{\gamma}} - u^{i\bar{l}} u^{k\bar{j}} u_{i\bar{j}\gamma} u_{k\bar{l}\bar{\gamma}} = (\log f)_{\gamma\bar{\gamma}}.$$

The second term in the left-hand side is nonnegative. Hence, we conclude

$$Lu_{\gamma\bar{\gamma}} \geq (\log f)_{\gamma\bar{\gamma}}.$$

In the following, we always choose a special class of γ. For real constants a_j and b_j, $j = 1, \ldots, n$, set

$$\gamma = \sum_{j=1}^{n} (a_j \partial_{x_j} + b_j \partial_{y_j}).$$

A simple calculation yields

$$\gamma = \sum_{j=1}^{n} (c_j \partial_j + \bar{c}_j \partial_{\bar{j}}),$$

where $c_j = a_j + \sqrt{-1} b_j$. Hence,

$$\sum_{j=1}^{n} |c_j|^2 = \sum_{j=1}^{n} (a_j^2 + b_j^2).$$

We now discuss the change of the complex Monge-Ampère operator under a holomorphic change of variables. Suppose $w = w(z)$ is a holomorphic change of variables. Then,

$$u_{z_i} = u_{w_k} \partial_{z_i} w_k,$$

and

$$u_{z_i \bar{z}_j} = u_{w_k \bar{w}_l} \partial_{\bar{z}_j} \bar{w}_l \partial_{z_i} w_k = \overline{\partial_{z_j} w_l} u_{w_k \bar{w}_l} \partial_{z_i} w_k.$$

Therefore, we obtain

$$\det(u_{z_i \bar{z}_j}) = |\det(w_z)|^2 \det(u_{w_k \bar{w}_l}).$$

Now, we prove a comparison principle for complex Monge-Ampère equations.

Lemma 7.1.2. *Let Ω be a bounded domain in $\mathbb{C}^n$. Suppose that u, v are pluri-subharmonic $C(\bar{\Omega}) \cap C^2(\Omega)$-functions and satisfy* $\det(u_{i\bar{j}}) \geq \det(v_{i\bar{j}})$ *in Ω and $u \leq v$ on $\partial\Omega$. Then, $u \leq v$ in Ω.*

Proof. First, we assume that u is strictly pluri-subharmonic. Then,

$$\begin{aligned} \det(u_{i\bar{j}}) - \det(v_{i\bar{j}}) &= \int_0^1 \frac{d}{dt} \det[(tu + (1-t)v)_{i\bar{j}}] dt \\ &= \int_0^1 B^{i\bar{j}}(t) dt (u - v)_{i\bar{j}}, \end{aligned}$$

where $\big(B^{i\bar{j}}(t)\big)$ is the cofactor matrix of $\big(tu_{i\bar{j}} + (1-t)v_{i\bar{j}}\big)$. Hence, $L(u-v) \geq 0$ in Ω for some linear uniformly elliptic operator L since u is strictly pluri-subharmonic. We have the desired result by the maximum principle.

If u is only pluri-subharmonic, we consider, for some positive constant ε,

$$u_\varepsilon = u + \varepsilon(|z|^2 - \max_{\partial\Omega} |z|^2).$$

Then, u_ε is strictly pluri-subharmonic. Also, $\det(\partial_{i\bar{j}} u_\varepsilon) \geq \det(v_{i\bar{j}})$ in Ω and $u_\varepsilon \leq v$ on $\partial\Omega$. By what we just proved, we have $u_\varepsilon \leq v$ in Ω. Hence, we get the desired result by letting $\varepsilon \to 0$. □

The higher regularity of solutions can also be established once they are known to be $C^{2,\alpha}$. The following results are simple corollaries of Proposition 5.1.9 and Proposition 5.1.10.

Proposition 7.1.3. *Let $\alpha \in (0,1)$ be a constant and Ω be a bounded domain in $\mathbb{C}^n$. Suppose that u is a strictly pluri-subharmonic $C^{2,\alpha}(\Omega)$-solution of*

$$\det(u_{i\bar{j}}) = f \quad \text{in } \Omega,$$

for some $f \in C^{\alpha}(\Omega)$, with $f > 0$ in Ω. For any integer $m \geq 1$, if $f \in C^{m,\alpha}(\Omega)$, then $u \in C^{m+2,\alpha}(\Omega)$. In particular, if $f \in C^{\infty}(\Omega)$, then $u \in C^{\infty}(\Omega)$.

Proposition 7.1.4. *Let $\alpha \in (0,1)$ be a constant and Ω be a bounded domain in $\mathbb{C}^n$ with a $C^{2,\alpha}$-boundary. Suppose that u is a strictly pluri-subharmonic $C^{2,\alpha}(\bar{\Omega})$-solution of*

$$\begin{aligned} \det(u_{i\bar{j}}) &= f \quad \text{in } \Omega, \\ u &= \varphi \quad \text{on } \partial\Omega, \end{aligned}$$

for some $f \in C^{\alpha}(\bar{\Omega})$ and $\varphi \in C^{2,\alpha}(\bar{\Omega})$, with $f > 0$ in $\bar{\Omega}$. For any integer $m \geq 1$, if $\partial\Omega \in C^{m+2,\alpha}$, $f \in C^{m,\alpha}(\bar{\Omega})$, and $\varphi \in C^{m+2,\alpha}(\bar{\Omega})$, then $u \in C^{m+2,\alpha}(\bar{\Omega})$. In particular, if $\partial\Omega \in C^{\infty}$, $f \in C^{\infty}(\bar{\Omega})$, and $\varphi \in C^{\infty}(\bar{\Omega})$, then $u \in C^{\infty}(\bar{\Omega})$.

7.2. Global C^2-Estimates

In this section, we discuss the Dirichlet problem for the complex Monge-Ampère equation. This section is similar to Section 6.2.

Let Ω be a bounded domain in $\mathbb{C}^n$ with a C^k-boundary, for some positive integer k. Then, $\Omega \subset \mathbb{C}^n$ is called *strongly pseudo-convex* if there exists a strictly pluri-subharmonic $C^k(\bar{\Omega})$-function σ such that $\sigma < 0$ in Ω and $\sigma = 0$, $\nabla\sigma \neq 0$ on $\partial\Omega$. Such a σ is called a *defining function* of Ω.

We now study the Dirichlet problem for the complex Monge-Ampère equation. Let Ω be a bounded strongly pseudo-convex domain in $\mathbb{C}^n$, f be a continuous function in $\bar{\Omega}$, and φ be a continuous function on $\partial\Omega$. We consider

$$\begin{aligned} \det(u_{i\bar{j}}) &= f \quad \text{in } \Omega, \\ u &= \varphi \quad \text{on } \partial\Omega. \end{aligned}$$

As in the real case, we reduce the solvability to $C^{2,\alpha}$-a priori estimates. We will derive a global C^2-estimate through a series of theorems.

We first derive the global C^1-estimate.

Theorem 7.2.1. *Let Ω be a bounded strongly pseudo-convex domain in $\mathbb{C}^n$ with a C^2-boundary. Suppose that u and u^0 are strictly pluri-subharmonic $C^1(\bar{\Omega}) \cap C^2(\Omega)$-functions and satisfy*

$$\det(u_{i\bar{j}}) = f \leq \det(u^0_{i\bar{j}}) \quad \text{in } \Omega,$$
$$u = u^0 = \varphi \quad \text{on } \partial\Omega,$$

for some $f \in C^1(\bar{\Omega})$ and $\varphi \in C^2(\bar{\Omega})$, with $f > 0$ in $\bar{\Omega}$. Then,

$$|u|_{C^1(\bar{\Omega})} \leq K,$$

where K is a positive constant depending only on Ω, the C^1-norm of u^0 in $\bar{\Omega}$, $\max_{\bar{\Omega}} f^{-1}$, the C^1-norm of f in $\bar{\Omega}$, and the C^2-norm of φ in $\bar{\Omega}$.

Proof. Note that the pluri-subharmonic function u is subharmonic. By the maximum principle, we have

$$u \leq \max_{\partial\Omega} \varphi.$$

Next, Lemma 7.1.2 implies

$$u \geq u^0 \quad \text{in } \Omega.$$

We conclude

$$|u|_{L^\infty(\Omega)} \leq K_0, \tag{1}$$

where K_0 is a positive constant depending only on $\max_{\partial\Omega} \varphi$ and $\min_{\bar{\Omega}} u^0$.

Let h be the harmonic function in Ω which equals φ on $\partial\Omega$. The maximum principle implies $u \leq h$ in Ω. Then,

$$u^0 \leq u \leq h \quad \text{in } \Omega.$$

Since these three functions have common values on $\partial\Omega$, then

$$\frac{\partial h}{\partial \nu} \leq \frac{\partial u}{\partial \nu} \leq \frac{\partial u^0}{\partial \nu} \quad \text{on } \partial\Omega,$$

where ν is the exterior unit normal to $\partial\Omega$. Therefore, we obtain

$$|\nabla u| \leq K_1 \quad \text{on } \partial\Omega, \tag{2}$$

where K_1 is a positive constant depending only on the $L^\infty(\partial\Omega)$-norms of ∇u^0 and ∇h, the latter of which depends on Ω and the C^2-norm of φ by Theorem 1.1.14.

We write the equation in the form

$$\log \det(u_{i\bar{j}}) = \log f.$$

Set

$$L = u^{i\bar{j}}\partial_{i\bar{j}},$$

where $(u^{i\bar{j}})$ is the inverse of the matrix $(u_{i\bar{j}})$. Hence, L is elliptic since u is strictly pluri-subharmonic.

By $\det(u^{i\bar{j}}) = f^{-1}$ and the inequality for arithmetic and geometric means, we get

$$\frac{1}{n}\sum_{j=1}^{n} u^{j\bar{j}} \geq f^{-\frac{1}{n}}.$$

Let a_j and b_j be real constants, $j = 1, \ldots, n$, with

$$\sum_{j=1}^{n}(a_j^2 + b_j^2) = 1,$$

and set

$$\gamma = \sum_{j=1}^{n}(a_j\partial_{x_j} + b_j\partial_{y_j}).$$

Then,

$$Lu_\gamma = (\log f)_\gamma.$$

For some positive constant μ to be determined, we consider the function

$$w = \pm u_\gamma + e^{\mu|z|^2}.$$

Note that

$$L(e^{\mu|z|^2}) = e^{\mu|z|^2}\left(\mu\sum_{j=1}^{n} u^{j\bar{j}} + \mu^2 u^{i\bar{j}}\bar{z}_i z_j\right) \geq n\mu e^{\mu|z|^2} f^{-\frac{1}{n}}.$$

Hence,

$$Lw \geq \pm(\log f)_\gamma + n\mu e^{\mu|z|^2} f^{-\frac{1}{n}} \geq 0$$

if we choose μ large enough. Then, the maximum principle implies

$$\sup_\Omega w \leq \max_{\partial\Omega} w,$$

and hence

$$\max_{\bar{\Omega}} |u_\gamma| \leq \max_{\partial\Omega} |u_\gamma| + C. \tag{3}$$

With (1), (2), and (3), we obtain

$$|u|_{C^1(\bar{\Omega})} \leq K.$$

This is the desired estimate. □

Next, we derive the fundamental boundary estimate of the second derivatives due to Caffarelli, Kohn, Nirenberg, and Spruck [**23**]. We present two proofs for the second normal derivatives, the first proof by Guan [**63**] and the second the original proof by Caffarelli, Kohn, Nirenberg, and Spruck.

Theorem 7.2.2. *Let Ω be a bounded strongly pseudo-convex domain in $\mathbb{C}^n$ with a C^4-boundary. Suppose that u is a strictly pluri-subharmonic $C^4(\bar{\Omega})$-solution of*

$$\begin{aligned} \det(u_{i\bar{j}}) &= f \quad \text{in } \Omega, \\ u &= \varphi \quad \text{on } \partial\Omega, \end{aligned}$$

for some $f \in C^2(\bar{\Omega})$ and $\varphi \in C^4(\bar{\Omega})$, with $f > 0$ in $\bar{\Omega}$. Then,

$$|\nabla^2 u|_{L^\infty(\partial\Omega)} \leq K,$$

where K is a positive constant depending only on Ω, the C^1-norm of u in $\bar{\Omega}$, $\max_{\bar{\Omega}} f^{-1}$, the C^2-norm of f in $\bar{\Omega}$, and the C^4-norm of φ in $\bar{\Omega}$.

Proof. In the following, a universal constant is a positive constant depending only on Ω, the C^1-norm of u in $\bar{\Omega}$, $\max_{\bar{\Omega}} f^{-1}$, the C^2-norm of f in $\bar{\Omega}$, and the C^4-norm of φ in $\bar{\Omega}$. As in the real case, we write the equation in the form

$$\log\det(u_{i\bar{j}}) = \log f.$$

Set

$$L = u^{i\bar{j}}\partial_{i\bar{j}},$$

where $(u^{i\bar{j}})$ is the inverse of the matrix $(u_{i\bar{j}})$. Then, L is elliptic since u is strictly pluri-subharmonic. By $\det(u^{i\bar{j}}) = f^{-1}$ and the inequality for arithmetic and geometric means, we get

$$\frac{1}{n}\sum_{i=1}^n u^{i\bar{i}} \geq f^{-\frac{1}{n}}. \tag{1}$$

Consider any boundary point; without loss of generality, we take it to be the origin and we take the x_n-axis to be the interior normal. Let σ be a strictly pluri-subharmonic C^4-defining function of Ω. Then, $\sigma_{z_j}(0) = 0$ for $j < n$ and $\sigma_{y_n}(0) = 0$. We also assume $\sigma_{x_n}(0) = -1$. In the following, we set $t' = (t_1, \ldots, t_{2n-1})$, with

$$\begin{aligned} &t_1 = x_1, \quad t_2 = y_1, \quad \ldots, \\ &t_{2n-3} = x_{n-1}, \quad t_{2n-2} = y_{n-1}, \quad t_{2n-1} = y_n. \end{aligned}$$

Writing the Taylor expansion of σ up to the second order, we obtain

$$\sigma = \operatorname{Re}\Big(- z_n + \sum_{i,j=1}^{n} a_{ij} z_i z_j \Big) + \sum_{i,j=1}^{n} b_{i\bar{j}} z_i \bar{z}_j + O(|z|^3).$$

This expression holds since σ is real-valued. Note that $\{a_{ij}\}$ has a uniform bound, independent of the points on $\partial\Omega$. Introducing new coordinates of the form

$$z'_k = z_k \quad \text{for } k = 1, \ldots, n-1,$$

$$z'_n = z_n - \sum_{i,j=1}^{n} a_{ij} z_i z_j,$$

we can write

$$\sigma = -\operatorname{Re}(z'_n) + \sum_{i,j=1}^{n} c_{i\bar{j}} z'_i \bar{z}'_j + O(|z'|^3).$$

Furthermore, at points where the Jacobian of the transform does not vanish, a function is pluri-subharmonic with respect to $\{z_j\}$ if and only if it is pluri-subharmonic with respect to $\{z'_j\}$, and $\det(u_{z_i\bar{z}_j})$ differs from $\det(u_{z'_i\bar{z}'_j})$ by a positive factor, given by the absolute value squared of the Jacobian. Thus, we drop the primes and assume that, in a neighborhood of 0, σ is of the form

$$\sigma = -\operatorname{Re}(z_n) + \sum_{i,j=1}^{n} c_{i\bar{j}} z_i \bar{z}_j + O(|z|^3). \tag{2}$$

Since σ is strictly pluri-subharmonic, $(c_{i\bar{j}})$ is positive definite. In a neighborhood of 0, $\partial\Omega$ can be expressed by

$$x_n = \rho(t') = \sum_{\alpha,\beta=1}^{2n-1} b_{\alpha\beta} t_\alpha t_\beta + O(|t'|^3), \tag{3}$$

where $(b_{\alpha\beta})$ is a (real) positive definite matrix. To check this, we simply note that

$$\sum_{\alpha,\beta=1}^{2n-1} b_{\alpha\beta} t_\alpha t_\beta = \sum_{i,j=1}^{n} c_{i\bar{j}} z_i \bar{z}_j,$$

for

$$z_1 = t_1 + \sqrt{-1}t_2, \ \ldots, \ z_{n-1} = t_{2n-3} + \sqrt{-1}t_{2n-2}, \ z_n = \sqrt{-1}t_{2n-1}.$$

Step 1. We first estimate $u_{t_\alpha t_\beta}(0)$ for $\alpha, \beta = 1, \ldots, 2n-1$. On $\partial\Omega$, we have

$$u - \varphi = 0,$$

or, for small t',

$$(u - \varphi)(t', \rho(t')) = 0.$$

Recall that φ is defined in $\bar{\Omega}$. By differentiating with respect to t_α and then t_β,

$$(\partial_{t_\alpha} + \rho_{t_\alpha}\partial_{x_n})(u - \varphi) = 0 \quad \text{on } \partial\Omega,$$

and

$$(\partial_{t_\beta} + \rho_{t_\beta}\partial_{x_n})(\partial_{t_\alpha} + \rho_{t_\alpha}\partial_{x_n})(u - \varphi) = 0 \quad \text{on } \partial\Omega.$$

Note that $\rho_{t_\alpha}(0) = 0$ and $\rho_{t_\alpha t_\beta}(0) = b_{\alpha\beta}$. Hence at 0, we obtain

$$(u - \varphi)_{t_\alpha t_\beta}(0) + b_{\alpha\beta}(u - \varphi)_{x_n}(0) = 0.$$

We have, for any $\alpha, \beta = 1, \ldots, 2n - 1$,

$$|u_{t_\alpha t_\beta}(0)| \le K_1, \tag{4}$$

for a universal constant K_1.

Step 2. Next, we fix an $\alpha = 1, \ldots, 2n - 1$ and estimate the mixed derivatives $u_{t_\alpha x_n}(0)$. We define T_α in a neighborhood of 0 by

$$T_\alpha = \partial_{t_\alpha} + \rho_{t_\alpha}\partial_{x_n}. \tag{5}$$

Note that T_α is a tangential vector to $\partial\Omega$. We first calculate $L(T_\alpha u)$. A straightforward calculation yields

$$\begin{aligned} L(T_\alpha u) &= u^{p\bar{q}}\partial_{p\bar{q}}(u_{t_\alpha} + \rho_{t_\alpha}u_{x_n}) \\ &= u^{p\bar{q}}\partial_{p\bar{q}}u_{t_\alpha} + \rho_{t_\alpha}u^{p\bar{q}}\partial_{p\bar{q}}u_{x_n} \\ &\quad + u^{p\bar{q}}\rho_{pt_\alpha}u_{x_n\bar{q}} + u^{p\bar{q}}\rho_{\bar{q}t_\alpha}u_{x_n p} + u^{p\bar{q}}\rho_{p\bar{q}t_\alpha}u_{x_n}. \end{aligned}$$

Hence,

$$L(T_\alpha u) = T_\alpha \log f + u^{p\bar{q}}\rho_{pt_\alpha}u_{x_n\bar{q}} + u^{p\bar{q}}\rho_{\bar{q}t_\alpha}u_{x_n p} + u^{p\bar{q}}\rho_{p\bar{q}t_\alpha}u_{x_n}. \tag{6}$$

Observe that $u^{p\bar{q}}u_{n\bar{q}} = \delta_n^p$ and

$$\partial_{z_n} = \frac{1}{2}\left(\partial_{x_n} - \sqrt{-1}\partial_{y_n}\right),$$

or

$$\partial_{x_n} = 2\partial_{z_n} + \sqrt{-1}\partial_{y_n},$$

so that

$$u_{x_n\bar{q}} = 2u_{n\bar{q}} + \sqrt{-1}u_{y_n\bar{q}}.$$

Thus, the second term in the right-hand side in (6) can be written as

$$\begin{aligned} u^{p\bar{q}}\rho_{pt_\alpha}u_{x_n\bar{q}} &= 2\rho_{pt_\alpha}u^{p\bar{q}}u_{n\bar{q}} + \sqrt{-1}u^{p\bar{q}}\rho_{pt_\alpha}u_{y_n\bar{q}} \\ &= 2\rho_{nt_\alpha} + \sqrt{-1}u^{p\bar{q}}\rho_{pt_\alpha}u_{y_n\bar{q}}. \end{aligned}$$

Here, the derivatives of ρ are bounded. We also have

$$|u^{p\bar{q}}\rho_{pt_\alpha}u_{y_n\bar{q}}| \le (u^{p\bar{q}}\rho_{pt_\alpha}\rho_{\bar{q}t_\alpha})^{\frac{1}{2}}(u^{p\bar{q}}u_{y_n p}u_{y_n\bar{q}})^{\frac{1}{2}}.$$

This follows from the Cauchy inequality and the fact that the positive matrix $(u^{p\bar{q}})$ induces an inner product in $\mathbb{C}^n$. Then,

$$|u^{p\bar{q}}\rho_{pt_\alpha}u_{x_n\bar{q}}| \le C + C\left(\sum_{i=1}^n u^{i\bar{i}}\right)^{\frac{1}{2}}(u^{p\bar{q}}u_{y_np}u_{y_n\bar{q}})^{\frac{1}{2}}.$$

A similar estimate holds for the third term in the right-hand side of (6), while the fourth term is bounded by a constant multiple of $\sum_{i=1}^n u^{i\bar{i}}$. Therefore, we have

$$|L(T_\alpha u)| \le C + C\sum_{i=1}^n u^{i\bar{i}} + C\left(\sum_{i=1}^n u^{i\bar{i}}\right)^{\frac{1}{2}}(u^{p\bar{q}}u_{y_np}u_{y_n\bar{q}})^{\frac{1}{2}},$$

and hence

$$|LT_\alpha(u-\varphi)| \le C + C\sum_{i=1}^n u^{i\bar{i}} + C\left(\sum_{i=1}^n u^{i\bar{i}}\right)^{\frac{1}{2}}(u^{p\bar{q}}u_{y_np}u_{y_n\bar{q}})^{\frac{1}{2}}.$$

To control the last term, we note that

$$\begin{aligned}
L(u_{y_n} - \varphi_{y_n})^2 &= u^{p\bar{q}}\partial_{p\bar{q}}(u_{y_n} - \varphi_{y_n})^2 \\
&= 2u^{p\bar{q}}\partial_p\big((u_{y_n} - \varphi_{y_n})(u_{y_n\bar{q}} - \varphi_{y_n\bar{q}})\big) \\
&= 2u^{p\bar{q}}(u_{y_np} - \varphi_{y_np})(u_{y_n\bar{q}} - \varphi_{y_n\bar{q}}) + 2u^{p\bar{q}}(u_{y_n} - \varphi_{y_n})(u_{y_np\bar{q}} - \varphi_{y_np\bar{q}}) \\
&= 2(I + II).
\end{aligned}$$

We expand I to get

$$I = u^{p\bar{q}}u_{y_np}u_{y_n\bar{q}} - u^{p\bar{q}}u_{y_n\bar{q}}\varphi_{y_np} - u^{p\bar{q}}u_{y_np}\varphi_{y_n\bar{q}} + u^{p\bar{q}}\varphi_{y_np}\varphi_{y_n\bar{q}}.$$

For the second term, we apply the Cauchy inequality as before to get

$$|u^{p\bar{q}}u_{y_n\bar{q}}\varphi_{y_np}| \le (u^{p\bar{q}}\varphi_{y_np}\varphi_{y_n\bar{q}})^{\frac{1}{2}}(u^{p\bar{q}}u_{y_n\bar{q}}u_{y_np})^{\frac{1}{2}}.$$

The third term is handled in a similar way. For II, we write

$$\begin{aligned}
II &= (u_{y_n} - \varphi_{y_n})(u^{p\bar{q}}u_{y_np\bar{q}} - u^{p\bar{q}}\varphi_{y_np\bar{q}}) \\
&= (u_{y_n} - \varphi_{y_n})\big((\log f)_{y_n} - u^{p\bar{q}}\varphi_{y_np\bar{q}}\big).
\end{aligned}$$

Therefore, we obtain

$$L(u_{y_n} - \varphi_{y_n})^2 \ge 2u^{p\bar{q}}u_{y_np}u_{y_n\bar{q}} - C\left(\sum_{i=1}^n u^{i\bar{i}}\right)^{\frac{1}{2}}(u^{p\bar{q}}u_{y_np}u_{y_n\bar{q}})^{\frac{1}{2}} - C\sum_{i=1}^n u^{i\bar{i}} - C.$$

To proceed, we set

$$w = \pm T_\alpha(u - \varphi) + (u_{y_n} - \varphi_{y_n})^2.$$

Then, we conclude, by the Cauchy inequality,

$$\begin{aligned} Lw &= \pm LT_\alpha(u-\varphi) + L(u_{y_n} - \varphi_{y_n})^2 \\ &\geq 2u^{p\bar{q}}u_{y_n p}u_{y_n \bar{q}} - C\left(\sum_{i=1}^n u^{i\bar{i}}\right)^{\frac{1}{2}} (u^{p\bar{q}}u_{y_n p}u_{y_n \bar{q}})^{\frac{1}{2}} - C\sum_{i=1}^n u^{i\bar{i}} - C \\ &\geq -C\sum_{i=1}^n u^{i\bar{i}} - C. \end{aligned}$$

We now claim, for some small constant $\varepsilon > 0$,

(i) $Lw \geq L(Ax_n - B|z|^2)$ in $\Omega \cap B_\varepsilon$;

(ii) $w \leq Ax_n - B|z|^2$ on $\partial(\Omega \cap B_\varepsilon)$,

for some universal constants A and B sufficiently large. We first note that

$$L(Ax_n - B|z|^2) = -Bu^{p\bar{q}}\partial_{p\bar{q}}|z|^2 = -B\sum_{i=1}^n u^{i\bar{i}}.$$

To prove (i), we need

$$(B - C)\sum_{i=1}^n u^{i\bar{i}} \geq C.$$

This can be achieved by (1) and choosing B sufficiently large. To prove (ii), consider first $\partial\Omega \cap B_\varepsilon$. Note that $T_\alpha(u - \varphi) = 0$ on $\partial\Omega \cap B_\varepsilon$. By (3), we have, for some constant $c_0 > 0$,

$$x_n \geq c_0|z|^2 \quad \text{on } \partial\Omega \cap B_\varepsilon.$$

Since $u(t', \rho(t')) = \varphi(t', \rho(t'))$, we obtain, by a differentiation with respect to y_n,

$$(u_{y_n} - \varphi_{y_n})^2 = \rho_{y_n}^2(u_{x_n} - \varphi_{x_n})^2 \leq C|t'|^2 \leq C|z|^2 \leq Cx_n.$$

Taking A sufficiently large, we obtain (ii) on $\partial\Omega \cap B_\varepsilon$. Note that $x_n \geq c_0'$ on $\Omega \cap \partial B_\varepsilon$, for some constant $c_0' > 0$. We choose A large further so that (ii) holds on $\Omega \cap \partial B_\varepsilon$. This finishes the proof of (i) and (ii). By the maximum principle, we obtain

$$\pm T_\alpha(u - \varphi) + (u_{y_n} - \varphi_{y_n})^2 \leq Ax_n - B|z|^2 \quad \text{in } \Omega \cap B_\varepsilon,$$

or

$$\pm T_\alpha(u - \varphi) \leq Ax_n - B|z|^2 \quad \text{in } \Omega \cap B_\varepsilon.$$

By letting $t' = 0$, dividing by x_n, and then letting $x_n \to 0$, we have

$$|(T_\alpha u)_{x_n}(0)| \leq A + |(T_\alpha \varphi)_{x_n}(0)|.$$

Note that

$$(T_\alpha u)_{x_n} = u_{t_\alpha x_n} + \rho_{t_\alpha x_n} u_{x_n} + \rho_{t_\alpha} u_{x_n x_n}.$$

Hence,

$$|u_{t_\alpha x_n}(0)| \le K_2, \tag{7}$$

for a universal constant K_2.

Step 3. Using the special coordinates above, we proceed to establish the estimate

$$|u_{x_n x_n}(0)| \le K_3.$$

Since we already proved, for $\alpha, \beta = 1, \ldots, 2n-1$,

$$|u_{t_\alpha t_\beta}(0)|, |u_{t_\alpha x_n}(0)| \le K,$$

it suffices to prove

$$u_{n\bar{n}}(0) \le C. \tag{8}$$

We can solve the equation $\det(u_{i\bar{j}}) = f$ for $u_{n\bar{n}}(0)$ and note that (8) follows provided the $(n-1)\times(n-1)$ matrix $\big(u_{i\bar{j}}(0)\big)_{1\le i,j\le n-1}$ satisfies

$$\big(u_{i\bar{j}}(0)\big)_{1\le i,j\le n-1} \ge c_1 I_{n-1},$$

for some positive constant $c_1 > 0$. In the following, we will prove, for any $\xi' = (\xi_1, \ldots, \xi_{n-1}) \in \mathbb{C}^{n-1}$,

$$\sum_{i,j=1}^{n-1} u_{i\bar{j}}(0)\xi_i\bar{\xi}_j \ge c_1|\xi'|^2. \tag{9}$$

After a subtraction of a linear function, we assume $\varphi_{t_\alpha}(0) = 0$, for $\alpha = 1, \ldots, 2n-1$.

Let σ be a pluri-subharmonic $C^4(\bar{\Omega})$-defining function of Ω introduced earlier; i.e.,

$$\Omega = \{\sigma < 0\}, \quad \partial\Omega = \{\sigma = 0\}, \quad \nabla\sigma|_{\partial\Omega} \ne 0.$$

For convenience, we assume $|\nabla\sigma| = 1$ on $\partial\Omega$. For any point $p \in \partial\Omega$, we define

$$\Sigma_p = \left\{\xi \in \mathbb{C}^n : \sum_{i=1}^n \sigma_i(p)\xi_i = 0\right\}.$$

Consider

$$m_0 = \min_{p\in\partial\Omega}\ \min_{\xi\in\Sigma_p, |\xi|=1} u_{i\bar{j}}(p)\xi_i\bar{\xi}_j.$$

We assume that m_0 is attained at $p = 0 \in \partial\Omega$ and $\xi = (1, 0, \ldots, 0) \in \Sigma_p$. We choose coordinates $z = (z_1, \ldots, z_n)$ as before so that the positive x_n-axis is the interior normal to $\partial\Omega$ at 0. We will prove

$$m_0 = u_{1\bar{1}}(0) \geq c_0, \tag{10}$$

for some universal constant c_0.

Without loss of generality, we assume that φ is strictly pluri-subharmonic in $\bar{\Omega}$; otherwise, we replace φ by $\varphi + t\sigma$ for sufficiently large constant t. There exists a function $h \in C^3(\bar{\Omega})$ such that, near $\partial\Omega$,

$$u - \varphi = h\sigma.$$

A simple differentiation yields

$$(u - \varphi)_{x_n} = h_{x_n}\sigma + h\sigma_{x_n}$$

and

$$(u - \varphi)_{i\bar{j}} = h_{i\bar{j}}\sigma + h_i\sigma_{\bar{j}} + h_{\bar{j}}\sigma_i + h\sigma_{i\bar{j}}.$$

Hence, for any $\xi \in \mathbb{C}^n$,

$$(u - \varphi)_{i\bar{j}}\xi_i\bar{\xi}_j = h_{i\bar{j}}\xi_i\bar{\xi}_j\sigma + h_i\sigma_{\bar{j}}\xi_i\bar{\xi}_j + h_{\bar{j}}\sigma_i\xi_i\bar{\xi}_j + h\sigma_{i\bar{j}}\xi_i\bar{\xi}_j.$$

First, by restricting to $\partial\Omega$, we have

$$(u - \varphi)_{x_n} = h\sigma_{x_n} \quad \text{on } \partial\Omega.$$

Next, we fix a point $p \in \partial\Omega$ and take $\xi \in \Sigma_p$. Then,

$$(u - \varphi)_{i\bar{j}}\xi_i\bar{\xi}_j = h\sigma_{i\bar{j}}\xi_i\bar{\xi}_j \quad \text{at } p,$$

and hence,

$$(u - \varphi)_{i\bar{j}}\xi_i\bar{\xi}_j = \frac{1}{\sigma_{x_n}}(u - \varphi)_{x_n}\sigma_{i\bar{j}}\xi_i\bar{\xi}_j \quad \text{at } p.$$

Moreover, at $p = 0$, we get, for any $i, j = 1, \ldots, n-1$,

$$(u - \varphi)_{i\bar{j}}(0) = -(u - \varphi)_{x_n}(0)\sigma_{i\bar{j}}(0),$$

where we used $\nabla_{\mathbb{C}}\sigma(0) = (0, \ldots, 0, -1)$. In particular,

$$u_{1\bar{1}}(0) = \varphi_{1\bar{1}}(0) - (u - \varphi)_{x_n}(0)\sigma_{1\bar{1}}(0).$$

We first consider $u_{1\bar{1}}(0) \geq \varphi_{1\bar{1}}(0)/2$. Then, (10) holds since φ is strictly pluri-subharmonic in $\bar{\Omega}$. Next, we assume $u_{1\bar{1}}(0) \leq \varphi_{1\bar{1}}(0)/2$. Then,

$$(u - \varphi)_{x_n}(0)\sigma_{1\bar{1}}(0) \geq \frac{1}{2}\varphi_{1\bar{1}}(0),$$

and hence

$$\sigma_{1\bar{1}}(0) \geq \frac{1}{2C}\varphi_{1\bar{1}}(0). \tag{11}$$

Let $\varepsilon > 0$ be a constant small enough so that

$$|\sigma_1|^2 + |\sigma_n|^2 > 0 \quad \text{in } \bar{\Omega} \cap \bar{B}_\varepsilon.$$

Define $\xi = (\xi_1, \ldots, \xi_n)$ by

$$\begin{aligned}
\xi_1 &= -\sigma_n(|\sigma_1|^2 + |\sigma_n|^2)^{-\frac{1}{2}}, \\
\xi_i &= 0 \quad \text{for } i = 2, \ldots, n-1, \\
\xi_n &= \sigma_1(|\sigma_1|^2 + |\sigma_n|^2)^{-\frac{1}{2}}.
\end{aligned}$$

Then, $\xi(p) \in \Sigma_p$ for any $p \in \partial\Omega \cap B_\varepsilon$ and $|\xi| = 1$, with $\xi(0) = (1, 0, \ldots, 0)$. Set

$$w = \varphi_{i\bar{j}}\xi_i\bar{\xi}_j + \frac{1}{\sigma_{x_n}}(u - \varphi)_{x_n}\sigma_{i\bar{j}}\xi_i\bar{\xi}_j - u_{1\bar{1}}(0) \quad \text{in } \Omega \cap B_\varepsilon.$$

Then,

$$w = u_{i\bar{j}}\xi_i\bar{\xi}_j - u_{1\bar{1}}(0) \quad \text{on } \partial\Omega \cap B_\varepsilon,$$

and hence, $w(0) = 0$ and

$$w \geq 0 \quad \text{on } \partial\Omega \cap B_\varepsilon.$$

Next, we calculate Lw. It is easy to check that

$$L\left(\varphi_{i\bar{j}}\xi_i\bar{\xi}_j - \frac{1}{\sigma_{x_n}}\varphi_{x_n}\sigma_{i\bar{j}}\xi_i\bar{\xi}_j\right) \leq C\sum_{i=1}^n u^{i\bar{i}}.$$

We set

$$v = \frac{1}{\sigma_{x_n}}\sigma_{i\bar{j}}\xi_i\bar{\xi}_j.$$

Then,

$$\begin{aligned}
L(u_{x_n}v) &= u_{x_n}Lv + vLu_{x_n} + u^{\bar{p}q}(u_{\bar{p}x_n}v_q + u_{qx_n}v_{\bar{p}}) \\
&= u_{x_n}Lv + v(\log f)_{x_n} + u^{\bar{p}q}(u_{\bar{p}x_n}v_q + u_{qx_n}v_{\bar{p}}).
\end{aligned}$$

With

$$u_{\bar{p}x_n} = 2u_{\bar{p}n} + \sqrt{-1}u_{\bar{p}y_n}, \quad u_{qx_n} = 2u_{q\bar{n}} - \sqrt{-1}u_{qy_n},$$

we have

$$u^{\bar{p}q}(u_{\bar{p}x_n}v_q + u_{qx_n}v_{\bar{p}}) = 2(v_n + v_{\bar{n}}) + \sqrt{-1}(u^{\bar{p}q}u_{\bar{p}y_n}v_q - u^{\bar{p}q}u_{qy_n}v_{\bar{p}}).$$

Therefore,

$$Lw \leq C + C\sum_{i=1}^n u^{i\bar{i}} + C\left(\sum_{i=1}^n u^{i\bar{i}}\right)^{\frac{1}{2}}\left(u^{\bar{p}q}u_{\bar{p}y_n}u_{qy_n}\right)^{\frac{1}{2}} \quad \text{in } \Omega \cap B_\varepsilon,$$

and

$$w \geq 0 \quad \text{on } \partial\Omega \cap B_\varepsilon.$$

Arguing as in Step 2, we get

$$-w + (u_{y_n} - \varphi_{y_n})^2 \leq Ax_n - B|z|^2 \quad \text{in } \Omega \cap B_\varepsilon,$$

for some universal constants A and B. Hence,

$$-w \leq Ax_n - B|z|^2 \quad \text{in } \Omega \cap B_\varepsilon,$$

and then,

$$-w_{x_n}(0) \leq A.$$

This implies, with $\sigma_{x_n}(0) = -1$,

$$(u - \varphi)_{x_n x_n}(0)\sigma_{1\bar{1}}(0) \leq C,$$

and hence, by (11),

$$u_{x_n x_n}(0) \leq C. \tag{12}$$

In view of (4), (7), and (12), we have an a priori upper bound for all eigenvalues of the Hessian matrix $(u_{i\bar{j}}(0))$. Since $\det(u_{i\bar{j}}(0)) = f(0) > 0$, the eigenvalues of $(u_{i\bar{j}}(0))$ also admit a positive lower bound; i.e.,

$$\min_{\xi \in \mathbb{C}^n, |\xi|=1} u_{i\bar{j}}(0)\xi_i\bar{\xi}_j \geq c_0.$$

Hence,

$$m_0 = \min_{\xi \in \Sigma_0, |\xi|=1} u_{i\bar{j}}(0)\xi_i\bar{\xi}_j \geq \min_{\xi \in \mathbb{C}^n, |\xi|=1} u_{i\bar{j}}(0)\xi_i\bar{\xi}_j \geq c_0.$$

This establishes (10). □

We note that (9) is proved by a global argument. In fact, we can prove (9) at each point of the boundary. We now provide such a proof.

Alternative Proof of Theorem 7.2.2. We provide an alternative proof of (9). Without loss of generality, we assume $\xi' = (1, 0, \ldots, 0) \in \mathbb{C}^{n-1}$ and prove

$$u_{1\bar{1}}(0) \geq c_1. \tag{9'}$$

After a subtraction of a linear function, we assume $\varphi_{t_\alpha}(0) = 0$, for $\alpha = 1, \ldots, 2n-1$.

Let $\widetilde{u} = u - \lambda x_n$, with λ chosen so that

$$\partial_{1\bar{1}}\widetilde{u}(t', \rho(t')) = 0 \quad \text{at } 0.$$

Note that

$$\partial_{1\bar{1}}\widetilde{u} = \widetilde{u}_{1\bar{1}} + \widetilde{u}_{1x_n}\rho_{\bar{1}} + \widetilde{u}_{\bar{1}x_n}\rho_1 + \widetilde{u}_{x_n x_n}\rho_1\rho_{\bar{1}} + \widetilde{u}_{x_n}\rho_{1\bar{1}}.$$

This implies at 0

$$u_{1\bar{1}}(0) + \widetilde{u}_{x_n}(0)\rho_{1\bar{1}}(0) = u_{1\bar{1}}(0) + (u_{x_n}(0) - \lambda)\rho_{1\bar{1}}(0) = 0,$$

and hence

$$u_{1\bar{1}}(0) = -\widetilde{u}_{x_n}(0)\rho_{1\bar{1}}(0). \tag{13}$$

In the following, we will estimate $\widetilde{u}_{x_n}(0)$ from above by a negative constant. We first claim, for some small constant $\varepsilon > 0$,

$$\widetilde{u} \le \operatorname{Re}[p(z)] + \operatorname{Re}\sum_{j=2}^{n} a_j z_1 \bar{z}_j + C\sum_{j=2}^{n} |z_j|^2 \quad \text{on } \partial\Omega \cap B_\varepsilon, \tag{14}$$

where p is a holomorphic cubic polynomial without linear terms, $a_2, \dots, a_n$ are complex numbers bounded by a universal constant, and C is a universal constant.

To prove (14), we consider the Taylor expansion of $\widetilde{u}\big(t', \rho(t')\big)$, given by

$$\widetilde{u}(t', \rho(t')) = u(t', \rho(t')) - \lambda\rho(t') = \varphi(t', \rho(t')) - \lambda\rho(t').$$

Obviously, there are no linear terms. For the quadratic terms, we consider $t_\alpha t_\beta$, $1 \le \alpha, \beta \le 2$, and $t_\alpha t_\beta$, $1 \le \alpha \le 2, 3 \le \beta \le 2n-1$. Since

$$\partial_{1\bar{1}}\widetilde{u}(0) = 0,$$

the linear combination of $t_\alpha t_\beta$, $1 \le \alpha, \beta \le 2$, can be written as

$$\operatorname{Re}[(a + \sqrt{-1}b)z_1^2],$$

for some real numbers a and b. Next, a linear combination of $t_1 t_{2\beta-1}, t_2 t_{2\beta-1}$, $t_1 t_{2\beta}, t_2 t_{2\beta}$, $2 \le \beta \le n-1$, can be written as

$$\operatorname{Re}\big((a + \sqrt{-1}b)z_1 z_\beta\big) + \operatorname{Re}\big((c + \sqrt{-1}d)z_1 \bar{z}_\beta\big).$$

In fact, we have

$$\begin{aligned}
\operatorname{Re}((a + \sqrt{-1}b)z_1 z_\beta) &= a(t_1 t_{2\beta-1} - t_2 t_{2\beta}) - b(t_1 t_{2\beta} + t_2 t_{2\beta-1}),\\
\operatorname{Re}((c + \sqrt{-1}d)z_1 \bar{z}_\beta) &= c(t_1 t_{2\beta-1} + t_2 t_{2\beta}) + d(t_1 t_{2\beta} - t_2 t_{2\beta-1}).
\end{aligned}$$

Last, a linear combination of $t_1 y_n$ and $t_2 y_n$ can be written as

$$\operatorname{Re}\big((a + \sqrt{-1}b)z_1 y_n\big),$$

for some real numbers a and b. For cubic terms, we first consider those which are cubic in (t_1, t_2). Note that any real homogeneous cubic polynomial in (t_1, t_2) admits a unique decomposition

$$\begin{aligned}
&\operatorname{Re}\big(a(t_1 + \sqrt{-1}t_2)^3 + b(t_1 - \sqrt{-1}t_2)(t_1 + \sqrt{-1}t_2)^2\big)\\
&\qquad = \operatorname{Re}(a z_1^3 + b z_1 |z_1|^2),
\end{aligned} \tag{15}$$

for some complex numbers a and b. This can be seen easily by expanding (15) in terms of t_1, t_2 and noting that complex numbers a and b provide four real parameters and any real homogeneous cubic polynomial in (t_1, t_2) has at most four coefficients. Hence,

$$\begin{aligned}\widetilde{u}|_{\partial\Omega} \leq \operatorname{Re}\sum_{j=2}^{n-1} a_j z_1 \bar{z}_j + \operatorname{Re}(c_1 z_1 y_n) + \operatorname{Re}(c_2 z_1 |z_1|^2) + \operatorname{Re} p(z_1, \ldots, z_n)\\ + \text{quadratic terms in } t_3, t_4, \ldots, t_{2n-1}\\ + \text{cubic terms where } (t_1, t_2) \text{ appears at most quadratically}\\ + \text{fourth-order terms,}\end{aligned} \tag{16}$$

for some complex numbers c_1, c_2, and $a_2, \ldots, a_{n-1}$. Now we focus on $z_1|z_1|^2$. By (3), it holds true that, for some positive constant ε,

$$x_n = \sum_{\alpha,\beta=1}^{2n-1} b_{\alpha\beta} t_\alpha t_\beta + O(|t'|^3) \quad \text{on } \partial\Omega \cap B_\varepsilon.$$

We assume $(b_{\alpha\beta})_{\alpha,\beta \leq 2n-2}$ is diagonal. Note that the coefficients of t_1^2 and t_2^2 are the same and are given by $\rho_{1\bar{1}}(0) > 0$. Hence,

$$|z_1|^2 = t_1^2 + t_2^2 = a x_n + \text{quadratic of } t' + O(|t'|^3) \quad \text{on } \partial\Omega \cap B_\varepsilon, \tag{17}$$

where there is no quadratic expression of (t_1, t_2) in the quadratic part of t'. By adjusting c_2 and p in (16) appropriately, we get

$$\begin{aligned}\widetilde{u}|_{\partial\Omega} \leq \operatorname{Re}\sum_{j=2}^{n-1} a_j z_1 \bar{z}_j + \operatorname{Re}(c_1 z_1 y_n) + \operatorname{Re}(c_2 z_1 x_n) + \operatorname{Re} p(z_1, \ldots, z_{n-1})\\ + \text{quadratic terms in } t_3, t_4, \ldots, t_{2n-1}\\ + \text{cubic terms where } (t_1, t_2) \text{ appears at most quadratically}\\ + \text{fourth-order terms.}\end{aligned}$$

Obviously,

$$\operatorname{Re}(c_1 z_1 y_n + c_2 z_1 x_n) = \operatorname{Re}(a z_1 z_n + b z_1 \bar{z}_n),$$

for some complex numbers a and b. Applying the Cauchy inequality to cubic terms where (t_1, t_2) appears at most quadratically, we obtain

$$\widetilde{u} \leq \operatorname{Re} p(z) + \operatorname{Re}\sum_{j=2}^{n} a_j z_1 \bar{z}_j + C\sum_{j=2}^{n} |z_j|^2 + C|z|^4 \quad \text{on } \partial\Omega \cap B_\varepsilon.$$

With (17), it is easy to check that

$$|z|^4 \leq C\sum_{j=2}^{n} |z_j|^2.$$

Hence, (14) is proved.

Set $\widehat{u} = \widetilde{u} - \mathrm{Re}[p(z)]$ and note that

$$\det \widehat{u}_{i\bar{j}} = \det u_{i\bar{j}} = f.$$

For some positive constants δ_0, δ_1, and μ to be determined, we set

$$\begin{aligned} h(z) &= -\delta_0 x_n + \delta_1 |z|^2 + \frac{1}{2\mu}\sum_{j=2}^n |a_j z_1 + \mu z_j|^2 \\ &= -\delta_0 x_n + \delta_1 |z|^2 + \frac{1}{2\mu}\sum_{j=2}^n |a_j|^2 |z_1|^2 + \mathrm{Re}\sum_{j=2}^n a_j z_1 \bar{z}_j + \frac{\mu}{2}\sum_{j=2}^n |z_j|^2, \end{aligned}$$

where $a_2, \ldots, a_n$ are constants as in (14). We claim, by choosing δ_0, δ_1, and μ appropriately,

$$\begin{aligned} \det(\widehat{u}_{i\bar{j}}) &\geq \det(h_{i\bar{j}}) \quad \text{in } \Omega \cap B_\varepsilon, \\ \widehat{u} &\leq h \quad \text{on } \partial(\Omega \cap B_\varepsilon). \end{aligned} \tag{18}$$

First, we note that $|z_n| \geq \varepsilon_0$ on $\partial B_\varepsilon \cap \Omega$, for some positive constant ε_0, and

$$h \geq -\delta_0 x_n + \frac{1}{2\mu}|a_n z_1 + \mu z_n|^2 \geq \frac{\mu}{2}|z_n|^2 + \mathrm{Re}(a_n z_1 \bar{z}_n) - \delta_0 x_n.$$

For any $\delta_0, \delta_1 \in (0,1)$, by choosing μ sufficiently large, we have

$$\widehat{u} \leq h \quad \text{on } \partial B_\varepsilon \cap \Omega.$$

By choosing also $\mu \geq 2C$ for the constant C in (14), we have by (14)

$$\widehat{u} \leq \frac{1}{2\mu}\sum_{j=2}^n |a_j z_1 + \mu z_j|^2 \quad \text{on } \partial\Omega \cap B_\varepsilon.$$

Next, we calculate $\det(h_{i\bar{j}})$. A straightforward calculation yields

$$(h_{i\bar{j}}) = \begin{pmatrix} \delta_1 + \frac{1}{2\mu}\sum_{j=2}^n |a_j|^2 & \frac{1}{2}a_2 & \cdots & \frac{1}{2}a_n \\ \frac{1}{2}\bar{a}_2 & \delta_1 + \frac{1}{2}\mu & & \\ \vdots & & \ddots & \\ \frac{1}{2}\bar{a}_n & & & \delta_1 + \frac{1}{2}\mu \end{pmatrix}.$$

The eigenvalues of $(h_{i\bar{j}})$ are given by

$$\delta_1,\ \delta_1 + \frac{1}{2}\mu + \frac{1}{2\mu}\sum_{j=2}^n |a_j|^2,\ \delta_1 + \frac{1}{2}\mu, \ldots, \delta_1 + \frac{1}{2}\mu.$$

This implies that h is strictly pluri-subharmonic and

$$\det(h_{i\bar{j}}) = \delta_1\Big(\delta_1 + \frac{1}{2}\mu\Big)^{n-2}\Big(\delta_1 + \frac{1}{2}\mu + \frac{1}{2\mu}\sum_{j=2}^n |a_j|^2\Big).$$

Therefore, by choosing δ_1 small, we have

$$\det(h_{i\bar{j}}) \leq f \quad \text{in } \Omega.$$

Last, on $\partial\Omega \cap B_\varepsilon$, we require

$$-\delta_0 x_n + \delta_1 |z|^2 \geq 0,$$

or

$$x_n \leq \frac{\delta_1}{\delta_0}|z|^2.$$

This can be achieved by taking δ_0 relatively small compared with δ_1. Then,

$$\widehat{u} \leq h \quad \text{on } \partial\Omega \cap B_\varepsilon.$$

This finishes the proof of (18). By Lemma 7.1.2, we obtain

$$\widehat{u} \leq h \quad \text{in } \Omega \cap B_\varepsilon.$$

Since $\widehat{u}(0) = h(0) = 0$, we get

$$\widehat{u}_{x_n}(0) \leq h_{x_n}(0) = -\delta_0.$$

Then, (13) implies

$$u_{1\bar{1}}(0) = -\widehat{u}_{x_n}(0)\rho_{1\bar{1}}(0) \geq \delta_0 \rho_{1\bar{1}}(0).$$

This is the desired estimate (9′). □

Next, we derive a global estimate of the second derivatives.

Theorem 7.2.3. *Let Ω be a bounded strongly pseudo-convex domain in $\mathbb{C}^n$ with a C^2-boundary. Suppose that u is a strictly pluri-subharmonic $C^2(\bar{\Omega})$-solution of*

$$\begin{aligned} \det(u_{i\bar{j}}) &= f \quad \text{in } \Omega, \\ u &= \varphi \quad \text{on } \partial\Omega, \end{aligned}$$

for some $f \in C^2(\bar{\Omega})$ and $\varphi \in C^2(\bar{\Omega})$, with $f > 0$ in $\bar{\Omega}$. Then,

$$|\nabla^2 u|_{L^\infty(\Omega)} \leq K,$$

where K is a positive constant depending only on Ω, the C^1-norm of u in $\bar{\Omega}$, the L^∞-norm of $\nabla^2 u$ on $\partial\Omega$, $\max_{\bar{\Omega}} f^{-1}$, and the C^2-norm of f in $\bar{\Omega}$.

Proof. For any real constants a_j and b_j, $j = 1, \ldots, n$, with

$$\sum_{j=1}^n (a_j^2 + b_j^2) = 1,$$

set

$$\gamma = \sum_{j=1}^n (a_j \partial_{x_j} + b_j \partial_{y_j}).$$

Then,

$$Lu_{\gamma\bar{\gamma}} \geq (\log f)_{\gamma\bar{\gamma}} \geq -C.$$

As in the proof of Theorem 7.2.1, we obtain, for large λ,

$$L(u_{\gamma\bar{\gamma}} + e^{\lambda|z|^2}) \geq 0 \quad \text{in } \Omega.$$

By the maximum principle, we conclude

$$\max_{\bar{\Omega}} u_{\gamma\bar{\gamma}} \leq \max_{\partial\Omega} u_{\gamma\bar{\gamma}} + C \leq K.$$

With the upper bound for every $u_{\gamma\bar{\gamma}}$ and the lower bound $u_{x_i x_i} + u_{y_i y_i} \geq 0$, we proceed to estimate second derivatives.

First, by taking $\gamma = \partial_{x_i}$ and $\gamma = \partial_{y_i}$, we have

$$u_{x_i x_i} \leq K, \quad u_{y_i y_i} \leq K.$$

With $u_{x_i x_i} + u_{y_i y_i} \geq 0$, we obtain

$$|u_{x_i x_i}|, \ |u_{y_i y_i}| \leq K.$$

Next, by taking $\gamma = \frac{1}{\sqrt{2}}(\partial_{x_i} \pm \partial_{y_i})$, we get

$$u_{\gamma\bar{\gamma}} = \frac{1}{2}(u_{x_i x_i} \pm 2u_{x_i y_i} + u_{y_i y_i}) \leq K,$$

and hence

$$|u_{x_i y_i}| \leq K.$$

Last, for $i \neq j$, by taking $\gamma = \frac{1}{\sqrt{2}}(\partial_{x_i} \pm \partial_{y_j})$ and $\gamma = \frac{1}{\sqrt{2}}(\partial_{x_j} \pm \partial_{y_i})$, we have

$$\frac{1}{2}(u_{x_i x_i} \pm 2u_{x_i y_j} + u_{y_j y_j}) \leq K,$$
$$\frac{1}{2}(u_{x_j x_j} \pm 2u_{x_j y_i} + u_{y_i y_i}) \leq K.$$

With lower bounds of $u_{x_i x_i}$, $u_{x_j x_j}$, $u_{y_i y_i}$, and $u_{y_j y_j}$ by $-K$, we obtain

$$|u_{x_i y_j}|, \ |u_{x_j y_i}| \leq 2K.$$

This finishes the proof. □

Now we can prove the main estimate in this section.

Theorem 7.2.4. *Let Ω be a bounded strongly pseudo-convex domain in $\mathbb{C}^n$ with a C^4-boundary. Suppose that u is a strictly pluri-subharmonic $C^4(\bar{\Omega})$-solution of*

$$\det(u_{i\bar{j}}) = f \quad \text{in } \Omega,$$
$$u = \varphi \quad \text{on } \partial\Omega,$$

for some $f \in C^2(\bar{\Omega})$ and $\varphi \in C^4(\bar{\Omega})$, with $f > 0$ in $\bar{\Omega}$. Then,

$$|u|_{C^2(\bar{\Omega})} \leq K,$$

where K is a positive constant depending only on Ω, $\max_{\bar{\Omega}} f^{-1}$, the C^2-norm of f in $\bar{\Omega}$, and the C^4-norm of φ in $\bar{\Omega}$.

Proof. Let u^0 be a strictly pluri-subharmonic $C^4(\bar{\Omega})$-function satisfying

$$\begin{aligned} \det(u^0_{i\bar{j}}) &\geq f \quad \text{in } \Omega, \\ u^0 &= \varphi \quad \text{on } \partial\Omega. \end{aligned}$$

To construct u^0, we let σ be a strictly pluri-subharmonic $C^4(\bar{\Omega})$-defining function of Ω; i.e., $\sigma < 0$ in Ω, and $\sigma = 0$ and $\nabla\sigma \neq 0$ on $\partial\Omega$. Then, we take $u^0 = \varphi + C\sigma$ for sufficiently large constant C. We have the desired estimate by combining Theorems 7.2.1, 7.2.2, and 7.2.3. □

We can solve the Dirichlet problem in strongly pseudo-convex domains.

Theorem 7.2.5. *Let $\alpha \in (0,1)$ be a constant and Ω be a bounded strongly pseudo-convex domain in $\mathbb{C}^n$ with a $C^{4,\alpha}$-boundary. Then for any $f \in C^{2,\alpha}(\bar{\Omega})$ with $f > 0$ in $\bar{\Omega}$ and any $\varphi \in C^{4,\alpha}(\bar{\Omega})$, there exists a unique strictly pluri-subharmonic $C^{4,\alpha}(\bar{\Omega})$-solution u of*

$$\begin{aligned} \det(u_{i\bar{j}}) &= f \quad \text{in } \Omega, \\ u &= \varphi \quad \text{on } \partial\Omega. \end{aligned}$$

The proof is similar to that of Theorem 6.2.6 and is omitted.

Chapter 8

Generalized Solutions of Monge-Ampère Equations

In this chapter, we discuss generalized solutions of Monge-Ampère equations. Such solutions are defined only for convex functions, which are not assumed to be C^2 to begin with. We will prove various regularity results under appropriate assumptions on the corresponding Monge-Ampère measures.

In Section 8.1, we discuss basic properties of convex functions and introduce several important concepts. Among those concepts, normal mappings play a role of central importance, based on which we introduce Monge-Ampère measures and generalized solutions of the Monge-Ampère equation.

In Section 8.2, we first derive the comparison principle for generalized solutions of the Monge-Ampère equation. Then, we discuss the existence of generalized solutions of the Dirichlet problem. We solve the Dirichlet problem in convex domains with convex boundary values and in strictly convex domains with arbitrary continuous boundary values.

In Section 8.3, we prove a normalization lemma due to John, which plays an important role in studying the Monge-Ampère equation. We also derive a global Hölder estimate for solutions of the Dirichlet problem in convex domains.

In Section 8.4, we discuss the strict convexity and the interior $C^{1,\alpha}$-regularity for solutions of the Monge-Ampère equation under the condition that the Monge-Ampère measures satisfy the doubling condition. The discussion is based on the level set approach.

In Section 8.5, we discuss the interior $C^{2,\alpha}$-regularity for solutions of the Monge-Ampère equation under the condition that the Monge-Ampère measures are induced by positive Hölder continuous functions.

Main regularity results in this chapter are due to Caffarelli [**14**], [**15**], [**16**]. For our presentation, we follow the survey paper [**154**] by Trudinger and Wang and also the book [**74**] by Gutiérrez.

8.1. Monge-Ampère Measures

In this section, we discuss basic properties of convex functions and introduce several important concepts. Among those concepts, normal mappings play a role of central importance, based on which we introduce the Monge-Ampère measures and the generalized solutions of the Monge-Ampère equation. In our definition of convexity, we do not assume the continuity. All functions are required to be pointwise defined.

We first introduce the notion of the supporting functions.

Definition 8.1.1. Let Ω be a domain in $\mathbb{R}^n$ and u be a function in Ω. For any point $x_0 \in \Omega$, a *supporting function* of u at x_0 is an affine function l with the properties $u(x_0) = l(x_0)$ and $u(x) \geq l(x)$ for any $x \in \Omega$.

We point out that any affine function l with $u(x_0) = l(x_0)$ can be written as $l(x) = u(x_0) + p \cdot (x - x_0)$ for some $p \in \mathbb{R}^n$. The vector p is the *gradient* of l.

Now, we introduce various notions of convexity.

Definition 8.1.2. Let Ω be a domain in $\mathbb{R}^n$, u be a function in Ω, and x_0 be a point in Ω.

(i) u is *locally convex* at x_0 if there exists a supporting function of $u|_{B_r(x_0)}$ at x_0, for some ball $B_r(x_0) \subset \Omega$.

(ii) u is *convex* at x_0 if there exists a supporting function of u at x_0.

(iii) u is *strictly convex* at x_0 if there exists a supporting function l of u at x_0 such that, for any $x \in \Omega \setminus \{x_0\}$,

$$u(x) > l(x).$$

Moreover, u is *locally convex*, *convex*, or *strictly convex* in Ω if u is locally convex, convex, or strictly convex at each point $x_0 \in \Omega$, respectively.

If u is locally convex at x_0, then

$$\liminf_{x \to x_0} u(x) \geq u(x_0).$$

In other words, locally convex functions are lower semi-continuous.

Next, we note that convex functions are subharmonic. To verify this, we let u be a convex function in some domain $\Omega \subset \mathbb{R}^n$. Take any $x_0 \in \Omega$. Then, there exists a $p \in \mathbb{R}^n$ such that, for any $x \in \Omega$,

$$u(x) \geq u(x_0) + p \cdot (x - x_0).$$

For any $B_r(x_0) \subset \Omega$, an integration with respect to x over $B_r(x_0)$ implies

$$u(x_0) \leq \frac{1}{\omega_n r^n} \int_{B_r(x_0)} u \, dx,$$

where ω_n is the volume of the unit ball in $\mathbb{R}^n$.

We have the following simple result concerning the relation between the local convexity and convexity. Recall that a domain Ω in $\mathbb{R}^n$ is *convex* if, for any $x_1, x_2 \in \Omega$, the line segment $\overline{x_1 x_2} \subset \Omega$.

Lemma 8.1.3. *Let Ω be a convex domain in $\mathbb{R}^n$ and u be a continuous function in Ω. If u is locally convex in Ω, then u is convex in Ω.*

Proof. Fix a point $x_0 \in \Omega$. By the assumption, there exist a ball $B_r(x_0) \subset \Omega$ and an affine function l such that $u(x_0) = l(x_0)$ and, for any $x \in B_r(x_0)$,

$$u(x) \geq l(x). \tag{1}$$

We need to prove that (1) holds for any $x \in \Omega$. To this end, we consider the restriction of $u - l$ to the line segment $\overline{x_0 x}$ and employ the assumption that $u - l$ is locally convex at each point on $\overline{x_0 x}$. We omit the details. □

Although our main object in this chapter is the collection of convex functions in convex domains, we will introduce several notations for functions in general domains. Among those notations, normal mappings play a role of central importance.

Definition 8.1.4. Let Ω be a domain in $\mathbb{R}^n$ and u be a function in Ω. For any x_0, we set

$$\partial u(x_0) = \{p \in \mathbb{R}^n : u(x) \geq u(x_0) + p \cdot (x - x_0) \text{ for any } x \in \Omega\}.$$

In other words, $p \in \partial u(x_0)$ if and only if p is the gradient of some supporting function of u at x_0. For any subset $E \subset \Omega$, we set

$$\partial u(E) = \bigcup_{x \in E} \partial u(x).$$

The ∂u, viewed as a set-valued function in Ω, is called the *normal mapping* of u.

The normal mapping has the following properties.

Lemma 8.1.5. *Let Ω be a domain in $\mathbb{R}^n$, u be a function in Ω, and x_0 be a point in Ω.*

(i) *u is convex at x_0 if and only if $\partial u(x_0) \neq \emptyset$.*

(ii) *The set $\partial u(x_0)$ is convex and closed.*

(iii) *If u is convex at x_0 and $\nabla u(x_0)$ exists, then there exists a unique supporting function of u at x_0 and $\partial u(x_0) = \{\nabla u(x_0)\}$.*

Proof. The proofs of (i) and (ii) are simple. To prove (iii), we take any $p \in \partial u(x_0)$ and fix an $i = 1, \ldots, n$. By the convexity of u at x_0, we have, for all small t,

$$u(x_0 + te_i) \geq u(x_0) + tp \cdot e_i,$$

or

$$u(x_0 + te_i) - u(x_0) \geq tp_i.$$

If $t > 0$, we obtain

$$\frac{1}{t}\big[u(x_0 + te_i) - u(x_0)\big] \geq p_i,$$

and hence $u_i(x_0) \geq p_i$ by letting $t \to 0^+$. If $t < 0$, then

$$\frac{1}{t}\big[u(x_0 + te_i) - u(x_0)\big] \leq p_i,$$

and hence $u_i(x_0) \leq p_i$. Therefore, $u_i(x_0) = p_i$ for each $i = 1, \ldots, n$. □

We point out that the set $\partial u(K)$ is not necessarily convex in $\mathbb{R}^n$ even if u is convex in Ω and K is a convex subset of Ω. For example, we consider

$$u(x) = e^{|x|^2} \quad \text{for any } x \in \mathbb{R}^n$$

and

$$K = \{x \in \mathbb{R}^n : |x_i| \leq 1 \text{ for } i = 1, \ldots, n\}.$$

We note that u is a convex function in $\mathbb{R}^n$ and K is a convex set in $\mathbb{R}^n$. The set $\partial u(K)$ is a star-shaped symmetric set about the origin that is not convex.

We now consider a useful example.

Example 8.1.6. Let $B_R(x_0)$ be a ball in $\mathbb{R}^n$ and set, for some $h > 0$,

$$u(x) = \frac{h}{R}|x - x_0| \quad \text{for any } x \in B_R(x_0).$$

The graph of u is a cone in $\mathbb{R}^{n+1}$ with its vertex $(x_0, 0)$ and $u = h$ on $\partial B_R(x_0)$. Since u is C^1 in $B_R(x_0) \setminus \{x_0\}$, then, for any $x \in B_R(x_0) \setminus \{x_0\}$,

$$\partial u(x) = \left\{\frac{h}{R}\frac{x - x_0}{|x - x_0|}\right\}.$$

Next, we claim

$$\partial u(x_0) = \overline{B_{h/R}}.$$

To prove this, we note that, by the definition of the normal mapping, $p \in \partial u(x_0)$ if and only if, for any $x \in B_R(x_0)$,

$$\frac{h}{R}|x - x_0| \geq p \cdot (x - x_0). \tag{1}$$

If p satisfies (1) and $p \neq 0$, we take $x = x_0 + Rp/|p|$ in (1) and get $|p| \leq h/R$. On the other hand, if $|p| \leq h/R$, then (1) holds.

We now estimate gradients of convex functions in terms of distances to the boundary. Compare the following result with Lemma 6.4.2.

Lemma 8.1.7. *Let Ω be a domain in $\mathbb{R}^n$ and u be a bounded function in Ω. Then, for any $x \in \Omega$ and any $p \in \partial u(x)$,*

$$|p| \leq \frac{1}{\operatorname{dist}(x, \partial\Omega)} \left(\sup_{\Omega} u - u(x) \right).$$

Proof. We consider the case $p \neq 0$. The definition of $\partial u(x)$ implies, for any $y \in \Omega$,

$$u(y) \geq u(x) + p \cdot (y - x).$$

For any $0 < r < \operatorname{dist}(x, \partial\Omega)$, take

$$y = x + r\frac{p}{|p|} \in \Omega.$$

Then,

$$u(y) \geq u(x) + r|p|.$$

We have the desired result by letting $r \to \operatorname{dist}(x, \partial\Omega)$. □

We now relate the convexity and the sign of the Hessian matrices under appropriate regularity assumptions.

Lemma 8.1.8. *Let Ω be a domain in $\mathbb{R}^n$, u be a function in Ω, and x_0 be a point in Ω. If u is convex and differentiable up to the second order at x_0, then $\nabla^2 u(x_0) \geq 0$.*

Proof. We first note that ∇u is defined in a neighborhood of x_0. Fix a vector $\xi \in \mathbb{R}^n$ with $|\xi| = 1$ and restrict u to $\{x_0 + t\xi : t \text{ sufficiently small}\}$. By the convexity of u at x_0, we have, for any t small,

$$u(x_0 + t\xi) \geq u(x_0) + \nabla u(x_0) \cdot (t\xi),$$
$$u(x_0 - t\xi) \geq u(x_0) + \nabla u(x_0) \cdot (-t\xi).$$

A simple addition yields

$$u(x_0 + t\xi) + u(x_0 - t\xi) - 2u(x_0) \geq 0.$$

Hence,

$$u_{\xi\xi}(x_0) = \lim_{t\to 0} \frac{1}{t^2}\big(u(x_0 + t\xi) + u(x_0 - t\xi) - 2u(x_0)\big) \geq 0.$$

This implies the desired result. □

Next, we prove a basic property of convex functions.

Theorem 8.1.9. *Let Ω be a domain in $\mathbb{R}^n$ and $u \in L^1(\Omega)$ be a convex function in Ω. Then, u is locally Lipschitz in Ω and, for any $\Omega' \Subset \Omega$,*

$$|u|_{C^{0,1}(\Omega')} \leq C\|u\|_{L^1(\Omega)},$$

where C is a positive constant depending only on n, Ω', and $\operatorname{dist}(\Omega', \partial\Omega)$. Moreover, u is differentiable a.e. in Ω.

Proof. We take an arbitrary ball $B_r(x_0) \subset \Omega$ and claim

$$|u(x_0)| \leq \frac{C}{r^n}\|u\|_{L^1(B_r(x_0))} \tag{1}$$

and, for any $p \in \partial u(x_0)$,

$$|p| \leq \frac{C}{r^{n+1}}\|u\|_{L^1(B_r(x_0))}, \tag{2}$$

where C is a positive constant depending only on n.

We note that (1) implies that u is locally bounded in Ω. Under the assumptions of (1) and (2), we prove that u is locally Lipschitz. To this end, take an arbitrary subdomain $\Omega' \Subset \Omega$ and a positive constant r such that $r < \operatorname{dist}(\Omega', \partial\Omega)$. For any $x_0 \in \Omega'$ and any $p \in \partial u(x_0)$, we have, for any $x \in \Omega$,

$$u(x) \geq u(x_0) + p\cdot(x - x_0). \tag{3}$$

Then, (2) implies, for any $x \in \Omega'$,

$$u(x) \geq u(x_0) - |p||x - x_0| \geq u(x_0) - \frac{C}{r^{n+1}}\|u\|_{L^1(\Omega)}|x - x_0|.$$

Since this holds for any $x, x_0 \in \Omega'$, we can reverse the role of x and x_0. Hence, for any $x, x_0 \in \Omega'$,

$$|u(x) - u(x_0)| \leq \frac{C}{r^{n+1}}\|u\|_{L^1(\Omega)}|x - x_0|.$$

Then, we conclude the first assertion that u is locally Lipschitz. The second assertion follows from Rademacher's theorem that Lipschitz functions are differentiable almost everywhere. Refer to page 81 of [**53**] for details.

Now we proceed to prove (1) and (2). We assume $p \neq 0$. Set

$$B_r^+(x_0) = \{x \in B_r(x_0) : p \cdot (x - x_0) > 0\}.$$

In other words, $B_r^+(x_0)$ is the half-ball of $B_r(x_0)$ in the direction of p. Note that

$$\frac{1}{r^{n+1}} \int_{B_r^+(x_0)} \frac{p}{|p|} \cdot (x - x_0)\, dx$$

is a positive constant depending only on n, independent of x_0, r, and p. In fact, this constant can be calculated explicitly. By integrating (3) with respect to x in $B_r^+(x_0)$, we have

$$u(x_0) + cr|p| \leq \frac{2}{\omega_n r^n} \int_{B_r^+(x_0)} u\, dx, \tag{4}$$

where c is a positive constant depending only on n.

If $u \geq 0$ in Ω, then (4) implies (1) and (2). As a consequence, u is locally Lipschitz.

Next, we consider the general case. We take a supporting function l_{x_0} at x_0. Then, $u - l_{x_0}$ is a nonnegative convex function in Ω. By what we just proved, u is locally Lipchitz in Ω and in particular continuous in Ω. Then, there exists a point $\overline{x} \in B_{r/2}(x_0)$ such that

$$u(\overline{x}) = \frac{2^n}{\omega_n r^n} \int_{B_{r/2}(x_0)} u\, dx.$$

Hence,

$$|u(\overline{x})| \leq \frac{C}{r^n} \int_{B_r(x_0)} |u|\, dx. \tag{5}$$

Next, take any $\overline{p} \in \partial u(\overline{x})$. Note that $B_{r/2}(\overline{x}) \subset B_r(x_0)$. By applying (4) to u in $B_{r/2}(\overline{x})$, we obtain

$$\frac{cr}{2}|\overline{p}| \leq \frac{2 \cdot 2^n}{\omega_n r^n} \int_{B_{r/2}^+(\overline{x})} u\, dx - u(\overline{x}),$$

and hence

$$|\overline{p}| \leq \frac{C}{r^{n+1}} \int_{B_r(x_0)} |u|\, dx. \tag{6}$$

Set

$$\overline{u}(x) = u(x) - u(\overline{x}) - \overline{p} \cdot (x - \overline{x}).$$

Then, $\overline{u}$ is a nonnegative convex function in Ω. Therefore, (1) and (2) hold for $\overline{u}$ at x_0 by what we proved in the special case and hence also for u at x_0 in view of (5), (6), and the fact that $\partial \overline{u}(x_0) = \partial u(x_0) - \{\overline{p}\}$. □

Next, we prove that convex functions are differentiable at points where supporting functions are unique.

Lemma 8.1.10. *Let Ω be a domain in $\mathbb{R}^n$, $u \in C(\Omega)$ be a convex function, and $x_0 \in \Omega$ be a point. If u has a unique supporting function at x_0, then u is differentiable at x_0; namely, there exists an affine function l such that, for any $x \in \Omega$,*

$$u(x) = l(x) + o(|x - x_0|) \quad \text{as } x \to x_0.$$

Proof. Without loss of generality, we assume $x_0 = 0$ and $u \geq u(0) = 0$; otherwise, we consider $u(x + x_0) - l_{x_0}(x + x_0)$ instead, for some supporting function l_{x_0} of u at x_0. Then, $l = 0$ is a supporting function of u at 0. We claim

$$u(x) = o(|x|) \quad \text{as } x \to 0. \tag{1}$$

Since (1) involves behaviors of u near 0, we assume, without loss of generality, that Ω is a bounded domain and u is a bounded function. Take any sequence $x_i \in \Omega$ with $x_i \to 0$ and any sequence $p_i \in \partial u(x_i)$. Then, for any $x \in \Omega$,

$$u(x) \geq u(x_i) + p_i \cdot (x - x_i). \tag{2}$$

Lemma 8.1.7 implies that $\{p_i\}$ is a bounded sequence. Let p_{i_k} be a convergent subsequence with $p_{i_k} \to p_0$, for some $p_0 \in \mathbb{R}^n$. By the continuity of u in Ω and taking the limit in (2), we obtain, for any $x \in \Omega$,

$$u(x) \geq u(0) + p_0 \cdot (x - 0).$$

Hence, $p_0 \in \partial u(0)$. Since u has a unique supporting function at 0, then $p_0 = 0$. We note that any convergent subsequence of p_i has the same limit 0. Then, $p_i \to 0$. Taking $x = 0$ in (2), we get

$$0 \geq u(x_i) - |p_i||x_i|.$$

This implies

$$0 \leq \frac{u(x_i)}{|x_i|} \leq |p_i|,$$

and hence

$$\frac{u(x_i)}{|x_i|} \to 0 \quad \text{as } i \to \infty.$$

We conclude the proof of (1). □

We note that Lemma 8.1.10 is the converse of Lemma 8.1.5(iii).

A deep result due to Alexandrov asserts that convex functions have second-order derivatives almost everywhere. Refer to page 242 of [**53**] for the details.

In many cases, we need to approximate convex functions by smooth convex functions. Consider a function $\eta \in C_0^\infty(B_1)$, with $\eta \geq 0$ in $\mathbb{R}^n$ and $\int_{\mathbb{R}^n} \eta \, dx = 1$. We may further assume that η depends only on $|x|$. Such a function η is called a *mollifier* in $\mathbb{R}^n$. For any positive ε, we set

$$\eta_\varepsilon(x) = \frac{1}{\varepsilon^n}\eta\left(\frac{x}{\varepsilon}\right).$$

Let Ω be a domain in $\mathbb{R}^n$. For each constant $\varepsilon > 0$, set

$$\Omega_\varepsilon = \{x \in \Omega : \operatorname{dist}(x, \partial\Omega) > \varepsilon\}.$$

Consider $u \in L^1_{\mathrm{loc}}(\Omega)$. Define, for any $x \in \Omega_\varepsilon$,

$$u^\varepsilon(x) = \eta_\varepsilon * u(x) = \int_{\mathbb{R}^n} u(y)\eta_\varepsilon(y - x)\, dy.$$

Then, $u^\varepsilon \in C^\infty(\Omega_\varepsilon)$ and $u^\varepsilon \to u$ in $L^1_{\mathrm{loc}}(\Omega)$ as $\varepsilon \to 0$. Moreover, if $u \in C(\Omega)$, then $u^\varepsilon \to u$ locally uniformly in Ω as $\varepsilon \to 0$.

We now characterize the convexity in an equivalent way.

Lemma 8.1.11. *Let Ω be a convex domain in $\mathbb{R}^n$ and let $u \in L^1(\Omega)$. Then, u is convex in Ω as in Definition* 8.1.2(ii) *if and only if, for any $x_1, x_2 \in \Omega$ and any $t \in (0, 1)$,*

$$u(tx_1 + (1-t)x_2) \leq tu(x_1) + (1-t)u(x_2). \tag{1}$$

Proof. We first prove the *only if* part. Suppose that u satisfies Definition 8.1.2(ii). For any fixed $x_1, x_2 \in \Omega$ and $t \in (0,1)$, we write $x_0 = tx_1 + (1-t)x_2$ and take a supporting function l of u at x_0. Then, $l(x_0) = u(x_0)$ and $l(x_i) \leq u(x_i)$ for $i = 1, 2$. Since l is affine, we have

$$\begin{aligned} u(x_0) = l(x_0) &= tl(x_1) + (1-t)l(x_2) \\ &\leq tu(x_1) + (1-t)u(x_2). \end{aligned}$$

We now prove the *if* part. We assume (1) and prove that u satisfies Definition 8.1.2(ii). We first consider a special case where $u \in C^1(\Omega)$. We fix an $x_0 \in \Omega$. Then, for any $x \in \Omega$ and any $t \in (0, 1)$,

$$u(x_0 + t(x - x_0)) \leq u(x_0) + t\big(u(x) - u(x_0)\big),$$

and hence

$$\frac{1}{t}\big[u(x_0 + t(x - x_0)) - u(x_0)\big] \leq u(x) - u(x_0).$$

Letting $t \to 0$, we obtain, for any $x \in \Omega$.

$$u(x) \geq u(x_0) + \nabla u(x_0) \cdot (x - x_0).$$

Next, we consider the general case. Let η be a mollifier and define $u^\varepsilon = \eta_\varepsilon * u$, for any positive ε. Fix a convex subdomain $\Omega' \Subset \Omega$ and consider $\varepsilon < \operatorname{dist}(\Omega', \partial\Omega)$. We claim, for any $x_1, x_2 \in \Omega'$ and any $t \in [0,1]$,

$$u^\varepsilon(tx_1 + (1-t)x_2) \le tu^\varepsilon(x_1) + (1-t)u^\varepsilon(x_2). \tag{2}$$

To prove (2), we fix $x_1, x_2 \in \Omega'$ and $t \in (0,1)$. Then, for each $y \in B_\varepsilon$,

$$\begin{aligned} u\big(y + (tx_1 + (1-t)x_2)\big) &= u\big(t(y+x_1) + (1-t)(y+x_2)\big) \\ &\le tu(y+x_1) + (1-t)u(y+x_2). \end{aligned}$$

By multiplying the above inequality by $\eta_\varepsilon(y) \ge 0$ and integrating over B_ε, we have

$$\begin{aligned} &u^\varepsilon(tx_1 + (1-t)x_2) \\ &\quad = \int u(y + (tx_1 + (1-t)x_2))\eta_\varepsilon(y)\,dy \\ &\quad \le t\int u(y+x_1)\eta_\varepsilon(y)\,dy + (1-t)\int u(y+x_2)\eta_\varepsilon(y)\,dy \\ &\quad = tu^\varepsilon(x_1) + (1-t)u^\varepsilon(x_2). \end{aligned}$$

This ends the proof of (2). Since u^ε is smooth, we can apply what we proved in the special case to u^ε and conclude that u^ε is convex in Ω' as in Definition 8.1.2(ii). Fix an $x_0 \in \Omega'$. We have, for any $x \in \Omega'$,

$$u^\varepsilon(x) \ge u^\varepsilon(x_0) + \nabla u^\varepsilon(x_0) \cdot (x - x_0). \tag{3}$$

For any compact subset $K \subset \Omega$, we take a convex subdomain Ω' such that $K \subset \Omega' \Subset \Omega$. By applying Theorem 8.1.9 to u^ε in Ω', we obtain

$$\sup_K(|u^\varepsilon| + |\nabla u^\varepsilon|) \le C\|u^\varepsilon\|_{L^1(\Omega')},$$

where C is a positive constant depending only on n and $\operatorname{dist}(K, \partial\Omega')$. With $\varepsilon < \operatorname{dist}(\Omega', \partial\Omega)$, we have

$$\sup_K(|u^\varepsilon| + |\nabla u^\varepsilon|) \le C\|u\|_{L^1(\Omega)}.$$

Take any sequence $\varepsilon_i \to 0$. Then, $\{u^{\varepsilon_i}\}$ is a sequence of bounded and equicontinuous functions in K. By the Arzela-Ascoli theorem, there exists a uniformly convergent subsequence of $\{u^{\varepsilon_i}\}$ in K. We now take a sequence of increasing compact subsets $K_i \subset \Omega$ such that $\bigcup_i K_i = \Omega$ and, by a diagonalization process, we have a subsequence of $\{u^{\varepsilon_i}\}$ uniformly convergent in any compact subset $K \subset \Omega$. The limit $\widetilde{u}$ is a continuous function in Ω. Since $u^{\varepsilon_i} \to u$ in L^1 locally in Ω, then $\widetilde{u} = u$, which is independent of the sequence

$\{\varepsilon_i\}$. For each fixed $x_0 \in \Omega$, we have the convergence $\nabla u^{\varepsilon_i}(x_0) \to p$ for some $p \in \mathbb{R}^n$. Letting $\varepsilon_i \to 0$ in (3), we obtain, for any $x \in \Omega'$,

$$u(x) \geq u(x_0) + p \cdot (x - x_0).$$

Since Ω' is arbitrary, the above inequality holds for any $x \in \Omega$. Hence, u is convex at x_0 as in Definition 8.1.2(ii). □

In fact, we have the following convergence result from the proof of Lemma 8.1.11.

Lemma 8.1.12. *Let Ω be a domain in $\mathbb{R}^n$, $u \in L^1(\Omega)$ be convex, and $u^\varepsilon = \eta_\varepsilon * u$, for a mollifier η. Then, for any convex subdomain $\Omega' \Subset \Omega$ and any positive constant $\varepsilon < \operatorname{dist}(\Omega', \partial\Omega)$, u^ε is convex in Ω' and*

$$\sup_{\Omega'}(|u^\varepsilon| + |\nabla u^\varepsilon|) \leq C\|u\|_{L^1(\Omega)},$$

where C is a positive constant depending only on n and $\operatorname{dist}(\Omega', \partial\Omega)$. Moreover, u^ε converges to u uniformly in any compact subset of Ω.

In many cases, it is convenient to employ the equivalent description (1) of the convexity in Lemma 8.1.11. For example, it is easy to prove by (1) that the supremum of certain linear functions is convex.

We now introduce the notion of Legendre transforms.

Definition 8.1.13. Let Ω and Ω^* be domains in $\mathbb{R}^n$ and u be a continuous function in Ω. For any $p \in \Omega^*$, define

$$u^*(p) = \sup\{x \cdot p - u(x) : x \in \Omega\}.$$

The function u^* is called the *Legendre transform* of u.

Under appropriate assumptions of Ω and u, u^* is a well-defined function.

Lemma 8.1.14. *Let Ω be a bounded domain in $\mathbb{R}^n$, u be a bounded continuous function in Ω, and u^* be the Legendre transform of u defined in some domain $\Omega^* \subset \mathbb{R}^n$. Then, u^* is a well-defined convex function in Ω^*.*

Proof. Since Ω is bounded and u is bounded, then u^* is a well-defined function taking a finite value at each point $p \in \Omega^*$. In fact, $u^* \in L^\infty_{\mathrm{loc}}(\Omega^*)$. Since u^* is the supremum of linear functions, u^* is convex by Lemma 8.1.11. We now provide an alternative proof by using Definition 8.1.2(ii). Take an arbitrary point $p_0 \in \Omega^*$. Then, there exists a sequence $\{x_m\} \subset \Omega$ such that

$$p_0 \cdot x_m - u(x_m) \to u^*(p_0).$$

By passing to subsequences, we assume $x_m \to x_0$ and $u(x_m) \to a$ for some point $x_0 \in \overline{\Omega}$ and some constant a. Hence,

$$u^*(p_0) = p_0 \cdot x_0 - a. \tag{1}$$

Now, we take an arbitrary $p \in \Omega^*$. By the definition of $u^*(p)$, we have, for any m,

$$u^*(p) \geq p \cdot x_m - u(x_m).$$

Taking the limit $m \to \infty$, we get

$$u^*(p) \geq p \cdot x_0 - a. \tag{2}$$

By subtracting (1) from (2), we obtain, for any $p \in \Omega^*$,

$$u^*(p) \geq u^*(p_0) + x_0 \cdot (p - p_0).$$

This implies that u^* is convex at p_0. □

In the proof above, if $x_0 \in \Omega$, then $a = u(x_0)$ by the continuity of u, and $p_0 \in \partial u(x_0)$. In other words, the affine function $u(x_0) + p_0 \cdot (x - x_0)$ is a supporting function of u at x_0.

By Definition 8.1.13, we have, for any $x \in \Omega$ and any $p \in \Omega^*$,

$$u(x) + u^*(p) \geq x \cdot p.$$

For given Ω and u, we usually take Ω^* to be a domain containing $\partial u(\Omega)$, where ∂u is the normal mapping of u as in Definition 8.1.4. The simplest choice is $\Omega^* = \partial u(\Omega)$ or $\Omega^* = \mathbb{R}^n$.

Lemma 8.1.15. *Let Ω be a domain in $\mathbb{R}^n$, u be a continuous function in Ω, and u^* be its Legendre transform defined in some domain $\Omega^* \subset \mathbb{R}^n$ containing $\partial u(\Omega)$. Assume $x \in \Omega$ is a point with $\partial u(x) \neq \emptyset$. Then, for any $p \in \partial u(x)$,*

$$u(x) + u^*(p) = x \cdot p, \tag{1}$$

and $x \in \partial u^*(p)$. Moreover, $(u^*)^*(x) = u(x)$.

Proof. Take any $x_0 \in \Omega$ and any $p_0 \in \partial u(x_0)$. Then, for any $x \in \Omega$,

$$u(x) \geq u(x_0) + p_0 \cdot (x - x_0),$$

or

$$p_0 \cdot x_0 - u(x_0) \geq p_0 \cdot x - u(x).$$

Hence, $\sup\{p_0 \cdot x - u(x) : x \in \Omega\}$ is attained at x_0, and then

$$u^*(p_0) = p_0 \cdot x_0 - u(x_0). \tag{2}$$

This yields (1), with x and p there replaced by x_0 and p_0.

Next, we take any $p \in \Omega^*$. The definition of $u^*(p)$ implies

$$u^*(p) \geq p \cdot x_0 - u(x_0). \tag{3}$$

By subtracting (2) from (3), we obtain

$$u^*(p) \geq u^*(p_0) + x_0 \cdot (p - p_0).$$

Therefore, $x_0 \in \partial u^*(p_0)$.

We rewrite (2) and (3) as

$$u(x_0) = x_0 \cdot p_0 - u^*(p_0)$$

and, for any $p \in \Omega^*$,

$$u(x_0) \geq x_0 \cdot p - u^*(p).$$

Then,

$$u(x_0) = \sup\{x_0 \cdot p - u^*(p) : p \in \Omega^*\}.$$

Therefore, $(u^*)^*(x_0) = u(x_0)$. □

Legendre transforms play important roles in the study of convex functions. We now discuss one application.

Lemma 8.1.16. *Let Ω be a domain in $\mathbb{R}^n$ and u be a continuous function in Ω. Set*

$$S = \{p \in \mathbb{R}^n : p \in \partial u(x_1) \cap \partial u(x_2) \text{ for some } x_1, x_2 \in \Omega \text{ with } x_1 \neq x_2\}.$$

Then, S has Lebesgue measure zero.

We note that S is the collection of points in $\mathbb{R}^n$ that belong to the image by the normal mapping of more than one point of Ω.

Proof. We first consider the case where Ω is a bounded domain and u is a bounded function in Ω. Take a domain Ω^* in $\mathbb{R}^n$ such that $\partial u(\Omega) \subset \Omega^*$ and let u^* be the Legendre transform of u defined in Ω^*. Lemma 8.1.14 implies that u^* is convex in Ω^*. Then, u^* is differentiable a.e. in Ω^* by Theorem 8.1.9. We now claim that u^* is not differentiable at any $p \in S$. Then, S has measure zero. To prove the claim, we take $p \in S$ such that $p \in \partial u(x_1) \cap \partial u(x_2)$ for some $x_1, x_2 \in \Omega$ with $x_1 \neq x_2$. By Lemma 8.1.15, $x_i \in \partial u^*(p)$, for $i = 1, 2$. In other words, u^* has two different supporting functions at p given by $l_i(q) = u^*(p) + x_i \cdot (q - p)$, for $i = 1, 2$. Hence, u^* is not differentiable at p.

In the general case, we write

$$\Omega = \bigcup_{m=1}^{\infty} \Omega_m,$$

where $\{\Omega_m\}$ is a sequence of increasing bounded domains in Ω. Hence, u is bounded in each Ω_m. Take any $p \in S$. There exist some points $x_1, x_2 \in \Omega$, with $x_1 \neq x_2$, such that, for any $x \in \Omega$ and any $i = 1, 2$,

$$u(x) \geq u(x_i) + p \cdot (x - x_i). \tag{1}$$

Since Ω_m increases, we have $x_1, x_2 \in \Omega_m$ for m sufficiently large, and (1) obviously holds for any $x \in \Omega_m$. Now, we set

$$S_m = \{p \in \mathbb{R}^n : p \in \partial(u|_{\Omega_m})(x_1) \cap \partial(u|_{\Omega_m})(x_2)$$
$$\text{for some } x_1, x_2 \in \Omega_m \text{ with } x_1 \neq x_2\}.$$

We have proved

$$S \subset \bigcup_{m=1}^{\infty} S_m.$$

By applying what we proved in the special case to $u|_{\Omega_m}$, we conclude that S_m has measure zero. □

Now, we prove an important result concerning normal mappings.

Theorem 8.1.17. *Let Ω be a domain in $\mathbb{R}^n$ and u be a continuous function in Ω.*

(i) *If $K \subset \Omega$ is compact, then $\partial u(K)$ is compact in $\mathbb{R}^n$.*

(ii) *If $E \subset \Omega$ is a Borel set, then $\partial u(E)$ is Lebesgue measurable in $\mathbb{R}^n$.*

(iii) *If $\{E_m\}$ is a sequence of disjoint Borel sets in Ω, then*

$$\left|\partial u\left(\bigcup_m E_m\right)\right| = \sum_m |\partial u(E_m)|.$$

Proof. (i) We consider a sequence $\{p_m\} \subset \partial u(K)$. For each m, there exists a point $x_m \in K$ such that, for any $x \in \Omega$,

$$u(x) \geq u(x_m) + p_m \cdot (x - x_m). \tag{1}$$

Since K is compact in Ω, we assume, by passing through a subsequence, $x_m \to x_0$ for some $x_0 \in K$. Next, we take a subdomain Ω' such that $K \subset \Omega' \Subset \Omega$. Then, $\operatorname{dist}(K, \partial\Omega') > r$ for some positive constant r and u is bounded in Ω'. By applying Lemma 8.1.7 to u in Ω', we obtain

$$|p_m| \leq \frac{1}{r}\left(\sup_{\Omega'} u - \inf_{\Omega'} u\right).$$

Hence, $\{p_m\}$ is a bounded sequence. We assume, by passing through a subsequence, $p_m \to p_0$ for some $p_0 \in \mathbb{R}^n$. By letting $m \to \infty$ in (1) and the continuity of u, we conclude, for any $x \in \Omega$,

$$u(x) \geq u(x_0) + p_0 \cdot (x - x_0).$$

This implies $p_0 \in \partial u(x_0)$, and hence $p_0 \in \partial u(K)$.

(ii) We set

$$\mathcal{S} = \{E \subset \Omega : \partial u(E) \text{ is Lebesgue measurable}\}.$$

By (i), the class $\mathcal{S}$ contains all compact subsets of Ω. Next, we note that if $\{E_m\}$ is a sequence of subsets of Ω, then

$$\partial u\Big(\bigcup_m E_m\Big) = \bigcup_m \partial u(E_m). \tag{2}$$

Hence, if $E_m \in \mathcal{S}$ for $m = 1, 2, \ldots$, then $\bigcup_m E_m \in \mathcal{S}$. In particular, by writing

$$\Omega = \bigcup_m K_m,$$

for a sequence of compact subsets $\{K_m\}$, we conclude $\Omega \in \mathcal{S}$. Next, we note that, for any subset $E \subset \Omega$,

$$\partial u(\Omega \setminus E) = \big(\partial u(\Omega) \setminus \partial u(E)\big) \cup \big(\partial u(\Omega \setminus E) \cap \partial u(E)\big).$$

By Lemma 8.1.16, we have $|\partial u(\Omega \setminus E) \cap \partial u(E)| = 0$ for any subset $E \subset \Omega$. We then obtain $\Omega \setminus E \in \mathcal{S}$ if $E \in \mathcal{S}$. In conclusion, the class $\mathcal{S}$ is a σ-algebra containing all Borel subsets of Ω.

(iii) In view of (2), we will prove

$$\Big|\bigcup_m \partial u(E_m)\Big| = \sum_m |\partial u(E_m)|. \tag{3}$$

To this end, we set $F_m = \partial u(E_m)$ and write

$$\bigcup_m F_m = F_1 \bigcup (F_2 \setminus F_1) \bigcup (F_3 \setminus (F_1 \cup F_2))$$
$$\bigcup (F_4 \setminus (F_1 \cup F_2 \cup F_3)) \bigcup \cdots,$$

where the sets in the right-hand side are disjoint. On the other hand, we have

$$F_m = \big(F_m \cap (F_{m-1} \cup \cdots \cup F_1)\big) \bigcup \big(F_m \setminus (F_{m-1} \cup \cdots \cup F_1)\big).$$

Since $E_i \cap E_j = \emptyset$ for $i \neq j$, we have, by Lemma 8.1.16,

$$\big|F_m \cap (F_{m-1} \cup \cdots \cup F_1)\big| = 0,$$

and hence

$$|F_m| = \big|F_m \setminus (F_{m-1} \cup \cdots \cup F_1)\big|.$$

Therefore,

$$\Big|\bigcup_m F_m\Big| = \sum_m |F_m|,$$

which is (3). $\square$

Next, we introduce the important Monge-Ampère measures.

Definition 8.1.18. Let Ω be a domain in $\mathbb{R}^n$ and u be a continuous function in Ω. Define, for any Borel set $E \subset \Omega$,

$$Mu(E) = \big|\partial u(E)\big|.$$

We call Mu the *Monge-Ampère measure* associated with the function u.

Theorem 8.1.17 asserts that Mu is a locally finite Borel measure in Ω.

Now, we calculate Monge-Ampère measures associated with C^2-convex functions.

Proposition 8.1.19. *Let Ω be a domain in $\mathbb{R}^n$ and $u \in C^2(\Omega)$ be a convex function. Then,*

$$Mu = \det \nabla^2 u \, dx;$$

namely, for any Borel subset $E \subset \Omega$,

$$Mu(E) = \int_E \det \nabla^2 u \, dx.$$

We point out that the domain Ω is not assumed to be convex.

Proof. Since $u \in C^2(\Omega)$ is convex, then $\nabla^2 u \geq 0$ in Ω by Lemma 8.1.8. Set

$$A = \{x \in \Omega : \nabla^2 u(x) > 0\}.$$

We claim that ∇u is one-to-one in A. To prove this, we take $x_1, x_2 \in A$ with $\nabla u(x_1) = \nabla u(x_2)$. The convexity implies, for any $x \in \Omega$ and any $i = 1, 2$,

$$u(x) \geq u(x_i) + \nabla u(x_i) \cdot (x - x_i).$$

By taking $i = 1$ and $x = x_2$ and then $i = 2$ and $x = x_1$, we have

$$u(x_2) - u(x_1) = \nabla u(x_1) \cdot (x_2 - x_1).$$

If Ω is convex, we have, by the Taylor expansion,

$$\begin{aligned} u(x_2) = {} & u(x_1) + \nabla u(x_1) \cdot (x_2 - x_1) \\ & + \int_0^1 t(x_2 - x_1)^T \nabla^2 u(x_1 + t(x_2 - x_1))(x_2 - x_1) \, dt. \end{aligned}$$

Then, the integral is zero and the integrand vanishes for all $t \in [0, 1]$. Since $x_1 \in A$, then $x_1 + t(x_2 - x_1) \in A$ for t sufficiently small and hence $x_1 = x_2$. If Ω is not necessarily convex, we proceed as follows. Set

$$v(x) = u(x) - u(x_1) - \nabla u(x_1) \cdot (x - x_1).$$

Then, $v \in C^2(\Omega)$ is a convex function, with $v(x_1) = v(x_2) = 0$, $\nabla v(x_1) = \nabla v(x_2) = 0$, and $\nabla^2 u = \nabla^2 v$. In particular, $v \geq 0$ in Ω and $l = 0$ is the supporting function of v at both x_1 and x_2. If $x_1 \neq x_2$, we assume, without loss of generality, that x_1 and x_2 are on the x_n-axis with $(x_2 - x_1) \cdot e_n > 0$. Take an arbitrary point $x_0 \in \overline{x_1 x_2} \cap \Omega$. Then, for any $x \in \Omega$,

$$v(x) \geq v(x_0) + \nabla v(x_0) \cdot (x - x_0).$$

Since v is C^2 and $\nabla^2 v(x_1) > 0$, by taking x_0 sufficiently close to x_1, we have $v(x_0) > 0$ and $\nabla v(x_0) \cdot e_n > 0$. Hence,

$$v(x_2) \geq v(x_0) + |x_2 - x_0| \nabla v(x_0) \cdot e_n > 0,$$

which leads to a contradiction. Therefore, $x_1 = x_2$.

Since $u \in C^2(\Omega)$, then $\nabla u \in C^1(\Omega)$ and $Mu(E) = |\nabla u(E)|$ for any Borel set $E \subset \Omega$. We write

$$\nabla u(E) = \nabla u(E \cap A) \bigcup \nabla u(E \setminus A).$$

Since $\Omega \setminus A = \{x \in \Omega : \det \nabla^2 u(x) = 0\}$, then $|\nabla u(\Omega \setminus A)| = 0$ by Sard's theorem. By a change of variables, we obtain

$$\begin{aligned} Mu(E) &= Mu(E \cap A) + Mu(E \setminus A) = |\nabla u(E \cap A)| \\ &= \int_{E \cap A} \det \nabla^2 u \, dx = \int_E \det \nabla^2 u \, dx. \end{aligned}$$

This is the desired result. □

We now consider an example.

Example 8.1.20. Let $B_R(x_0)$ be a ball in $\mathbb{R}^n$ and set, for some $h > 0$,

$$u(x) = \frac{h}{R}|x - x_0|.$$

This is the function in Example 8.1.6. Then, it is easy to check that

$$Mu = |B_{h/R}| \delta_{x_0},$$

where δ_{x_0} is the Dirac measure at x_0.

Next, we discuss the convergence of Monge-Ampère measures.

Lemma 8.1.21. *Let Ω be a domain in $\mathbb{R}^n$ and $u_m, u \in C(\Omega)$ be convex functions. Assume $u_m \to u$ locally uniformly in Ω. Then, for any compact subset $K \subset \Omega$ and any open subset $U \Subset \Omega$,*

$$\limsup_{m \to \infty} Mu_m(K) \leq Mu(K),$$

and

$$Mu(U) \leq \liminf_{m \to \infty} Mu_m(U).$$

Proof. (i) We first prove

$$\limsup_{m\to\infty} \partial u_m(K) \subset \partial u(K). \tag{1}$$

Then,

$$\limsup_{m\to\infty} \chi_{\partial u_m(K)} \leq \chi_{\partial u(K)},$$

and the first desired result follows from the dominated convergence theorem. To prove (1), we take any $p \in \limsup_{m\to\infty} \partial u_m(K)$. Then, there exists a subsequence $m' \to \infty$ such that $p_{m'} \to p$ for some $p_{m'} \in \partial u_{m'}(K)$; namely, $p_{m'} \in \partial u_{m'}(x_{m'})$ for some $x_{m'} \in K$. By passing to a subsequence if necessary, we assume $x_{m'} \to x_0$ for some $x_0 \in K$. We note that u_m is Lipschitz in K with a uniform Lipschitz norm by Lemma 8.1.7. Hence, $u_{m'}(x_{m'}) \to u(x_0)$. The definition of $\partial u_{m'}(x_{m'})$ implies, for any $x \in \Omega$,

$$u_{m'}(x) \geq u_{m'}(x_{m'}) + p_{m'} \cdot (x - x_{m'}).$$

By letting $m' \to \infty$, we obtain

$$u(x) \geq u(x_0) + p \cdot (x - x_0).$$

Hence, $p \in \partial u(x_0)$, which implies (1).

(ii) Set

$$S = \{p \in \mathbb{R}^n : p \in \partial u(x_1) \cap \partial u(x_2) \text{ for some } x_1, x_2 \in \Omega \text{ with } x_1 \neq x_2\}.$$

By Lemma 8.1.16, $|S| = 0$. We now prove, for any compact subset $K \subset U$,

$$\partial u(K) \setminus S \subset \liminf_{m\to\infty} \partial u_m(U). \tag{2}$$

Then,

$$\chi_{\partial u(K)} \leq \liminf_{m\to\infty} \chi_{\partial u_m(U)} \quad \text{a.e. in } \mathbb{R}^n,$$

and the Fatou lemma implies

$$Mu(K) \leq \liminf_{m\to\infty} Mu_m(U).$$

Since K is an arbitrary compact subset in U, we have the second desired result. To prove (2), we take any $p \in \partial u(K) \setminus S$. Then, there exists a point $x_0 \in K$ such that $p \in \partial u(x_0)$ and $p \notin \partial u(x_1)$ for any $x_1 \in \Omega \setminus \{x_0\}$. Hence, for any $x \in \Omega$,

$$u(x) \geq u(x_0) + p \cdot (x - x_0).$$

We also note that, for any $x \in \Omega \setminus \{x_0\}$,

$$u(x) > u(x_0) + p \cdot (x - x_0). \tag{3}$$

If (3) does not hold, there exists an $x_1 \in \Omega \setminus \{x_0\}$ such that

$$u(x_1) = u(x_0) + p \cdot (x_1 - x_0).$$

Then, for any $x \in \Omega$,

$$\begin{aligned} u(x) &\geq u(x_0) + p \cdot (x - x_0) \\ &= u(x_1) + p \cdot (x - x_1). \end{aligned}$$

This implies $p \in \partial u(x_1)$, which leads to a contradiction. Therefore, (3) holds for any $x \in \Omega \setminus \{x_0\}$, and in particular for any $x \in \partial U$. By the compactness of ∂U and the uniform convergence of $u_m \to u$ in compact sets, there exist some constant $\varepsilon > 0$ and some integer m_0 such that, for any $m \geq m_0$ and any $x \in \partial U$,

$$u_m(x) \geq u_m(x_0) + p \cdot (x - x_0) + \varepsilon. \tag{4}$$

Next, we set

$$\delta_m = \min\{u_m(x) - u_m(x_0) - p \cdot (x - x_0) : x \in \overline{U}\}.$$

By evaluating at x_0, we have $\delta_m \leq 0$. Assume δ_m is attained at some $x_m \in \overline{U}$. Then, (4) implies $x_m \in U$. By the definitions of δ_m and x_m, we obtain, for any $x \in \overline{U}$,

$$u_m(x) \geq u_m(x_0) + p \cdot (x - x_0) + \delta_m$$

and

$$u_m(x_m) = u_m(x_0) + p \cdot (x_m - x_0) + \delta_m.$$

A simple subtraction implies, for any $x \in \bar{U}$,

$$u_m(x) \geq u_m(x_m) + p \cdot (x - x_m). \tag{5}$$

Since u_m is convex in Ω, U is open, and $x_m \in U$, (5) holds for any $x \in \Omega$. This follows easily from Lemma 8.1.3. Therefore, $p \in \partial u_m(x_m)$ for any $m \geq m_0$, which implies (2). □

We then conclude the weak convergence of Monge-Ampère measures.

Theorem 8.1.22. *Let Ω be a domain in $\mathbb{R}^n$ and $u_m, u \in C(\Omega)$ be convex functions. Assume $u_m \to u$ locally uniformly in Ω. Then, Mu_m converges weakly as measures to Mu in Ω; namely, for any $f \in C_0(\Omega)$,*

$$\int_\Omega f\, dMu_m(x) \to \int_\Omega f\, dMu(x).$$

Refer to page 54 of [**53**] for a proof.

Next, we introduce the notion of generalized solutions.

Definition 8.1.23. Let Ω be a domain in $\mathbb{R}^n$, ν be a Borel measure in Ω, and u be a continuous function in Ω. Then, u is a *generalized solution*, in the sense of Alexandrov, of the Monge-Ampère equation

$$\det \nabla^2 u = \nu$$

if u is locally convex and $Mu = \nu$. Here, Mu is the Monge-Ampère measure associated with u defined as in Definition 8.1.18.

In Definition 8.1.23, we regard $\det \nabla^2 u$ as a measure when u is a generalized solution. However, when u is C^2, we also regard $\det \nabla^2 u$ as a function, as in the usual sense.

Theorem 8.1.22 asserts that the class of generalized solutions is closed under the uniform convergence. Specifically, if $\{u_m\}$ is a sequence of generalized solutions of $\det \nabla^2 u = \nu$ in Ω and $u_m \to u$ locally uniformly in Ω, then u is also a generalized solution of $\det \nabla^2 u = \nu$ in Ω.

Theorem 8.1.22 has many important applications. We now prove a result concerning the sum of two convex functions.

Lemma 8.1.24. *Let Ω be a bounded convex domain in $\mathbb{R}^n$ and $u, v \in C(\bar{\Omega})$ be convex functions in Ω. Then,*

$$M(u+v) \geq Mu + Mv;$$

i.e., for any Borel subset $\omega \subset \Omega$,

$$M(u+v)(\omega) \geq Mu(\omega) + Mv(\omega).$$

Proof. We first consider the case $u, v \in C^2(\Omega)$. We note that if A and B are symmetric and nonnegative definite matrices, then

$$\det(A+B) \geq \det A + \det B.$$

Hence,

$$\det(\nabla^2 u + \nabla^2 v) \geq \det \nabla^2 u + \det \nabla^2 v \quad \text{in } \Omega.$$

Proposition 8.1.19 implies the desired result.

If at least one of u and v is not C^2, we can approximate u or v by a sequence of convex $C^2(\Omega)$-functions converging locally uniformly in Ω. This can be achieved by Lemma 8.1.12. Then, Theorem 8.1.22 implies the desired result. □

We now construct a convex generalized solution of the Monge-Ampère equation which is not necessarily C^2, even if the corresponding Monge-Ampère measure is given by a nice function. As we will see, the singular set, where the solution fails to be C^2, consists of a line segment which extends to the boundary of the maximal region of the existence. Such an example is due to Pogorelov [**122**].

Theorem 8.1.25. *For $n \geq 3$, there exists a convex generalized solution $u \in C^{1,1-\frac{2}{n}}(B_R)$ of*

$$\det \nabla^2 u = 1 \quad \text{in } B_R,$$

for some positive constant R depending only on n. Moreover, such a u is smooth except along a line through the origin.

Proof. In $\mathbb{R}^n$, we set $x = (x', x_n)$, with $x' = (x_1, \ldots, x_{n-1}) \in \mathbb{R}^{n-1}$, and $\rho = |x'|$.

Step 1. We first consider the function of the form $u = u(\rho, x_n)$ and calculate $\det \nabla^2 u$. We have, for any $\alpha, \beta = 1, \ldots, n-1$,

$$\begin{aligned}
u_\alpha &= u_\rho \frac{x_\alpha}{\rho}, \\
u_{\alpha\beta} &= u_{\rho\rho} \frac{x_\alpha x_\beta}{\rho^2} + u_\rho \left(\frac{\delta_{\alpha\beta}}{\rho} - \frac{x_\alpha x_\beta}{\rho^3} \right), \\
u_{\alpha n} &= u_{\rho x_n} \frac{x_\alpha}{\rho}.
\end{aligned}$$

A straightforward calculation yields

$$\det \nabla^2 u = (\rho^{-1} u_\rho)^{n-2} (u_{\rho\rho} u_{x_n x_n} - u_{\rho x_n}^2). \tag{1}$$

In fact, we can employ the following trick. By an appropriate rotation in $\mathbb{R}^{n-1}$, we calculate $\det \nabla^2 u$ at $(0, \ldots, 0, \rho, x_n)$. At such a point, we have

$$\nabla^2 u = \begin{pmatrix} \rho^{-1} u_\rho & & & & \\ & \ddots & & & \\ & & \rho^{-1} u_\rho & & \\ & & & u_{\rho\rho} & u_{\rho x_n} \\ & & & u_{\rho x_n} & u_{x_n x_n} \end{pmatrix},$$

where all the other elements are zero. This implies (1).

Step 2. Next, we consider

$$u = \rho^a v(x_n).$$

We will find a constant a and a function v such that u is convex and

$$\det \nabla^2 u = 1.$$

A simple calculation, with the help of (1), yields

$$\det \nabla^2 u = a^{n-1} \rho^{na-2n+2} v^{n-2} \left((a-1) v v'' - a v'^2 \right).$$

Note that $\rho = 0$ at the origin. In order to have a positive $\det \nabla^2 u$, we require $na = 2n - 2$, or

$$a = 2 - \frac{2}{n}. \tag{2}$$

In the following, a is fixed by (2). Next, we find v such that

$$a^{n-1}v^{n-2}\Big((a-1)vv'' - av'^2\Big) = 1. \tag{3}$$

This is a nonlinear ordinary differential equation. We prescribe

$$v(0) = 1, \quad v'(0) = 0.$$

The existence of v in $(-c, c) \subset \mathbb{R}$, for some $c > 0$, follows from the standard theory of ordinary differential equations. It is easy to see that v is smooth, positive, and convex in $(-c, c)$.

Step 3. In summary, we set

$$u(x', x_n) = |x'|^{2-\frac{2}{n}} v(x_n),$$

where v is a positive and convex solution of (3) in $(-c, c)$, for some $c > 0$. Then,

$$u \text{ is convex and } C^{1,1-\frac{2}{n}} \text{ in } \mathbb{R}^{n-1} \times (-c, c),$$
$$u \text{ is smooth in } \mathbb{R}^{n-1} \setminus \{0\} \times (-c, c),$$

and

$$\det \nabla^2 u = 1 \quad \text{in } \mathbb{R}^{n-1} \setminus \{0\} \times (-c, c). \tag{4}$$

Next, we prove that u is a generalized solution of $\det \nabla^2 u = 1$ in $\mathbb{R}^{n-1} \times (-c, c)$. To this end, we claim, for any domain $\Omega \subset \mathbb{R}^{n-1} \times (-c, c)$,

$$|\partial u(\Omega)| = |\Omega|. \tag{5}$$

First, (5) holds for any domain $\Omega \subset \mathbb{R}^{n-1} \setminus \{0\} \times (-c, c)$, by (4) and Proposition 8.1.19. Next, by the explicit expression of ∇u, we have, for any fixed $\tau \in (0, c)$,

$$|\partial u(B'_\varepsilon \times (-\tau, \tau))| \to 0 \quad \text{as } \varepsilon \to 0.$$

As a result, (4) holds for any domain $\Omega \subset \mathbb{R}^{n-1} \times (-c, c)$. □

Next, we introduce a class of convex sets which play a role of significant importance in the study of Monge-Ampère equations. Let Ω be a convex domain in $\mathbb{R}^n$ and $u \in C(\Omega)$ be a convex function in Ω. For any $x_0 \in \mathbb{R}^n$, any supporting function l_{x_0} of u at x_0, and any $h > 0$, we define

$$S_{u,h}(x_0) = \{x \in \Omega : \ u(x) < l_{x_0}(x) + h\}.$$

These sets are called *sections* of u at x_0. We point out that sections also depend on supporting functions. When it is clear from the context, we usually write $S_h(x_0)$. Moreover, if u assumes its minimum at x_0, we simply write S_h.

To end this section, we discuss briefly how solutions change by affine transforms. Let Ω be a domain in $\mathbb{R}^n$ and $u \in C(\Omega)$ be a convex function in Ω. Consider an invertible affine transform T in $\mathbb{R}^n$ given by $Tx = Ax + y_*$, for some point $y_* \in \mathbb{R}^n$ and some invertible matrix A. For some positive constant μ, define $v = \mu u \circ T^{-1}$; namely, for any $y \in T(\Omega)$,

$$v(y) = \mu u(T^{-1}y).$$

It is straightforward to check that, for any $x_0 \in \Omega$, the function $l(x) = u(x_0) + p \cdot (x - x_0)$ is a supporting function of u at x_0 if and only if $\widetilde{l}(y) = v(y_0) + \mu A^{-T} p \cdot (y - y_0)$ is a supporting function of v at $y_0 = Tx_0$. In other words, $p \in \partial u(x_0)$ if and only if $\mu A^{-T} p \in \partial v(Tx_0)$. Hence, for any subset $\omega \subset \Omega$, $p \in \partial u(w)$ if and only if $\mu A^{-T} p \in \partial v(T(\omega))$. Therefore,

$$\partial v(T(\omega)) = \mu A^{-T} \partial u(\omega).$$

This implies, for any subset $\omega \subset \Omega$,

$$|\partial v(T(\omega))| = \mu^n |\det A|^{-1} |\partial u(\omega)|,$$

or

$$Mv(T(\omega)) = \mu^n |\det A|^{-1} Mu(\omega).$$

Now, we assume ν is a finite measure in Ω and $u \in C(\Omega)$ is a convex generalized solution of

$$\det \nabla^2 u = \nu \quad \text{in } \Omega.$$

Then, $v \in C(T(\Omega))$ is a generalized solution of

$$\det \nabla^2 v = \widetilde{\nu} \quad \text{in } T(\Omega),$$

where $\widetilde{\nu}$ is a finite measure in $T(\Omega)$ such that, for any Borel set $\omega \subset \Omega$,

$$\widetilde{\nu}\big(T(\omega)\big) = \mu^n |\det A|^{-1} \nu(\omega).$$

Also, the sections of u and v are related by

$$T(S_{u,h}(x_0)) = S_{v,\mu h}(Tx_0).$$

In practice, we choose μ such that the new solution v or its corresponding Monge-Ampère measure has nice properties. For example, if we take

$$\mu = \left(\frac{|\det A|}{\nu(\Omega)}\right)^{\frac{1}{n}},$$

then the corresponding measure $\widetilde{\nu}$ is given by, for any $\omega \subset \Omega$,

$$\widetilde{\nu}(T(\omega)) = \frac{1}{\nu(\Omega)}\nu(\omega).$$

In particular, $\widetilde{\nu}(T(\Omega)) = 1$.

8.2. Dirichlet Problems

In this section, we first derive the comparison principle for generalized solutions of the Monge-Ampère equation. Then, we discuss the existence of generalized solutions of the Dirichlet problem. We solve the Dirichlet problem in convex domains with convex boundary values and in strictly convex domains with arbitrary continuous boundary values. We begin with the following simple lemma.

Lemma 8.2.1. *Let Ω be a bounded domain in $\mathbb{R}^n$ and u, v be continuous functions in $\bar{\Omega}$. If $u = v$ on $\partial\Omega$ and $u \leq v$ in Ω, then $\partial v(\Omega) \subset \partial u(\Omega)$.*

Proof. Take any $p \in \partial v(\Omega)$. There exists a point $x_0 \in \Omega$ such that $p \in \partial v(x_0)$; namely, for any $x \in \Omega$,

$$v(x) \geq v(x_0) + p \cdot (x - x_0).$$

Set

$$a = \sup\{v(x_0) + p \cdot (x - x_0) - u(x) : x \in \Omega\}.$$

Then, $a \geq 0$ since $v(x_0) \geq u(x_0)$. We assume that a is attained at some point $x_1 \in \bar{\Omega}$. Then,

$$a = v(x_0) + p \cdot (x_1 - x_0) - u(x_1),$$

and, for any $x \in \Omega$,

$$u(x) \geq v(x_0) + p \cdot (x - x_0) - a.$$

A simple substitution yields, for any $x \in \Omega$,

$$u(x) \geq u(x_1) + p \cdot (x - x_1)$$

and

$$v(x_1) \geq v(x_0) + p \cdot (x_1 - x_0) = u(x_1) + a.$$

If $a > 0$, then $x_1 \notin \partial\Omega$ and hence $p \in \partial u(x_1)$. If $a = 0$, then, for any $x \in \Omega$,

$$u(x) \geq v(x_0) + p \cdot (x - x_0) \geq u(x_0) + p \cdot (x - x_0),$$

and consequently $p \in \partial u(x_0)$. □

Lemma 8.2.1 has a geometric proof. Take any supporting hyperplane of the graph of v at some point in Ω. This hyperplane has a nonempty intersection with the graph of u since $v \geq u$. Now, we lower this hyperplane vertically until it touches the graph of u for the last time. The new hyperplane provides a supporting hyperplane of the graph of u somewhere in Ω.

Next, we prove a comparison principle for generalized solutions.

Theorem 8.2.2. *Let Ω be a bounded domain in $\mathbb{R}^n$ and u, v be locally convex $C(\bar{\Omega})$-functions. Assume $Mu \geq Mv$ in Ω and $u \leq v$ on $\partial\Omega$. Then, $u \leq v$ in Ω.*

Here, $Mu \geq Mv$ in Ω means $Mu(\omega) \geq Mv(\omega)$ for any Borel set $\omega \subset \Omega$.

Proof. We prove by contradiction and assume $\{x \in \Omega : u > v\} \neq \emptyset$. Without loss of generality, we assume $0 \in \Omega$ and $\Omega \Subset B_R$ for some constant $R > 0$. Set, for some constant $\varepsilon > 0$,

$$u_\varepsilon(x) = u(x) + \varepsilon(|x|^2 - R^2).$$

Then, $u_\varepsilon < u$ on $\partial\Omega$. By taking ε sufficiently small, we assume $G = \{x \in \Omega : u_\varepsilon > v\} \neq \emptyset$. Then, $G \Subset \Omega$ and $u_\varepsilon = v$ on ∂G. By Lemma 8.2.1, we have $\partial u_\varepsilon(G) \subset \partial v(G)$, and hence

$$Mu_\varepsilon(G) \leq Mv(G). \tag{1}$$

By Lemma 8.1.24 and Proposition 8.1.19, we obtain

$$Mu_\varepsilon(G) \geq Mu(G) + (2\varepsilon)^n |G|. \tag{2}$$

By combining (1) and (2), we conclude $Mv(G) > Mu(G)$, which contradicts the assumption $Mu \geq Mv$ in Ω. □

A consequence of Theorem 8.2.2 is the uniqueness of the generalized solutions of the Dirichlet boundary-value problem.

Corollary 8.2.3. *Let Ω be a bounded domain in $\mathbb{R}^n$. Assume ν is a locally finite measure in Ω and $\varphi \in C(\partial\Omega)$. Then, there exists at most one generalized solution $u \in C(\bar{\Omega})$ of the problem*

$$\begin{aligned} \det \nabla^2 u &= \nu \quad \text{in } \Omega, \\ u &= \varphi \quad \text{on } \partial\Omega. \end{aligned}$$

Next, we prove a fundamental estimate due to Alexandrov.

Theorem 8.2.4. *Let Ω be a bounded convex domain in $\mathbb{R}^n$ and u be a convex $C(\bar{\Omega})$-function with $u = 0$ on $\partial\Omega$. Then, for any $x \in \Omega$,*

$$|u(x)|^n \leq C(\operatorname{diam}(\Omega))^{n-1} \operatorname{dist}(x, \partial\Omega) |\partial u(\Omega)|,$$

where C is a positive constant depending only on n.

Proof. By the convexity, we have $u \le 0$ in Ω. We fix a point $x_0 \in \Omega$ and assume $u(x_0) < 0$. Let v be the convex function satisfying $v = 0$ on $\partial\Omega$ and the graph of v is the convex cone in $\mathbb{R}^n \times \mathbb{R}$ with its vertex $(x_0, u(x_0))$. In particular, $v(x_0) = u(x_0)$. Then, $u = v = 0$ on $\partial\Omega$ and $u \le v$ in Ω by the convexity of u. By Lemma 8.2.1, we have

$$\partial v(\Omega) \subset \partial u(\Omega). \tag{1}$$

Set $D = \operatorname{diam}(\Omega)$ and $d = \operatorname{dist}(x_0, \partial\Omega)$. We claim

$$\left(\frac{|v(x_0)|}{D}\right)^{n-1} \cdot \frac{|v(x_0)|}{d} \le C|\partial v(\Omega)|, \tag{2}$$

where C is a positive constant depending only on n. Then, the desired result follows from (1) and (2).

We now prove (2). First, by the definition of v, we have, for any $p \in B_{|v(x_0)|/D}$ and any $x \in \Omega$,

$$v(x) \ge v(x_0) + p \cdot (x - x_0).$$

Hence,

$$B_{|v(x_0)|/D} \subset \partial v(\Omega). \tag{3}$$

Next, we note that

$$\text{there exists a } p_0 \in \partial v(\Omega) \text{ such that } |p_0| = |v(x_0)|/d. \tag{4}$$

To check this, we take a point $\widetilde{x} \in \partial\Omega$ such that $|\widetilde{x} - x_0| = d$ and H is a supporting hyperplane of Ω at $\widetilde{x}$ in $\mathbb{R}^n$. The existence of such an H follows from the convexity of Ω. The hyperplane in $\mathbb{R}^{n+1}$ generated by H and the point $(x_0, v(x_0))$ is a supporting hyperplane of v and its gradient p_0 satisfies $|p_0| = |v(x_0)|/d$. Last, since the graph of v is a cone with the vertex $(x_0, v(x_0))$, then $\partial v(\Omega) = \partial v(x_0)$ and hence $\partial v(\Omega)$ is a convex set. By combining (3), (4), and $|p_0| \ge |v(x_0)|/D$, we conclude that $\partial v(\Omega)$ contains the convex hull of $B_{|v(x_0)|/D}$ and p_0. A geometric argument yields

$$C\left(\frac{|v(x_0)|}{D}\right)^{n-1} \cdot |p_0| \le |\partial v(\Omega)|.$$

This proves (2). □

In the rest of this section, we discuss the existence of generalized solutions of the Dirichlet problem. We first consider the case that the Monge-Ampère measures are discrete.

Lemma 8.2.5. *Let Ω be a bounded convex domain in $\mathbb{R}^n$, ν be a discrete measure given by, for some $c_i > 0$ and $x_i \in \Omega$, $i = 1, \ldots, N$,*

$$\nu = \sum_{i=1}^{N} c_i \delta_{x_i},$$

and $\varphi \in C(\bar{\Omega})$ be a convex function in Ω. Then, there exists a unique generalized convex solution $u \in C(\bar{\Omega})$ of the Dirichlet problem

$$\begin{aligned} \det \nabla^2 u &= \nu \quad \text{in } \Omega, \\ u &= \varphi \quad \text{on } \partial\Omega. \end{aligned}$$

Proof. The uniqueness follows from Corollary 8.2.3. We now prove the existence.

Step 1. We first prove the existence of subsolutions. For each $i = 1, \ldots, N$ and each $t_i < 0$, let $\underline{u}_{t_i} \in C(\bar{\Omega})$ be the convex function satisfying $\underline{u}_{t_i} = 0$ on $\partial\Omega$ and the graph of $\underline{u}_{t_i}$ is the convex cone in $\mathbb{R}^n \times \mathbb{R}$ with its vertex (x_i, t_i). Then, $M\underline{u}_{t_i} = 0$ in $\Omega \setminus \{x_i\}$ and $\underline{u}_{t_i} < \underline{u}_{t_i'}$ in Ω for any $t_i < t_i' < 0$. Moreover, for each fixed $t_i^0 < 0$, $\underline{u}_{t_i} \to \underline{u}_{t_i^0}$ uniformly in Ω as $t_i \to t_i^0$. Theorem 8.1.22 implies $M\underline{u}_{t_i}(\{x_i\}) \to M\underline{u}_{t_i^0}(\{x_i\})$ as $t_i \to t_i^0$. In other words, $M\underline{u}_{t_i}(\{x_i\})$ is a continuous function of $t_i \in (-\infty, 0)$. In fact, $M\underline{u}_{t_i}(\{x_i\})$ is a decreasing function of $t_i \in (-\infty, 0)$ with $M\underline{u}_{t_i}(\{x_i\}) \to \infty$ as $t_i \to -\infty$ and $M\underline{u}_{t_i}(\{x_i\}) \to 0$ as $t_i \to 0^-$. By taking t_i appropriately, we assume that $\underline{u}_{t_i}$ is the generalized solution of

$$\begin{aligned} \det \nabla^2 \underline{u}_{t_i} &= c_i \delta_{x_i} \quad \text{in } \Omega, \\ \underline{u}_{t_i} &= 0 \quad \text{on } \partial\Omega. \end{aligned}$$

Set

$$\underline{u} = \varphi + \sum_{i=1}^{N} \underline{u}_{t_i}.$$

Since φ is convex in Ω, then $\underline{u} \in C(\bar{\Omega})$ is a convex function in Ω, with $\underline{u} = \varphi$ on $\partial\Omega$, and, by Lemma 8.1.24,

$$M\underline{u} \geq M\varphi + \sum_{i=1}^{N} M\underline{u}_{t_i} \geq \sum_{i=1}^{N} c_i \delta_{x_i} = \nu.$$

We point out that the boundary condition $\underline{u} = \varphi$ on $\partial\Omega$ is important for the rest of the proof.

Step 2. We now modify $\underline{u}$ to get a subsolution whose Monge-Ampère measure concentrates on $x_1, \ldots, x_N$. Set $\underline{b}_i = \underline{u}(x_i)$, for $i = 1, \ldots, N$, and

$$\underline{\mathcal{L}} = \{l : l \text{ is affine}, l(x_i) \leq \underline{b}_i \text{ for } i = 1, \ldots, N, l \leq \varphi \text{ on } \partial\Omega\}.$$

We note that all supporting functions of $\underline{u}$ are in $\underline{\mathcal{L}}$. Hence, $\underline{\mathcal{L}}$ is nonempty. Consider the harmonic function $h \in C(\bar{\Omega})$ with $h = \varphi$ on $\partial\Omega$. By the comparison principle (for harmonic functions), we have, for any $l \in \underline{\mathcal{L}}$,

$$l \leq h \quad \text{in } \Omega.$$

Now we set, for any $x \in \Omega$,

$$u_{(\underline{b}_1,\ldots,\underline{b}_N)}(x) = \sup\{l(x) : l \in \underline{\mathcal{L}}\}.$$

Then, $u_{(\underline{b}_1,\ldots,\underline{b}_N)}$ is well-defined in Ω and satisfies

$$u_{(\underline{b}_1,\ldots,\underline{b}_N)} \leq h \quad \text{in } \Omega, \tag{1}$$

and, for $i = 1, \ldots, N$,

$$u_{(\underline{b}_1,\ldots,\underline{b}_N)}(x_i) \leq \underline{b}_i. \tag{2}$$

Moreover, $u_{(\underline{b}_1,\ldots,\underline{b}_N)}$ is convex in Ω since it is the supremum of linear functions. Here, we employed the equivalent description of the convexity in Lemma 8.1.11.

For any $x_0 \in \Omega$, take a supporting function l of $\underline{u}$ at x_0. Then, $l \in \underline{\mathcal{L}}$ and $l(x_0) = \underline{u}(x_0)$. The definition of $u_{(\underline{b}_1,\ldots,\underline{b}_N)}$ at x_0 implies $u_{(\underline{b}_1,\ldots,\underline{b}_N)}(x_0) \geq \underline{u}(x_0)$. Hence,

$$u_{(\underline{b}_1,\ldots,\underline{b}_N)} \geq \underline{u} \quad \text{in } \Omega. \tag{3}$$

Evaluating (3) at x_i and combining with (2), we have, for $i = 1, \ldots, N$,

$$u_{(\underline{b}_1,\ldots,\underline{b}_N)}(x_i) = \underline{b}_i. \tag{4}$$

Next, for any $x_0 \in \partial\Omega$, we evaluate (1) and (3) at $x \in \Omega$ and then let $x \to x_0$. With $h = \underline{u} = \varphi$ on $\partial\Omega$, we can extend $u_{(\underline{b}_1,\ldots,\underline{b}_N)}$ continuously to $\partial\Omega$ and

$$u_{(\underline{b}_1,\ldots,\underline{b}_N)} = \underline{u} \quad \text{on } \partial\Omega. \tag{5}$$

By combining (3), (4), and (5), we have

$$Mu_{(\underline{b}_1,\ldots,\underline{b}_N)}(\{x_i\}) \geq M\underline{u}(\{x_i\}) \geq c_i.$$

Next, we set

$$S = \{p \in \mathbb{R}^n : p \in \partial u_{(\underline{b}_1,\ldots,\underline{b}_N)}(x') \cap \partial u_{(\underline{b}_1,\ldots,\underline{b}_N)}(x'') \quad \text{for some } x', x'' \in \Omega \text{ with } x' \neq x''\}.$$

We claim, for any $x_0 \in \Omega \setminus \{x_1, \ldots, x_N\}$,

$$\partial u_{(\underline{b}_1,\ldots,\underline{b}_N)}(x_0) \subset S. \tag{6}$$

To prove this, take any supporting function l of $u_{(\underline{b}_1,\ldots,\underline{b}_N)}$ at x_0. Then, $l \in \underline{\mathcal{L}}$ and $l(x_0) = u_{(\underline{b}_1,\ldots,\underline{b}_N)}(x_0)$. We note that either $l(x_i) = \underline{b}_i$ for some $i \in \{1, \ldots, N\}$ or $l(x_*) = \varphi(x_*)$ for some $x_* \in \partial\Omega$. If this is not true, then $l(x_i) < \underline{b}_i$ for any $i = 1, \ldots, N$ and $l(x) < \varphi(x)$ for any $x \in \partial\Omega$. By the continuity of φ, we have $l + \varepsilon \in \underline{\mathcal{L}}$ for some positive constant ε.

Then, the definition of $u_{(\underline{b}_1,\dots,\underline{b}_N)}$ implies $u_{(\underline{b}_1,\dots,\underline{b}_N)} \geq l+\varepsilon$, and in particular, $u_{(\underline{b}_1,\dots,\underline{b}_N)}(x_0) \geq l(x_0)+\varepsilon$, which leads to a contradiction. Let x_* be one of $x_1,\dots,x_N$ or the point on $\partial\Omega$ such that $u_{(\underline{b}_1,\dots,\underline{b}_N)}(x_*) = l(x_*)$. By the convexity of $u_{(\underline{b}_1,\dots,\underline{b}_N)}$, we obtain $u_{(\underline{b}_1,\dots,\underline{b}_N)} = l$ on $\overline{x_0x_*}$. Hence, l is the supporting function of $u_{(\underline{b}_1,\dots,\underline{b}_N)}$ at any point on $\overline{x_0x_*}\cap\Omega$. This proves (6). By Lemma 8.1.16, we have $|S|=0$, and hence

$$Mu_{(\underline{b}_1,\dots,\underline{b}_N)} = 0 \quad \text{in } \Omega\setminus\{x_1,\dots,x_N\}.$$

Step 3. We now lift $u_{(\underline{b}_1,\dots,\underline{b}_N)}$ and decrease its Monge-Ampère measures. Suppose $Mu_{(\underline{b}_1,\dots,\underline{b}_N)} > \nu$. Then, there exists an $i_0\in\{1,\dots,N\}$, say $i_0=1$, such that

$$Mu_{(\underline{b}_1,\dots,\underline{b}_N)}(\{x_1\}) > c_1. \tag{7}$$

We now discard $i=1$ from $\{1,\dots,N\}$ and consider only $\{2,\dots,N\}$. As in Step 2, we set

$$\overline{\mathcal{L}} = \{l : l \text{ is affine, } l(x_i)\leq \underline{b}_i \text{ for } i=2,\dots,N,\ l\leq\varphi \text{ on } \partial\Omega\}$$

and, for any $x\in\Omega$,

$$\overline{u}_{(\underline{b}_2,\dots,\underline{b}_N)}(x) = \sup\{l(x) : l\in\overline{\mathcal{L}}\}.$$

Proceeding as in Step 2, we conclude that $\overline{u}_{(\underline{b}_2,\dots,\underline{b}_N)}$ can be extended as a convex $C(\bar{\Omega})$-function satisfying

$$\begin{aligned} \overline{u}_{(\underline{b}_2,\dots,\underline{b}_N)} &= \varphi \quad \text{on } \partial\Omega,\\ \overline{u}_{(\underline{b}_2,\dots,\underline{b}_N)}(x_i) &= \underline{b}_i \quad \text{for } i=2,\dots,N, \end{aligned}$$

and

$$M\overline{u}_{(\underline{b}_2,\dots,\underline{b}_N)} = 0 \quad \text{in } \Omega\setminus\{x_2,\dots,x_N\}.$$

In particular,

$$M\overline{u}_{(\underline{b}_2,\dots,\underline{b}_N)}(\{x_1\}) = 0. \tag{8}$$

Moreover, we have $\underline{\mathcal{L}}\subset\overline{\mathcal{L}}$, and hence

$$u_{(\underline{b}_1,\dots,\underline{b}_N)} \leq \overline{u}_{(\underline{b}_2,\dots,\underline{b}_N)} \quad \text{in } \Omega.$$

Set $\bar{b}_1 = \overline{u}_{(\underline{b}_2,\dots,\underline{b}_N)}(x_1)$. Then, $\underline{b}_1 < \bar{b}_1$.

Take an arbitrary $b_1\in(\underline{b}_1,\bar{b}_1)$. As in Step 2, we set

$$\mathcal{L} = \{l : l \text{ is affine, } l(x_1)\leq b_1, l(x_i)\leq \underline{b}_i \text{ for } i=2,\dots,N,\ l\leq\varphi \text{ on } \partial\Omega\}$$

and, for any $x\in\Omega$,

$$u_{(b_1,\underline{b}_2,\dots,\underline{b}_N)}(x) = \sup\{l(x) : l\in\mathcal{L}\}.$$

Proceeding as in Step 2, we conclude that $u_{(b_1,\underline{b}_2,\ldots,\underline{b}_N)}$ can be extended as a convex $C(\bar{\Omega})$-function satisfying

$$\begin{aligned} u_{(b_1,\underline{b}_2,\ldots,\underline{b}_N)} &= \varphi \quad \text{on } \partial\Omega, \\ u_{(b_1,\underline{b}_2,\ldots,\underline{b}_N)}(x_1) &= b_1, \\ u_{(b_1,\underline{b}_2,\ldots,\underline{b}_N)}(x_i) &= \underline{b}_i \quad \text{for } i = 2, \ldots, N, \end{aligned}$$

and

$$Mu_{(b_1,\underline{b}_2,\ldots,\underline{b}_N)} = 0 \quad \text{in } \Omega \setminus \{x_1, x_2, \ldots, x_N\}.$$

For any other $b_1' \in (\underline{b}_1, \bar{b}_1)$, we can define $\mathcal{L}'$ and $u_{(b_1',\underline{b}_2,\ldots,\underline{b}_N)}$ with b_1' replacing b_1 in $\mathcal{L}$ and $u_{(b_1,\underline{b}_2,\ldots,\underline{b}_N)}$. We consider the case $b_1' > b_1$. Then, $l \in \mathcal{L}'$ for any $l \in \mathcal{L}$ and $l' - b_1' + b_1 \in \mathcal{L}$ for any $l' \in \mathcal{L}'$. This implies

$$u_{(b_1',\underline{b}_2,\ldots,\underline{b}_N)} - b_1' + b_1 \le u_{(b_1,\underline{b}_2,\ldots,\underline{b}_N)} \le u_{(b_1',\underline{b}_2,\ldots,\underline{b}_N)} \quad \text{in } \Omega.$$

If l is a supporting function of $u_{(b_1',\underline{b}_2,\ldots,\underline{b}_N)}$ at x_1, then $l - b_1' + b_1$ is a supporting function of $u_{(b_1,\underline{b}_2,\ldots,\underline{b}_N)}$ at x_1. Therefore, $\partial u_{(b_1',\underline{b}_2,\ldots,\underline{b}_N)}(x_1) \subset \partial u_{(b_1,\underline{b}_2,\ldots,\underline{b}_N)}(x_1)$ and then

$$Mu_{(b_1',\underline{b}_2,\ldots,\underline{b}_N)}(\{x_1\}) \le Mu_{(b_1,\underline{b}_2,\ldots,\underline{b}_N)}(\{x_1\}).$$

Moreover, for any fixed $b_1^0 \in [\underline{b}_1, \bar{b}_1]$,

$$u_{(b_1,\underline{b}_2,\ldots,\underline{b}_N)} \to u_{(b_1^0,\underline{b}_2,\ldots,\underline{b}_N)} \quad \text{uniformly in } \Omega \text{ as } b_1 \to b_1^0.$$

Theorem 8.1.22 implies

$$Mu_{(b_1,\underline{b}_2,\ldots,\underline{b}_N)}(\{x_1\}) \to Mu_{(b_1^0,\underline{b}_2,\ldots,\underline{b}_N)}(\{x_1\}).$$

Hence, $Mu_{(b_1,\underline{b}_2,\ldots,\underline{b}_N)}(\{x_1\})$ is a continuous and decreasing function of $b_1 \in [\underline{b}_1, \bar{b}_1]$. With (7) and (8), there exists a $b_1 \in (\underline{b}_1, \bar{b}_1)$ such that

$$Mu_{(b_1,\underline{b}_2,\ldots,\underline{b}_N)}(\{x_1\}) = c_1.$$

Note that

$$u_{(b_1,\underline{b}_2,\ldots,\underline{b}_N)} = u_{(\underline{b}_1,\underline{b}_2,\ldots,\underline{b}_N)} \quad \text{on } \partial\Omega$$

and

$$u_{(b_1,\underline{b}_2,\ldots,\underline{b}_N)} \ge u_{(\underline{b}_1,\underline{b}_2,\ldots,\underline{b}_N)} \quad \text{in } \Omega,$$

with strict inequality near x_1. Then,

$$Mu_{(b_1,\underline{b}_2,\ldots,\underline{b}_N)}(\Omega) \le Mu_{(\underline{b}_1,\underline{b}_2,\ldots,\underline{b}_N)}(\Omega), \tag{9}$$

and, for any $i = 2, \ldots, N$,

$$Mu_{(b_1,\underline{b}_2,\ldots,\underline{b}_N)}(\{x_i\}) \ge Mu_{(\underline{b}_1,\underline{b}_2,\ldots,\underline{b}_N)}(\{x_i\}). \tag{10}$$

We point out that the strict inequality in (9) may hold depending on the location of x_1 relative to $\partial\Omega$ and that the strict inequality in (10) may hold for some $i = 2, \ldots, N$, if $N \ge 2$.

Step 4. We discuss the limit of liftings. Set $u_0 = u_{(\underline{b}_1,\ldots,\underline{b}_N)}$, where $u_{(\underline{b}_1,\ldots,\underline{b}_N)}$ is the function constructed in Step 2. We arrange the N points $x_1, \ldots, x_N$ as an infinite sequence $\{x_j\}$ given by

$$x_1, x_2, \ldots, x_N, x_1, x_2, \ldots, x_N, \ldots.$$

Recall that $Mu_0(\{x_i\}) \geq c_i$ for any $i = 1, \ldots, N$. Let $j_1 \in \{1, \ldots, N\}$ be the first integer such that $Mu_0(\{x_{j_1}\}) > c_{j_1}$. Then, we lift $u_0(x_{j_1})$ to get u_1 such that $Mu_1(\{x_{j_1}\}) = c_{j_1}$. Note that $u_1 \in C(\bar{\Omega})$ is convex, $u_1 \geq u_0$ in Ω, $u_1 = u_0$ on $\partial\Omega$, $Mu_1(\Omega) \leq Mu_0(\Omega)$, $Mu_1(\{x_i\}) \geq c_i$ for any $i = 1, \ldots, N$, and $Mu_1 = 0$ in $\Omega \backslash \{x_1, \ldots, x_N\}$. If $Mu_1(\{x_i\}) = c_i$ for any $i = j_1, \ldots, j_1+N-1$, then u_1 is the desired generalized solution. Otherwise, we assume j_2 is the first integer after j_1 such that $Mu_1(\{x_{j_2}\}) > c_{j_2}$. We then lift $u_1(x_{j_2})$ to get u_2 such that $Mu_2(\{x_{j_2}\}) = c_{j_2}$. Note that $u_2 \in C(\bar{\Omega})$ is convex, $u_2 \geq u_1$ in Ω, $u_2 = u_1$ on $\partial\Omega$, $Mu_2(\Omega) \leq Mu_1(\Omega)$, $Mu_2(\{x_i\}) \geq c_i$ for any $i = 1, \ldots, N$, and $Mu_2 = 0$ in $\Omega \setminus \{x_1, \ldots, x_N\}$. We point out that it is possible that $Mu_2(\{x_{j_1}\}) > c_{j_1}$. If $Mu_2(\{x_i\}) = c_i$ for any $i = j_2, \ldots, j_2 + N - 1$, then u_2 is the desired generalized solution. Otherwise, we assume j_3 is the first integer after j_2 such that $Mu_2(\{x_{j_3}\}) > c_{j_3}$. By continuing this process, we either stop after finitely many steps or we can continue indefinitely. In the first case, we get a generalized solution u_k for some k. In the second case, we get a sequence of convex functions $u_k \in C(\bar{\Omega})$ such that $u_k = \varphi$ on $\partial\Omega$,

$$\begin{aligned} &u_k \leq u_{k+1} \leq h \quad \text{in } \Omega, \\ &Mu_k(\Omega) \geq Mu_{k+1}(\Omega) > \nu, \\ &Mu_k(\{x_i\}) \geq c_i \quad \text{for any } i = 1, \ldots, N, \end{aligned}$$

and

$$Mu_k = 0 \quad \text{in } \Omega \setminus \{x_1, \ldots, x_N\}.$$

Moreover, for each $i \in \{1, \ldots, N\}$, there exists a subsequence $k'(i)$ such that $Mu_{k'(i)}(\{x_i\}) = c_i$. Since u_k is increasing and bounded from above, we may assume

$$u_\infty = \lim_{k\to\infty} u_k \quad \text{in } \Omega.$$

By Lemma 8.1.7, u_k is locally uniformly Lipschitz in Ω and by the Arzela-Ascoli theorem the convergence above is locally uniformly in Ω. Hence, $u_\infty \in C(\Omega)$ and u_∞ is convex in Ω. Next, by $u_0 \leq u_\infty \leq h$ in Ω, we get $u \in C(\bar{\Omega})$ and $u_\infty = \varphi$ on $\partial\Omega$. By Theorem 8.1.22, we conclude $Mu_\infty = 0$ in $\Omega \setminus \{x_1, \ldots, x_N\}$ and $Mu_\infty(\{x_i\}) = c_i$ for each $i \in \{1, \ldots, N\}$. Hence, $Mu_\infty = \nu$ and then u_∞ is the desired generalized solution. □

We now prove the existence of generalized solutions of the Dirichlet problem for general measures and homogeneous boundary values.

Lemma 8.2.6. *Let Ω be a bounded convex domain in $\mathbb{R}^n$ and ν be a finite measure in Ω. Then, there exists a unique generalized convex solution $u \in C(\bar{\Omega})$ of the Dirichlet problem*

$$\begin{aligned} \det \nabla^2 u &= \nu \quad \text{in } \Omega, \\ u &= 0 \quad \text{on } \partial\Omega. \end{aligned}$$

Proof. As in the proof of Lemma 8.2.5, we only prove the existence. Take a sequence of discrete measures ν_k of the form, for some $x_{i,k} \in \Omega$ and $c_{i,k} > 0$,

$$\nu_k = \sum_{i=1}^{N_k} c_{i,k} \delta_{x_{i,k}},$$

such that $\nu_k \to \nu$ weakly in Ω. In particular, for some positive constant A, $\nu_k(\Omega) \leq A$ for any $k \geq 1$. By Lemma 8.2.5, let $u_k \in C(\bar{\Omega})$ be the generalized convex solution of

$$\begin{aligned} \det \nabla^2 u_k &= \nu_k \quad \text{in } \Omega, \\ u_k &= 0 \quad \text{on } \partial\Omega. \end{aligned}$$

By the convexity and Theorem 8.2.4, we have, for any $x \in \Omega$,

$$-C\big(\operatorname{dist}(x, \partial\Omega)\big)^{\frac{1}{n}} \leq u_k(x) \leq 0,$$

where C is a positive constant depending only on n, A, and $\operatorname{diam}(\Omega)$, independent of k. By Lemma 8.1.7, u_k is locally uniformly Lipschitz in Ω and by the Arzela-Ascoli theorem a subsequence of u_k converges to u locally uniformly in Ω, for some convex function $u \in C(\Omega)$. Moreover, for any $x \in \Omega$,

$$-C\big(\operatorname{dist}(x, \partial\Omega)\big)^{\frac{1}{n}} \leq u(x) \leq 0.$$

This implies $u \in C(\bar{\Omega})$ and $u = 0$ on $\partial\Omega$. By Theorem 8.1.22, we conclude $Mu = \nu$ in Ω and hence u is the desired generalized solution. □

We are ready to prove the existence of generalized solutions of the Dirichlet problem for general measures and general convex boundary values.

Theorem 8.2.7. *Let Ω be a bounded convex domain in $\mathbb{R}^n$, ν be a finite measure in Ω, and $\varphi \in C(\bar{\Omega})$ be a convex function in Ω. Then, there exists a unique generalized convex solution $u \in C(\bar{\Omega})$ of the Dirichlet problem*

$$\begin{aligned} \det \nabla^2 u &= \nu \quad \text{in } \Omega, \\ u &= \varphi \quad \text{on } \partial\Omega. \end{aligned} \tag{1}$$

Proof. We first derive an a priori estimate. Let $u \in C(\bar{\Omega})$ be a generalized convex solution of (1). Consider the harmonic function $h \in C(\bar{\Omega})$ with $h = \varphi$ on $\partial\Omega$. Since convex functions are subharmonic, the comparison principle implies

$$u \leq h \quad \text{in } \Omega.$$

Next, we consider the Dirichlet problems

$$\begin{aligned} \det \nabla^2 v &= 0 \quad \text{in } \Omega, \\ v &= \varphi \quad \text{on } \partial\Omega, \end{aligned}$$

and

$$\begin{aligned} \det \nabla^2 w &= \nu \quad \text{in } \Omega, \\ w &= 0 \quad \text{on } \partial\Omega. \end{aligned}$$

The existence of the generalized convex solution $v \in C(\bar{\Omega})$ is implied by Lemma 8.2.5 and the existence of the generalized convex solution $w \in C(\bar{\Omega})$ is implied by Lemma 8.2.6. By Lemma 8.1.24,

$$M(v + w) \geq Mv + Mw = Mw = \nu.$$

Hence,

$$\begin{aligned} M(v + w) &\geq M(u) \quad \text{in } \Omega, \\ v + w &= u \quad \text{on } \partial\Omega. \end{aligned}$$

By Theorem 8.2.2, we have

$$v + w \leq u \quad \text{in } \Omega.$$

Theorem 8.2.4 implies, for any $x \in \Omega$,

$$w(x) \geq -C(\operatorname{diam}(\Omega))^{\frac{n-1}{n}} \big(\nu(\Omega)\big)^{\frac{1}{n}} \big(\operatorname{dist}(x, \partial\Omega)\big)^{\frac{1}{n}}.$$

Hence, we obtain, for any $x \in \Omega$,

$$(2) \qquad v(x) - C(\operatorname{diam}(\Omega))^{\frac{n-1}{n}} \big(\nu(\Omega)\big)^{\frac{1}{n}} \big(\operatorname{dist}(x, \partial\Omega)\big)^{\frac{1}{n}} \leq u(x) \leq h(x).$$

We point out that the functions h and v depend only on φ, independent of ν.

Now, we prove the existence of the generalized convex solution $u \in C(\bar{\Omega})$ of (1). Take a sequence of discrete measures ν_k of the form, for some $x_{i,k} \in \Omega$ and $c_{i,k} > 0$,

$$\nu_k = \sum_{i=1}^{N_k} c_{i,k} \delta_{x_{i,k}},$$

such that $\nu_k \to \nu$ weakly in Ω. In particular, for some positive constant A, $\nu_k(\Omega) \leq A$ for any $k \geq 1$. By Lemma 8.2.5, let $u_k \in C(\bar{\Omega})$ be the generalized convex solution of

$$\begin{aligned} \det \nabla^2 u_k &= \nu_k \quad \text{in } \Omega, \\ u_k &= \varphi \quad \text{on } \partial\Omega. \end{aligned}$$

By applying (2) to u_k, we have, for any $x \in \Omega$,

$$v(x) - C\big(\operatorname{dist}(x, \partial\Omega)\big)^{\frac{1}{n}} \leq u_k(x) \leq h(x),$$

where C is a positive constant depending only on n, A, and $\operatorname{diam}(\Omega)$, independent of k. By Lemma 8.1.7, u_k is locally uniformly Lipschitz in Ω and, by the Arzela-Ascoli theorem, a subsequence of u_k converges to u locally uniformly in Ω, for some convex function $u \in C(\Omega)$. Moreover, for any $x \in \Omega$,

$$v(x) - C\big(\operatorname{dist}(x, \partial\Omega)\big)^{\frac{1}{n}} \leq u(x) \leq h(x).$$

This implies $u \in C(\bar{\Omega})$ and $u = \varphi$ on $\partial\Omega$. By Theorem 8.1.22, we conclude $Mu = \nu$ in Ω and hence u is the desired generalized solution. □

In Theorem 8.2.7, it is necessary to assume that the boundary values are convex. For example, we consider $\Omega = B_1^+$, the upper half unit ball in $\mathbb{R}^n$. Any convex solutions $u \in C(\bar{B}_1^+)$ of (1) in Theorem 8.2.7 induce convex functions φ in $\partial B_1^+ \cap \{x_n = 0\}$. However, if Ω is a strictly convex domain, it is not necessary to assume the convexity for boundary values. Recall that a domain $\Omega \subset \mathbb{R}^n$ is *strictly convex* if, for any $x, y \in \bar{\Omega}$, the open line segment joining x and y lies in Ω.

We first consider homogeneous Monge-Ampère equations in strictly convex domains.

Lemma 8.2.8. *Let Ω be a bounded strictly convex domain in $\mathbb{R}^n$ and $\varphi \in C(\partial\Omega)$. Then, there exists a unique generalized convex solution $u \in C(\bar{\Omega})$ of the Dirichlet problem*

$$\begin{aligned} \det \nabla^2 u &= 0 \quad \text{in } \Omega, \\ u &= \varphi \quad \text{on } \partial\Omega. \end{aligned}$$

Proof. We only prove the existence. Set

$$\mathcal{L} = \{l : l \text{ is affine, } l \leq \varphi \text{ on } \partial\Omega\}.$$

We first note that the constant function $\min_{\partial\Omega} \varphi \in \mathcal{L}$, and hence $\mathcal{L}$ is nonempty. Consider the harmonic function $h \in C(\bar{\Omega})$ with $h = \varphi$ on $\partial\Omega$. By the comparison principle (for harmonic functions), we have, for any $l \in \mathcal{L}$,

$$l \leq h \quad \text{in } \Omega.$$

Now we set, for any $x \in \Omega$,

$$u(x) = \sup\{l(x) : l \in \mathcal{L}\}.$$

Then, u is well-defined in Ω and satisfies

$$u \leq h \quad \text{in } \Omega. \tag{1}$$

Since u is the supremum of linear functions, u is convex in Ω and, in particular, u is continuous in Ω. Here, we employed the equivalent description of the convexity in Lemma 8.1.11.

For any $x_0 \in \partial\Omega$, we evaluate (1) at $x \in \Omega$ and then let $x \to x_0$. With $h = \varphi$ on $\partial\Omega$, we obtain

$$\limsup_{x \to x_0} u(x) \leq \varphi(x_0). \tag{2}$$

Now, we prove

$$\liminf_{x\to x_0} u(x) \geq \varphi(x_0). \tag{3}$$

To this end, we claim, for any $x_0 \in \partial\Omega$ and any $\varepsilon > 0$, there exists a function $l_{x_0,\varepsilon} \in \mathcal{L}$ such that

$$l_{x_0,\varepsilon}(x_0) \geq \varphi(x_0) - \varepsilon. \tag{4}$$

Then, the definition of u implies $u \geq l_{x_0,\varepsilon}$ in Ω, and hence

$$\liminf_{x\to x_0} u(x) \geq l_{x_0,\varepsilon}(x_0) \geq \varphi(x_0) - \varepsilon.$$

With $\varepsilon \to 0$, we obtain (3). We now prove (4). With $x_0 \in \partial\Omega$ fixed, for any $\varepsilon > 0$, there exists a $\delta > 0$ such that, for any $x \in \partial\Omega \cap B_\delta(x_0)$,

$$|\varphi(x) - \varphi(x_0)| < \varepsilon.$$

Let P be an affine function in $\mathbb{R}^n$ such that $\{P = 0\}$ is a supporting hyperplane to Ω at x_0 and $\Omega \subset \{P > 0\}$. For example, we can take $P(x) = \mathbf{n} \cdot (x - x_0)$ for some unit vector $\mathbf{n} \in \mathbb{R}^n$. Since Ω is strictly convex, there exists an $\eta > 0$ such that

$$D \equiv \{x \in \bar{\Omega} : P(x) \leq \eta\} \subset B_\delta(x_0).$$

Set

$$m = \min\{\varphi(x) : x \in \partial\Omega,\ P(x) \geq \eta\}$$

and

$$l(x) = \varphi(x_0) - \varepsilon - AP(x),$$

where A is a constant satisfying

$$A \geq \max\left\{\frac{1}{\eta}[\varphi(x_0) - \varepsilon - m], 0\right\}.$$

Now, we prove the claim. First, we have

$$l(x_0) = \varphi(x_0) - \varepsilon - AP(x_0) = \varphi(x_0) - \varepsilon.$$

Hence, l satisfies (4). For any $x \in \partial\Omega \cap D$, we have $P(x) \geq 0$ and hence

$$\varphi(x) \geq \varphi(x_0) - \varepsilon \geq \varphi(x_0) - \varepsilon - AP(x) = l(x).$$

For any $x \in \partial\Omega \setminus D$, we have $P(x) \geq \eta$ and hence

$$\begin{aligned}\varphi(x) \geq m &= l(x) + m - \varphi(x_0) + \varepsilon + AP(x)\\ &\geq l(x) + m - \varphi(x_0) + \varepsilon + A\eta \geq l(x).\end{aligned}$$

Therefore, $l \in \mathcal{L}$ and the claim holds. By combining (2) and (3), we conclude $u \in C(\bar{\Omega})$ and $u = \varphi$ on $\partial\Omega$.

Next, we set

$$S = \{p \in \mathbb{R}^n : p \in \partial u(x_1) \cap \partial u(x_2) \text{ for some } x_1, x_2 \in \Omega \text{ with } x_1 \neq x_2\}.$$

We claim

$$\partial u(\Omega) \subset S. \tag{5}$$

The proof is similar to that of (6) in the proof of Lemma 8.2.5. We simply sketch the proof. Take any $x_0 \in \Omega$ and any supporting function l of u at x_0. Then, $l \in \mathcal{L}$ and $l(x_0) = u(x_0)$. By the definition of u, we have $l(x_*) = \varphi(x_*)$ for some $x_* \in \partial\Omega$. By the convexity of u, we obtain $u = l$ on $\overline{x_0 x_*}$. Hence, l is the supporting function of u at any point on $\overline{x_0 x_*} \cap \Omega$. This proves (5). By Lemma 8.1.16, we have $|S| = 0$, and hence

$$Mu = 0 \quad \text{in } \Omega.$$

Therefore, u is the desired generalized solution. □

We now generalize Lemma 8.2.8 to discrete Monge-Ampère measures.

Lemma 8.2.9. *Let Ω be a bounded strictly convex domain in $\mathbb{R}^n$, ν be a discrete measure given by, for some $c_i > 0$ and $x_i \in \Omega$, $i = 1, \ldots, N$,*

$$\nu = \sum_{i=1}^{N} c_i \delta_{x_i},$$

and $\varphi \in C(\partial\Omega)$. Then, there exists a unique generalized convex solution $u \in C(\bar{\Omega})$ of the Dirichlet problem

$$\begin{aligned} \det \nabla^2 u &= \nu \quad \textit{in } \Omega, \\ u &= \varphi \quad \textit{on } \partial\Omega. \end{aligned}$$

Proof. We first prove the existence of subsolutions. For each $i = 1, \ldots, N$ and each $t_i < 0$, let $\underline{u}_{t_i} \in C(\bar{\Omega})$ be the convex function satisfying $\underline{u}_{t_i} = 0$ on $\partial\Omega$ and the graph of $\underline{u}_{t_i}$ is the convex cone in $\mathbb{R}^n \times \mathbb{R}$ with its vertex (x_i, t_i). By taking an appropriate t_i, we assume $\underline{u}_{t_i}$ is the generalized solution of

$$\begin{aligned} \det \nabla^2 \underline{u}_{t_i} &= c_i \delta_{x_i} \quad \text{in } \Omega, \\ \underline{u}_{t_i} &= 0 \quad \text{on } \partial\Omega. \end{aligned}$$

Next, consider

$$\begin{aligned} \det \nabla^2 v &= 0 \quad \text{in } \Omega, \\ v &= \varphi \quad \text{on } \partial\Omega. \end{aligned}$$

Lemma 8.2.8 implies the existence of a generalized convex solution $v \in C(\bar{\Omega})$. Set

$$\underline{u} = v + \sum_{i=1}^{N} \underline{u}_{t_i}.$$

Then, $\underline{u} \in C(\bar{\Omega})$ is a convex function in Ω, with $\underline{u} = \varphi$ on $\partial\Omega$, and, by Lemma 8.1.24,

$$M\underline{u} \geq Mv + \sum_{i=1}^{N} M\underline{u}_{t_i} \geq \sum_{i=1}^{N} c_i \delta_{x_i} = \nu.$$

The rest of the proof proceeds similarly to Steps 2–4 in the proof of Lemma 8.2.5 and is omitted. □

We now prove the existence of generalized solutions of the Dirichlet problem for general measures and general boundary values in strictly convex domains.

Theorem 8.2.10. *Let Ω be a bounded strictly convex domain in $\mathbb{R}^n$, ν be a finite measure in Ω, and $\varphi \in C(\partial\Omega)$. Then, there exists a unique generalized convex solution $u \in C(\bar{\Omega})$ of the Dirichlet problem*

$$\begin{aligned} \det \nabla^2 u &= \nu \quad \text{in } \Omega, \\ u &= \varphi \quad \text{on } \partial\Omega. \end{aligned}$$

Proof. The proof proceeds similarly to that of Theorem 8.2.7, with Lemma 8.2.9 replacing Lemma 8.2.5. □

8.3. Global Hölder Estimates

In this section, we discuss the global Hölder regularity of generalized convex solutions of the Dirichlet problem for Monge-Ampère equations.

First, we derive an estimate of the global Hölder norm of generalized solutions of the Dirichlet problem with the homogeneous boundary value.

Theorem 8.3.1. *Let Ω be a convex domain in $\mathbb{R}^n$ with $\Omega \subset B_1$ and ν be a finite measure in Ω. Suppose that $u \in C(\bar{\Omega})$ is a convex generalized solution of*

$$\begin{aligned} \det \nabla^2 u &= \nu \quad \text{in } \Omega, \\ u &= 0 \quad \text{on } \partial\Omega. \end{aligned}$$

Then, $u \in C^{1/n}(\bar{\Omega})$ and

$$|u|_{C^{1/n}(\bar{\Omega})} \leq C\{\nu(\Omega)\}^{\frac{1}{n}},$$

where C is a positive constant depending only on n.

Proof. For the estimate of the L^∞-norm, we simply apply Theorem 8.2.4 and obtain, for any $x \in \Omega$,

$$|u(x)|^n \leq C|\partial u(\Omega)| = C\nu(\Omega),$$

where C is a positive constant depending only on n. Hence,

$$|u|_{L^\infty(\Omega)} \le C\{\nu(\Omega)\}^{\frac{1}{n}}.$$

Next, we prove, for any $x_0, y_0 \in \Omega$,

$$(1) \qquad |u(x_0) - u(y_0)| \le C\{\nu(\Omega)\}^{\frac{1}{n}}|x_0 - y_0|^{\frac{1}{n}}.$$

Without loss of generality, we assume $u(x_0) < u(y_0)$. Set

$$G = \{x \in \Omega : u(x) < u(y_0)\}.$$

Then, G is a convex set, $x_0 \in G$, $y_0 \in \partial G$, and $u = u(y_0)$ on ∂G. Set $d = \text{dist}(x_0, \partial G)$ and take a point $z_0 \in \partial G$ such that $d = |x_0 - z_0|$. Then,

$$(2) \qquad u(y_0) = u(z_0), \quad |x_0 - z_0| \le |x_0 - y_0|.$$

By a translation and a rotation, we assume $z_0 = 0$ and $x_0 = de_n$. Then, $G \subset \{x_n > 0\}$. (We point out that it is important to take z_0 to be the closest point to x_0 on ∂G. If we simply put y_0 at the origin and x_0 on the positive x_n-axis, then G may not lie in the upper half-space.) Since $\Omega \subset B_1(x_*)$ for some $x_* \in B_1$, then

$$G \subset \widetilde{\Omega} \equiv \{x = (x', x_n) : |x'| < 2, \, 0 < x_n < 4\}.$$

Let v and w be the convex functions satisfying $v = u(y_0)$ on ∂G and $w = u(y_0)$ on $\partial\widetilde{\Omega}$ and the graphs of v and w are the convex cones in $\mathbb{R}^n \times \mathbb{R}$ with vertices $(x_0, u(x_0))$, respectively. Then, $v = u$ on ∂G and $v \ge u$ in G by the convexity of u. Hence,

$$\partial w(x_0) \subset \partial v(x_0) = \partial v(G) \subset \partial u(G),$$

where we used Lemma 8.2.1 for the last inclusion. Since w is convex, then $\partial w(x_0)$ is convex. With $d < 2$, it is straightforward to verify

$$|\partial w(x_0)| \ge c\,\frac{|u(x_0) - u(z_0)|}{d} \cdot \left(\frac{|u(x_0) - u(z_0)|}{2}\right)^{n-1}.$$

Hence,

$$|u(x_0) - u(z_0)|^n \le Cd|\partial w(x_0)| \le Cd|\partial u(G)| \le C\nu(\Omega)|x_0 - z_0|.$$

This implies (1) with the help of (2). □

Next, we derive an estimate of the global Hölder norm of generalized solutions of the Dirichlet problem for Hölder continuous boundary values.

Theorem 8.3.2. *Let Ω be a bounded convex domain in $\mathbb{R}^n$ with $\Omega \subset B_1$, ν be a finite measure in Ω, and $\varphi \in C^\alpha(\bar{\Omega})$ be a convex function in Ω, for some $\alpha \in (0,1)$. Suppose that $u \in C(\bar{\Omega})$ is a convex generalized solution of*

$$\det \nabla^2 u = \nu \quad \text{in } \Omega,$$
$$u = \varphi \quad \text{on } \partial\Omega.$$

Then, $u \in C^{\bar{\alpha}}(\bar{\Omega})$, for $\bar{\alpha} = \min\{\alpha, 1/n\}$, and

$$|u|_{L^\infty(\Omega)} \leq C\{\nu(\Omega)\}^{\frac{1}{n}} + |\varphi|_{L^\infty(\Omega)},$$

$$[u]_{C^{\bar{\alpha}}(\bar{\Omega})} \leq C\big(\{\nu(\Omega)\}^{\frac{1}{n}} + [\varphi]_{C^\alpha(\bar{\Omega})}\big),$$

where C is a positive constant depending only on n.

Proof. We first derive the L^∞-bound of u. The convexity of u implies

(1) $$u \leq \sup_\Omega \varphi \quad \text{in } \Omega.$$

Let $v \in C(\bar{\Omega})$ be the generalized convex solution of the Dirichlet problem

$$\det \nabla^2 v = \nu \quad \text{in } \Omega,$$
$$v = 0 \quad \text{on } \partial\Omega.$$

The existence of v is implied by Lemma 8.2.6. By Theorem 8.3.1, we obtain

(2) $$|v| \leq C\{\nu(\Omega)\}^{\frac{1}{n}} \quad \text{in } \Omega$$

and, for any $x, y \in \Omega$,

(3) $$|v(x) - v(y)| \leq C\{\nu(\Omega)\}^{\frac{1}{n}}|x - y|^{\frac{1}{n}},$$

where C is a positive constant depending only on n. Next, Lemma 8.1.24 implies

$$M(v + \varphi) \geq Mv + M\varphi \geq Mv = Mu.$$

Hence,

$$M(v + \varphi) \geq Mu \quad \text{in } \Omega,$$
$$v + \varphi = u \quad \text{on } \partial\Omega.$$

By Theorem 8.2.2, we have

(4) $$v + \varphi \leq u \quad \text{in } \Omega,$$

and hence, by (2),

(5) $$-C\{\nu(\Omega)\}^{\frac{1}{n}} + \inf_\Omega \varphi \leq u \quad \text{in } \Omega.$$

By combining (1) and (5), we have the desired L^∞-bound of u.

Next, we prove, for any $x_0, y_0 \in \bar{\Omega}$,

(6) $$|u(x_0) - u(y_0)| \leq C\{\nu(\Omega)\}^{\frac{1}{n}}|x_0 - y_0|^{\frac{1}{n}} + [\varphi]_{C^\alpha(\bar{\Omega})}|x_0 - y_0|^\alpha.$$

Let L be the straight line determined by x_0 and y_0. Assume

$$\max\{|u(x) - u(y)| : x, y \in L \cap \bar{\Omega},\ |x - y| = |x_0 - y_0|\}$$

is realized at $\bar{x}, \bar{y} \in \bar{\Omega}$. By the convexity of u, we conclude that one of $\bar{x}$ and $\bar{y}$ is on $\partial\Omega$. Therefore, it suffices to prove (6) if one of x_0 and y_0 is on $\partial\Omega$.

In the following, we assume $x_0 \in \Omega$ and $y_0 \in \partial\Omega$. We first consider the case $u(x_0) > u(y_0) = \varphi(y_0)$. As earlier, let L be the straight line passing through x_0 and y_0. Then, L intersects $\partial\Omega$ at y_0 and at another point, say z_0. We now consider u restricted to the line segment $\overline{y_0 z_0}$. Since u is convex and $u(x_0) > u(y_0)$, we have $u(z_0) = \varphi(z_0) \geq u(x_0)$ and

$$\frac{|u(x_0) - u(y_0)|}{|x_0 - y_0|} \leq \frac{|\varphi(z_0) - \varphi(y_0)|}{|z_0 - y_0|}.$$

Therefore,

$$\begin{aligned}|u(x_0) - u(y_0)| &\leq \frac{|\varphi(z_0) - \varphi(y_0)|}{|z_0 - y_0|} \cdot |x_0 - y_0| \\ &= \frac{|\varphi(z_0) - \varphi(y_0)|}{|z_0 - y_0|^\alpha} \cdot \frac{|x_0 - y_0|^{1-\alpha}}{|z_0 - y_0|^{1-\alpha}} \cdot |x_0 - y_0|^\alpha \\ &\leq [\varphi]_{C^\alpha(\bar{\Omega})} |x_0 - y_0|^\alpha.\end{aligned}$$

Next, we consider the case $u(x_0) \leq u(y_0) = \varphi(y_0)$. Then, by (4),

$$\begin{aligned}|u(x_0) - u(y_0)| &= u(y_0) - u(x_0) \leq (v + \varphi)(y_0) - (v + \varphi)(x_0) \\ &\leq |v(x_0) - v(y_0)| + |\varphi(x_0) - \varphi(y_0)| \\ &\leq C\{\nu(\Omega)\}^{\frac{1}{n}} |x_0 - y_0|^{\frac{1}{n}} + [\varphi]_{C^\alpha(\bar{\Omega})} |x_0 - y_0|^\alpha.\end{aligned}$$

By combining both cases, we have the desired estimate (6). □

In both Theorem 8.3.1 and Theorem 8.3.2, we assumed $\Omega \subset B_1$ so the estimates in both theorems have simple forms. Of course, the global regularity $u \in C^{1/n}(\bar{\Omega})$ does not depend on such an assumption.

In our future study, it is convenient to arrange domains in a good form. For this purpose, we prove a normalization lemma due to John [**91**], which plays an important role in studying Monge-Ampère equations. An ellipsoid in $\mathbb{R}^n$ centered at x_0 can be written as

$$E(A, x_0) = \{x \in \mathbb{R}^n : \langle A(x - x_0), x - x_0 \rangle < 1\},$$

for some positive definite matrix A. The volume of $E(A, x_0)$ is given by

$$|E(A, x_0)| = \frac{|B_1|}{\sqrt{\det(A)}}.$$

Lemma 8.3.3. *Let Ω be a bounded convex domain in $\mathbb{R}^n$. Then, there exists an ellipsoid E such that*

$$\frac{1}{n} E \subset \Omega \subset E.$$

Moreover, E can be taken as the ellipsoid of the minimum volume containing Ω.

Here, $\frac{1}{n}E$ is dilated with respect to the center of E; namely, if x_0 is the center of E, then,

$$\frac{1}{n}E = \left\{ x_0 + \frac{1}{n}(x - x_0) : x \in E \right\}.$$

Proof. *Step 1.* Let $\mathcal{E}$ be the collection of all ellipsoids E with $\Omega \subset E$, and set

$$V_0 = \inf_{E \in \mathcal{E}} |E|.$$

We claim that there exists an $E \in \mathcal{E}$ such that $V_0 = |E|$.

To prove the claim, we assume, without loss of generality, $0 \in \Omega$. We note that there exist constants r and R with $r < R$ such that, for any $E \in \mathcal{E}$ with $|E| \leq V_0 + 1$,

$$B_r \subset E \subset B_R. \tag{1}$$

To verify this, we first take $B_r \Subset \Omega$ for some constant $r > 0$. Then, $B_r \Subset E$ for any $E \in \mathcal{E}$. This implies that the least axis of E is greater than r. Next, for any $E \in \mathcal{E}$ with $|E| \leq V_0 + 1$, the maximal axis is less than R, for some positive constant R depending only on r and V_0. Then, (1) follows. Now, we take a sequence $\{E_m\} \subset \mathcal{E}$ such that $|E_m| \to V_0$. Then, for m sufficiently large, E_m satisfies (1). It is easy to conclude that $E_m \to E$ for some $E \in \mathcal{E}$ with $|E| = V_0$.

Step 2. Let E be the ellipsoid constructed in Step 1. By an affine transform, we assume that E is the unit ball B_1. We now prove $B_{1/n} \subset \Omega$ by a contradiction argument. Consider $x_0 \in \partial\Omega$ such that

$$|x_0| = \inf_{\partial\Omega} |x|.$$

Assume

$$|x_0| < \frac{1}{n}.$$

By a rotation, we assume $x_0 = (0, \ldots, 0, -h)$ for some $h \in (0, 1/n)$. Then, the hyperplane $\{x_n = -h\}$ is tangent to $\partial\Omega$ at x_0 and

$$\Omega \subset D \equiv B_1 \cap \{x_n > -h\}.$$

In the following, we write $x' = (x_1, \ldots, x_{n-1})$. For some constant $\delta > 0$ small, consider the ellipsoid E_δ defined by

$$\frac{1}{(1+\delta)^2}|x'|^2 + (1+\delta)^{2(n-1)}\left(x_n - 1 + \frac{1}{(1+\delta)^{n-1}}\right)^2 < 1.$$

Then, $|E_\delta| = |B_1|$. We now consider the intersections of ∂B_1 and ∂E_δ. To find these intersections, we need to solve

$$|x'|^2 + x_n^2 = 1$$

and

$$\frac{1}{(1+\delta)^2}|x'|^2 + (1+\delta)^{2(n-1)}\left(x_n - 1 + \frac{1}{(1+\delta)^{n-1}}\right)^2 = 1.$$

By eliminating x', we have

$$(1+\delta)^{2(n-1)}(x_n-1)^2 + 2(1+\delta)^{n-1}(x_n-1) + (1+\delta)^{-2}(1-x_n^2) = 0.$$

Hence, $x_n = 1$ or

$$(1+\delta)^{2(n-1)}(x_n-1) + 2(1+\delta)^{n-1} - (1+\delta)^{-2}(1+x_n) = 0.$$

Then,

$$\big((1+\delta)^{2(n-1)} - (1+\delta)^{-2}\big)x_n = (1+\delta)^{2(n-1)} - 2(1+\delta)^{n-1} + (1+\delta)^{-2}.$$

A simple expansion in δ yields

$$\big(2n\delta + O(\delta^2)\big)x_n = -2\delta + O(\delta^2),$$

or

$$x_n = -\frac{1}{n} + O(\delta).$$

Hence, for $\delta, \widetilde{\delta} > 0$ sufficiently small, $\partial(E_\delta + \widetilde{\delta} e_n)$ strictly contains D. Therefore, we can find an ellipsoid $\widetilde{E}$ such that $\Omega \subset \widetilde{E}$ and $|\widetilde{E}| < |E_\delta| = |B_1|$. This contradicts the choice of B_1. $\square$

In Lemma 6.4.1, we proved a similar result, requiring the center of the ellipsoid to be at the center of mass of the convex set. Such a requirement is not imposed in Lemma 8.3.3. In this chapter, we will use Lemma 8.3.3 instead of Lemma 6.4.1.

A convex domain Ω is *normalized* if

$$B_{1/n} \subset \Omega \subset B_1.$$

For any ellipsoid E, there is an affine transform T such that $T(E) = B_1$. By Lemma 8.3.3, for any bounded convex domain Ω, there exists an affine transform T such that

$$B_{1/n} \subset T(\Omega) \subset B_1.$$

The set $T(\Omega)$ is called a *normalization* of Ω, and T is called an affine transform that *normalizes* Ω. Lemma 8.3.3 simply asserts that any bounded convex domain can be transformed to a normalized one by an affine transform. We write $Tx = Ax + y_0$, for some point $y_0 \in \mathbb{R}^n$ and some invertible $n \times n$ matrix A. Then,

$$c_1 \le |T(\Omega)| = |\det A||\Omega| \le c_2,$$

where c_1 and c_2 are positive constants depending only on n.

Next, we prove a basic result concerning dilations of convex domains.

Lemma 8.3.4. *Let Ω be a bounded convex domain in $\mathbb{R}^n$ with $0 \in \Omega$ and r, R be positive constants.*

(i) *If $\Omega \subset B_R$, then, for any constant $\eta \in (0, R)$,*

$$(1+\eta/R)\Omega \setminus \overline{(1-\eta/R)\Omega} \subset \{x \in \mathbb{R}^n : \operatorname{dist}(x, \partial\Omega) < \eta\}.$$

(ii) *If $B_r \subset \Omega$, then, for any constant $\eta \in (0, r)$,*

$$\{x \in \mathbb{R}^n : \operatorname{dist}(x, \partial\Omega) < \eta\} \subset (1+\eta/r)\Omega \setminus \overline{(1-\eta/r)\Omega}.$$

Here, all dilations of Ω are with respect to the origin; namely, for any $\tau > 0$,

$$\tau\Omega = \{\tau x : x \in \Omega\}.$$

Proof. (i) Take an arbitrary $x \in (1+\eta/R)\Omega \setminus \overline{(1-\eta/R)\Omega}$. There exist a point $\bar{x} \in \partial\Omega$ and a constant $\lambda \in (1-\eta/R, 1+\eta/R)$ such that $x = \lambda\bar{x}$. By noting that $|\bar{x}| < R$, we have

$$|x - \bar{x}| = |\lambda - 1||\bar{x}| < \eta,$$

and hence $\operatorname{dist}(x, \partial\Omega) < \eta$.

(ii) Take any $x \in \Omega$ with $\operatorname{dist}(x, \partial\Omega) < \eta$. There exists a point $x_* \in \partial\Omega$ such that $|x - x_*| = \operatorname{dist}(x, \partial\Omega)$. Assume the ray from O to x intersects $\partial\Omega$ at some $z \in \partial\Omega$. We only consider the case $z \neq x_*$. Then, $|x - z| \geq |x - x_*|$ and the angle between the segments $\overline{zx}$ and $\overline{zx_*}$ is smaller than $\pi/2$. Let l be the line joining z and x_*. Since Ω is convex, the set $l \setminus \overline{zx_*}$ does not intersect Ω. Hence, $|y| > r$ for any $y \in l \setminus \overline{zx_*}$. Let l' be the ray from O that is parallel to the segment $\overline{xx_*}$ and lies in the plane containing z, x, and x_*. Let P be the intersection of l' and l. Then, $\triangle OPz$ is similar to $\triangle xx_*z$, and hence

$$\frac{|xz|}{|Oz|} = \frac{|\overline{xx_*}|}{|OP|} < \frac{\eta}{r}. \tag{1}$$

This implies $x \in \Omega \setminus \overline{(1-\eta/r)\Omega}$.

For $x \notin \bar{\Omega}$ with $\operatorname{dist}(x, \partial\Omega) < \eta$, we can proceed similarly and obtain (1). Then, $x \in (1+\eta/r)\Omega \setminus \bar{\Omega}$. □

For any bounded convex domain $\Omega \subset \mathbb{R}^n$, by Lemma 8.3.3, there exists a minimum ellipsoid E such that

$$\frac{1}{n}E \subset \Omega \subset E.$$

For any positive constant τ, we define $\tau\Omega$ to be the τ-dilation of the set Ω with respect to the center of the ellipsoid E.

In the next result, we derive a global estimate of generalized solutions of the Dirichlet problem.

Theorem 8.3.5. *Let Ω be a bounded convex domain in $\mathbb{R}^n$ and ν be a finite measure in Ω satisfying, for some $b > 0$,*

$$\nu(\Omega) \leq b\nu(2^{-1}\Omega).$$

Suppose that $u \in C(\bar{\Omega})$ is a convex generalized solution of

$$\begin{aligned} \det \nabla^2 u &= \nu \quad \text{in } \Omega, \\ u &= 0 \quad \text{on } \partial\Omega. \end{aligned}$$

Then,

$$C^{-1}\{|\Omega|\nu(\Omega)\}^{\frac{1}{n}} \leq \sup_{\Omega} |u| \leq C\{|\Omega|\nu(\Omega)\}^{\frac{1}{n}},$$

where C is a positive constant depending only on n and b.

Proof. If $\nu \equiv 0$ in Ω, then $u \equiv 0$ by Corollary 8.2.3. In the following, we consider the case that ν is a nontrivial measure.

Step 1. We first consider the special case $B_{1/n} \subset \Omega \subset B_1$ and $\nu(\Omega) = 1$ and proceed to prove

$$C^{-1} \leq \sup_{\Omega} |u| \leq C. \tag{1}$$

We already proved the upper bound in Theorem 8.3.1. In fact, by applying Theorem 8.2.4, we obtain, for any $x \in \Omega$,

$$|u(x)|^n \leq C|\partial u(\Omega)| = C\nu(\Omega) = C,$$

where C is a positive constant depending only on n. This proves the upper bound in (1).

For the lower bound, we first note that, for any $\alpha \in (0, 1)$,

$$\operatorname{dist}(\alpha\Omega, \partial\Omega) \geq \frac{1}{n}(1 - \alpha). \tag{2}$$

To prove this, we recall $B_{1/n} \subset \Omega$. By Lemma 8.3.4(ii), we have, for any constant $\eta \in (0, 1/n)$,

$$\{x \in \Omega : \operatorname{dist}(x, \partial\Omega) < \eta\} \subset \Omega \setminus \overline{(1 - n\eta)\Omega}.$$

This implies (2) if we take $\eta = (1 - \alpha)/n$. Next, we take $\alpha = 1/2$. By applying Lemma 8.1.7 to u and using (2), we have

$$\partial u(2^{-1}\Omega) \subset B_R,$$

where we can take

$$R = 2n \sup_{\Omega} |u|.$$

Then,

$$\nu(2^{-1}\Omega) = |\partial u(2^{-1}\Omega)| \leq C \sup_{\Omega} |u|^n.$$

This implies the lower bound in (1) since $\nu(2^{-1}\Omega) \geq 1/b$.

Step 2. For the general case, we consider an affine transform T such that $B_{1/n} \subset T(\Omega) \subset B_1$. We write $T(x) = Ax + y_0$, for some point $y_0 \in \mathbb{R}^n$ and some invertible $n \times n$ matrix A. Then,

$$c_1 \leq |T(\Omega)| = |\det A||\Omega| \leq c_2, \tag{3}$$

where c_1 and c_2 are positive constants depending only on n. Set, for any $y \in T(\Omega)$,

$$v(y) = \frac{|\det A|^{\frac{1}{n}}}{\nu(\Omega)^{\frac{1}{n}}} u(T^{-1}y).$$

Then,

$$\begin{aligned} \det \nabla^2 v &= \widetilde{\nu} \quad \text{in } T(\Omega), \\ v &= 0 \quad \text{on } \partial T(\Omega) = T(\partial\Omega), \end{aligned}$$

where $\widetilde{\nu}$ is a finite measure in $T(\Omega)$ such that, for any Borel set $\omega \subset \Omega$,

$$\widetilde{\nu}\big(T(\omega)\big) = \frac{1}{\nu(\Omega)} \nu(\omega).$$

Then,

$$\widetilde{\nu}\big(T(\Omega)\big) = 1,$$

and

$$\widetilde{\nu}(2^{-1}T(\Omega)) = \widetilde{\nu}(T(2^{-1}\Omega)) \geq \frac{1}{b},$$

where $2^{-1}T(\Omega)$ is dilated with respect to the origin and $2^{-1}\Omega$ with respect to $-A^{-1}y_0$. By what we proved in Step 1, we have

$$C^{-1} \leq \sup_{T(\Omega)} |v| \leq C,$$

and hence

$$C^{-1} \frac{\nu(\Omega)^{\frac{1}{n}}}{|\det A|^{\frac{1}{n}}} \leq \sup_{\Omega} |u| \leq C \frac{\nu(\Omega)^{\frac{1}{n}}}{|\det A|^{\frac{1}{n}}}.$$

This implies the desired result with the help of (3). □

Next, we characterize the distance of the minimal point to the boundary.

Corollary 8.3.6. *Let Ω be a bounded convex domain in $\mathbb{R}^n$ and ν be a nontrivial finite measure in Ω satisfying, for some $b > 0$,*

$$\nu(\Omega) \leq b\nu(2^{-1}\Omega).$$

Suppose that $u \in C(\bar{\Omega})$ is a convex generalized solution of

$$\det \nabla^2 u = \nu \quad \text{in } \Omega,$$
$$u = 0 \quad \text{on } \partial\Omega.$$

Assume l is a line segment in Ω with its two end points $x', x'' \in \partial\Omega$. If, for some $x \in l$,

$$u(x) \leq \frac{1}{2} \inf_{\Omega} u,$$

then

$$|x - x'| \geq c|x'' - x'|,$$

where c is a positive constant depending only on n and b.

Proof. We note that the ratio $|x - x'|/|x'' - x'|$ is invariant under affine transforms. Hence, by using an affine transform, we assume $B_{1/n} \subset \Omega \subset B_1$ and $\inf_\Omega u = -1$. By Theorem 8.3.5, we have $\nu(\Omega) \leq C$. Therefore, if $u(x) \leq -1/2$, then Theorem 8.3.1 implies, for any $y \in \partial\Omega$,

$$\frac{1}{2} \leq |u(x) - u(y)| \leq C|x - y|^{\frac{1}{n}},$$

and hence $\text{dist}(x, \partial\Omega) \geq c_0$, for some positive constant c_0 depending only on n and b. □

To end this section, we introduce the doubling condition and prove one of its consequences.

Definition 8.3.7. Let Ω be a domain in $\mathbb{R}^n$ and ν be a locally finite Borel measure. Then, ν satisfies the *doubling condition* if there exists a positive constant b such that, for any convex subdomain $\omega \subset \Omega$,

$$\nu(\omega) \leq b\nu\left(2^{-1}\omega\right).$$

The constant b is called the *doubling constant.*

Next, we prove that generalized solutions have unique supporting functions if they are strictly convex and the corresponding Monge-Ampère measures satisfy the doubling condition.

Lemma 8.3.8. *Let Ω be a convex domain in $\mathbb{R}^n$ and ν be a finite measure in Ω satisfying the doubling condition. Suppose that $u \in C(\Omega)$ is a convex generalized solution of*

$$\det \nabla^2 u = \nu \quad \text{in } \Omega.$$

If u is strictly convex at some $x_0 \in \Omega$, then u has a unique supporting function at x_0.

Proof. Without loss of generality, we assume $x_0 = 0$, Ω is a bounded domain, and u is a bounded function. We note that $\partial u(0)$ is a compact convex subset of $\mathbb{R}^n$. Suppose that $\partial u(0)$ contains more than one point. By the strict convexity of u at 0, we first take $p_1 \in \partial u(0)$ such that, for any $x \in \Omega \setminus \{0\}$,

$$u(x) > u(0) + p_1 \cdot x.$$

Next, we take $p_0 \in \partial u(0)$ such that

$$|p_0 - p_1| = \max\{|p - p_1| : p \in \partial u(0)\}.$$

We now claim

$$p \notin \partial u(0) \text{ if } \langle p - p_0, p_1 - p_0 \rangle < 0. \tag{1}$$

To verify this, we simply note that

$$\begin{aligned} |p - p_1|^2 &= |p - p_0 + p_0 - p_1|^2 \\ &= |p - p_0|^2 - 2\langle p - p_0, p_1 - p_0 \rangle + |p_0 - p_1|^2. \end{aligned}$$

If $\langle p - p_0, p_1 - p_0 \rangle < 0$, then $|p - p_1| > |p_0 - p_1|$. This proves (1). Next, by the convexity of $\partial u(0)$, we note that $(1 - \alpha)p_0 + \alpha p_1 \in \partial u(0)$ for any $\alpha \in [0, 1]$. Moreover, for any $\alpha \in (0, 1]$ and any $x \in \Omega \setminus \{0\}$,

$$u(x) > u(0) + \big[(1 - \alpha)p_0 + \alpha p_1\big] \cdot x. \tag{2}$$

We point out that the strict inequality may not hold for $\alpha = 0$.

Set, for any $x \in \Omega$,

$$v(x) = u(x) - u(0) - p_0 \cdot x.$$

Since $p_0, p_1 \in \partial u(0)$, then $l_0(x) = 0$ and $l_1(x) = (p_1 - p_0) \cdot x$ are two supporting functions of v at 0. Hence,

$$v(0) = 0, \quad v \geq 0 \quad \text{in } \Omega.$$

By a rotation in $\mathbb{R}^n$, we assume that the line segment $\overline{p_0 p_1}$ is on the positive x_n-axis, with $a = |p_1 - p_0|$. To be specific, we set $y = Ax$ for some orthogonal matrix A and $w(y) = v(x)$. Hence,

$$w(0) = 0, \quad w \geq 0 \quad \text{in } A(\Omega). \tag{3}$$

The supporting function l_1 in the new coordinates is given by

$$l_1(x) = (p_1 - p_0) \cdot x = (p_1 - p_0) \cdot A^{-1}y = A^{-T}(p_1 - p_0) \cdot y.$$

We choose A such that $A^{-T}(p_1 - p_0) = ae_n$. Then,

$$l_1(x) = ae_n \cdot y = ay_n,$$

and hence, for any $y \in A(\Omega)$,

$$w(y) \geq ay_n. \tag{4}$$

Moreover, (2) implies, for any $\tau \in (0, a]$ and any $y \in A(\Omega \setminus \{0\})$,

$$w(y) > \tau y_n. \tag{5}$$

By (1), we note that $q \notin \partial w(0)$ if $q \cdot e_n < 0$. Next, we claim

$$\frac{w(-te_n)}{t} \to 0 \quad \text{as } t \to 0^+. \tag{6}$$

To prove (6), we note that $w(te_n)$ is a convex function of $t \in (-\varepsilon, \varepsilon)$ for some constant $\varepsilon > 0$, with $w(0) = 0$ and $w(-te_n) \geq 0$ for any $t \in (0, \varepsilon)$ by (3). Then, $\frac{w(-te_n)}{t}$ is an increasing function of $t \in (0, \varepsilon)$. We assume, for some $\mu \geq 0$,

$$\frac{w(-te_n)}{t} \to \mu \quad \text{as } t \to 0^+.$$

Take an arbitrary sequence $t_i \to 0^+$ as $i \to \infty$ and a $q_i \in \partial w(-t_i e_n)$ for each i. Then, for any $y \in A(\Omega)$,

$$w(y) \geq w(-t_i e_n) + q_i \cdot (y + t_i e_n). \tag{7}$$

By Lemma 8.1.7, q_i is a bounded sequence. We assume, up to a subsequence, $q_i \to q_0$, for some $q_0 \in \mathbb{R}^n$. By taking $i \to \infty$ in (7), we obtain, for any $y \in A(\Omega)$,

$$w(y) \geq q_0 \cdot y.$$

This implies $q_0 \in \partial w(0)$ and hence $q_0 \cdot e_n \geq 0$. On the other hand, by taking $y = 0$ in (7), we have

$$0 \geq \frac{w(-t_i e_n)}{t_i} + q_i \cdot e_n.$$

By letting $i \to \infty$, we get $0 \geq \mu + q_0 \cdot e_n \geq \mu$, and hence $\mu = 0$. Therefore, (6) follows readily.

Next, fix a constant $\tau \in (0, a)$ and set $w_\tau = w - \tau y_n$. Then, w_τ satisfies

$$\det \nabla^2 w_\tau = \widetilde{\nu} \quad \text{in } A(\Omega),$$

where $\widetilde{\nu} = \nu \circ A^{-1}$. For any $h > 0$, define

$$S_h = \{y \in A(\Omega) : w(y) < \tau y_n + h\}.$$

By (5), there exists a constant $h(\tau) > 0$ such that $S_h \Subset A(\Omega)$ for any $h \in (0, h(\tau))$. For any such h, we assume ∂S_h intersects the y_n-axis at $(0, y_n^+(h))$ and $(0, -y_n^-(h))$, with $y_n^+(h), y_n^-(h) > 0$. Here, we view $y_n^+(h)$ and $y_n^-(h)$ as functions of h. Since w_τ attains its minimum at 0 by (5), Corollary 8.3.6 implies, for any $h \in (0, h(\tau))$,

$$c_1 \leq \frac{y_n^+(h)}{y_n^-(h)} \leq c_2, \tag{8}$$

where c_1 and c_2 are positive constants depending only on n and b. The definition of $y_n^+(h)$ and (4) imply

$$w(y_n^+(h)) = \tau y_n^+(h) + h \geq a y_n^+(h),$$

and hence

$$y_n^+(h) \le \frac{h}{a-\tau}.$$

By the definition of $y_n^-(h)$, we have

$$w(-y_n^-(h)e_n) + \tau y_n^-(h) = h.$$

Since $w \ge 0$, then $y_n^-(h) \to 0$ as $h \to 0^+$, and

$$\frac{y_n^+(h)}{y_n^-(h)} \le \frac{1}{a-\tau} \cdot \frac{h}{y_n^-(h)} = \frac{1}{a-\tau}\left(\tau + \frac{w(-y_n^-(h)e_n)}{y_n^-(h)}\right).$$

By (6), we obtain

$$\limsup_{h\to 0^+} \frac{y_n^+(h)}{y_n^-(h)} \le \frac{\tau}{a-\tau}.$$

This contradicts (8) if we choose τ sufficiently small. Therefore, $\partial u(0)$ consists of one point. □

8.4. Interior $C^{1,\alpha}$-Regularity

In this section, we discuss the strict convexity and the interior $C^{1,\alpha}$-regularity for solutions of the Monge-Ampère equation under the condition that the Monge-Ampère measures satisfy the doubling condition. These results are based on the level set approach and are due to Caffarelli [**14**], [**16**].

We first introduce the notion of Hausdorff metrics. Let A and B be two subsets in $\mathbb{R}^n$. Their *Hausdorff distance* $d_H(A, B)$ is defined by

$$d_H(A,B) = \max\left\{\sup_{x\in A} \operatorname{dist}(x,B), \sup_{x\in B} \operatorname{dist}(x,A)\right\}.$$

It is straightforward to verify that d_H is a metric on the collection of nonempty compact subsets in $\mathbb{R}^n$.

We now discuss the convergence of convex domains and convex functions.

Lemma 8.4.1. (i) *Suppose that $\{\Omega_i\}$ is a sequence of convex domains in $\mathbb{R}^n$ satisfying*

$$B_r \subset \Omega_i \subset B_R, \tag{1}$$

for some positive constants r and R. Then, up to a subsequence, $\partial\Omega_i \to \partial\Omega$ in the Hausdorff metric for some convex domain Ω in $\mathbb{R}^n$, with $B_r \subset \Omega \subset B_R$.

(ii) *Assume, in addition, $u_i \in C(\bar{\Omega}_i)$ is a convex function satisfying $u_i = 0$ on $\partial\Omega_i$ and*

$$\sup_i \left|\partial u_i(\Omega_i)\right| < \infty. \tag{2}$$

Then, up to a subsequence, $u_i \to u$ locally uniformly in Ω for some convex function $u \in C(\bar{\Omega})$, with $u = 0$ on $\partial\Omega$.

(iii) *Let δ be a positive constant. Assume, in addition, $x_i \in \Omega_i$ is a point with* $\text{dist}(x_i, \partial\Omega_i) \geq \delta$. *Then, up to a subsequence, $x_i \to x_0$ for some $x_0 \in \Omega$ with* $\text{dist}(x_0, \partial\Omega) \geq \delta$. *Moreover, if $\bar{x}_i \in \partial\Omega_i$, then, up to a subsequence, $\bar{x}_i \to \bar{x}$ for some $\bar{x} \in \partial\Omega$.*

(iv) *Assume, in addition, l_i is a supporting function of u_i in Ω_i at x_i. Then, up to a subsequence, $l_i \to l$ for some supporting function l of u at x_0.*

Proof. We proceed in several steps.

Step 1. We discuss the convergence of Ω_i. Let F_i be the convex function in $\mathbb{R}^n$ satisfying $F_i = 0$ on $\partial\Omega_i$ and the graph of F_i is the convex cone in $\mathbb{R}^n \times \mathbb{R}$ with its vertex $(0, -1)$. Then,

$$\Omega_i = \{x \in \mathbb{R}^n : F_i(x) < 0\}.$$

We first note that $\partial F_i(\Omega_i) = \partial F_i(0)$. By (1), we have

$$\partial F_i(\Omega_i) \subset \overline{B}_{1/r}, \tag{3}$$

and hence, $-1 \leq F_i \leq -1 + 2R/r$ in $\bar{B}_{2R}$. Therefore, F_i is a sequence of bounded and equicontinuous functions in $\bar{B}_{2R}$. By the Arzela-Ascoli theorem, we have, by passing to a subsequence if necessary,

$$F_i \to F \quad \text{uniformly in } \bar{B}_{2R},$$

for a convex function $F \in C(\bar{B}_{2R})$. It is easy to check that F is in fact a convex function in $\mathbb{R}^n$ and the graph of F is a convex cone in $\mathbb{R}^n \times \mathbb{R}$ with its vertex $(0, -1)$. Set

$$\Omega = \{x \in \mathbb{R}^n : F(x) < 0\}.$$

Then, Ω is a convex set in B_{2R}.

Next, we prove the following statement: Let $K \subset \Omega$ be a compact set. Then, $K \subset \Omega_i$ for i sufficiently large. To verify this, we first have $F \leq -\delta$ in K by the compactness of K. Next, the uniform convergence of F_i in B_{2R} implies $F_i \leq -\delta/2$ in K for i sufficiently large. Hence, $K \subset \Omega_i$ for such i.

We now note that, for each integer i and any $\delta \in (0, 1)$,

$$\{x \in B_{2R} : |F_i(x)| < \delta\} = (1+\delta)\Omega_i \setminus \overline{(1-\delta)\Omega_i}, \tag{4}$$

where dilations of Ω_i are with respect to the origin. To prove this, we consider an arbitrary point $x \in B_{2R}$. Assume the ray from 0 to x intercepts $\partial\Omega_i$ at some $\bar{x}_i \in \partial\Omega$. By the similarity of triangles, $F_i(0) = -1$, and $F_i(\bar{x}_i) = 0$, we have

$$|F_i(x)| = \frac{|x - \bar{x}_i|}{|\bar{x}_i|}. \tag{5}$$

By writing $x = \lambda \bar{x}_i$ for some constant λ, we have $|F_i(x)| = |\lambda - 1|$. Hence, $|F_i(x)| < \delta$ if and only if $|\lambda - 1| < \delta$. This proves (4). Similarly, we have, for any $\delta \in (0, 1)$,

(6) $$\{x \in B_{2R} : |F(x)| < \delta\} = (1 + \delta)\Omega \setminus \overline{(1 - \delta)\Omega}.$$

We now prove

$$B_r \subset \Omega \subset B_R.$$

If $x \in \Omega$, then $F(x) < 0$. Since $F_i(x) \to F(x)$, then $F_i(x) < 0$ for i sufficiently large; that is, $x \in \Omega_i \subset B_R$. Next, we take an arbitrary $x \in B_r$. Then, $x \in B_{r-\delta}$ for some constant $\delta > 0$, and hence $x \in \Omega_i$ for each i. Assume the ray from 0 to x intercepts $\partial\Omega_i$ at some $\bar{x}_i \in \partial\Omega$. Then, (5) implies, with $r \leq |\bar{x}_i| \leq R$,

$$|F_i(x)| = \frac{|x - \bar{x}_i|}{|\bar{x}_i|} \geq \frac{|\bar{x}_i| - |x|}{|\bar{x}_i|} \geq \frac{\delta}{R}.$$

So, $F_i(x) \leq -\delta/R$ for each i. Letting $i \to \infty$, we get $F(x) \leq -\delta/R$, and hence $x \in \Omega$.

Next, we prove that, for any $\eta \in (0, r/2)$, there exists a positive integer $i_0(\eta)$ such that, for any $i \geq i_0(\eta)$,

(7) $$\{x \in B_{2R} : \operatorname{dist}(x, \partial\Omega) < \eta\} \subset \{x \in B_{2R} : \operatorname{dist}(x, \partial\Omega_i) < 2R\eta/r\}$$

and

(8) $$\{x \in B_{2R} : \operatorname{dist}(x, \partial\Omega_i) < \eta\} \subset \{x \in B_{2R} : \operatorname{dist}(x, \partial\Omega) < 2R\eta/r\}.$$

In fact, for any $\eta > 0$, there exists a positive integer $i_0(\eta)$ such that, for any $i \geq i_0(\eta)$,

$$|F_i - F| < \frac{\eta}{r} \quad \text{in } B_{2R}.$$

To prove (7), we take an arbitrary point $x \in B_{2R}$ with $\operatorname{dist}(x, \partial\Omega) < \eta$. By Lemma 8.3.4(ii), we have $x \in (1 + \eta/r)\Omega \setminus \overline{(1 - \eta/r)\Omega}$. Then, (6) implies $|F(x)| < \eta/r$. Hence, for $i \geq i_0(\eta)$, $|F_i(x)| < 2\eta/r$. Then, (4) implies $x \in (1+2\eta/r)\Omega_i \setminus \overline{(1 - 2\eta/r)\Omega_i}$. By Lemma 8.3.4(i), we obtain $\operatorname{dist}(x, \partial\Omega_i) < 2R\eta/r$ and hence (7). The proof of (8) is similar and is omitted. We note that (7) and (8) imply the convergence $\partial\Omega_i \to \partial\Omega$ in the Hausdorff metric.

Step 2. We discuss the convergence of u_i. First, by Theorem 8.3.1 and (2), we have, for any i,

(9) $$|u_i| \leq C \quad \text{in } \Omega_i,$$

where C is a positive constant depending only on n, R, and $|\partial u_i(\Omega_i)|$. Here, we only need the estimate of the L^∞-norm in Theorem 8.3.1. Next, we take an arbitrary compact set $K \subset \Omega$. Then, $K \subset \Omega_i$ for i sufficiently large.

Moreover, take a positive constant $\eta(K)$ such that $\operatorname{dist}(x, \partial\Omega) \geq \eta(K)$ for any $x \in K$. By (8), we have, for any $x \in K$ and any i sufficiently large,

$$\operatorname{dist}(x, \partial\Omega_i) \geq \frac{r}{2R}\eta(K).$$

Lemma 8.1.7 implies, for any $p \in \partial u_i(K)$ and any i sufficiently large,

$$|p| \leq C(K),$$

where $C(K)$ is a positive constant depending only on n, r, R, $|\partial u_i(\Omega_i)|$, and K. Therefore, $\{u_i\}$ is a sequence of bounded and equicontinuous functions in K. By the Arzela-Ascoli theorem, there exists a uniformly convergent subsequence of $\{u_i\}$ in K. We now take a sequence of increasing compact subsets $K_i \subset \Omega$ such that $\bigcup_i K_i = \Omega$ and, by a diagonalization process, we have a subsequence of $\{u_i\}$ uniformly convergent in any compact subset $K \subset \Omega$.

We define, for any $x \in \Omega$,

$$u(x) = \lim_{i \to \infty} u_i(x).$$

Then, u is continuous and convex in Ω. For each $x \in \Omega$, since $u_i(x) \leq 0$ for i sufficiently large, then $u \leq 0$ in Ω. We point out that u is defined only in Ω at this time.

Next, we prove $u \in C(\bar{\Omega})$ and $u = 0$ on $\partial\Omega$. To this end, we take an arbitrary point $x \in \Omega$. Then, $x \in \Omega_i$ for i sufficiently large. By applying Theorem 8.2.4 to u_i in Ω_i, we have

$$|u_i(x)|^n \leq C \operatorname{dist}(x, \partial\Omega_i).$$

Take $y_0 \in \partial\Omega$ such that $|x - y_0| = \operatorname{dist}(x, \partial\Omega)$. By (7), for any constant $\varepsilon > 0$, we have $B_\varepsilon(y_0) \cap \partial\Omega_i \neq \emptyset$, for i sufficiently large. We take $y_{i,\varepsilon} \in \partial\Omega_i \cap B_\varepsilon(y_0)$. Then, for i sufficiently large,

$$|u_i(x)|^n \leq C|x - y_{i,\varepsilon}| \leq C(|x - y_0| + \varepsilon).$$

Letting $i \to \infty$ and then $\varepsilon \to 0$, we have, by the convergence of u_i to u,

$$|u(x)|^n \leq C|x - y_0|.$$

Therefore, we conclude, for any $x \in \Omega$,

$$|u(x)|^n \leq C \operatorname{dist}(x, \partial\Omega).$$

Hence, u can be continuously extended to $\bar{\Omega}$ and $u = 0$ on $\partial\Omega$.

Step 3. We discuss the convergence of x_i. For $x_i \in \Omega_i$ with $\operatorname{dist}(x_i, \partial\Omega_i) \geq \delta$, we have, by (7), $x_i \in \Omega$ with $\operatorname{dist}(x_i, \partial\Omega) \geq r\delta/(2R)$. We assume, by passing to a subsequence, $x_i \to x_0$ for some $x_0 \in \Omega$. We now prove $\operatorname{dist}(x_0, \partial\Omega) \geq \delta$. To this end, we take any $y \in \partial\Omega$. By (7), for any constant

$\varepsilon > 0$, we have $B_\varepsilon(y) \cap \partial\Omega_i \neq \emptyset$, for i sufficiently large. We take $y_{i,\varepsilon} \in \partial\Omega_i \cap B_\varepsilon(y)$. Then, for i sufficiently large,

$$\delta \leq |x_i - y_{i,\varepsilon}| \leq |x_i - y| + \varepsilon.$$

Letting $i \to \infty$ and then $\varepsilon \to 0$, we have $|x_0 - y| \geq \delta$. Since $y \in \partial\Omega$ is arbitrary, then $\text{dist}(x_0, \partial\Omega) \geq \delta$. Moreover, by (8), $\text{dist}(x_0, \partial\Omega_i) \geq r\delta/(2R)$ for i sufficiently large.

Next, we assume $\bar{x}_i \in \partial\Omega_i \subset B_R$ and $\bar{x}_i \to \bar{x}$ and proceed to prove $\bar{x} \in \partial\Omega$. To this end, let F_i and F be the convex functions introduced in Step 1. Then, $F_i(\bar{x}_i) = 0$. Since F_i is Lipschitz in B_{2R} with the Lipschitz constant independent of i, we have $F(\bar{x}) = 0$ and hence $\bar{x} \in \partial\Omega$.

Step 4. We discuss the convergence of $\{l_i\}$. For each i, we write

$$l_i(x) = u_i(x_i) + p_i \cdot (x - x_i),$$

for some $p_i \in \partial u_i(x_i)$. Then, for any $x \in \Omega_i$,

$$u_i(x) \geq u_i(x_i) + p_i \cdot (x - x_i). \tag{10}$$

By Step 3, we have $x_i \to x_0$ for some $x_0 \in \Omega$ with $\text{dist}(x_0, \partial\Omega) \geq \delta$. Note that, for any compact set $K \subset \Omega$ and i sufficiently large, u_i is Lipschitz in K with the Lipschitz constant independent of i. Then, $u_i(x_i) \to u(x_0)$. By (9) and Lemma 8.1.7, we have

$$\sup_i |p_i| < \infty.$$

We assume, by passing to a subsequence, $p_i \to p$ for some $p \in \mathbb{R}^n$. By letting $i \to \infty$ in (10), we get, for any $x \in \Omega$,

$$u(x) \geq u(x_0) + p \cdot (x - x_0).$$

Hence, $l(x) = u(x_0) + p \cdot (x - x_0)$ is a supporting function of u at x_0. □

Next, we discuss supporting functions of the generalized solutions.

Lemma 8.4.2. *Let Ω be a convex domain in $\mathbb{R}^n$ with $B_{1/n} \subset \Omega \subset B_1$ and ν be a nontrivial finite measure in Ω satisfying the doubling condition with the doubling constant b. Suppose that $u \in C(\bar{\Omega})$ is a convex generalized solution of*

$$\det \nabla^2 u = \nu \quad \text{in } \Omega,$$
$$u = 0 \quad \text{on } \partial\Omega.$$

Then, for any point $x_0 \in \Omega$ and any supporting function l_{x_0} of u at x_0,

$$\sup_{\partial\Omega} l_{x_0} < 0.$$

Proof. We take an arbitrarily fixed $x_0 \in \Omega$ and let l be a supporting function of u at x_0. Then, $l \le 0$ on $\partial\Omega$ since $u = 0$ on $\partial\Omega$. We now prove the desired result by a contradiction argument. Assume

$$\max_{\partial\Omega} l = 0.$$

For convenience, we assume $x_0 = 0$ by a translation. Hence, $0 \in \Omega$, and $B_{1/n}(x_*) \subset \Omega \subset B_1(x_*)$ for some $x_* \in B_1$. Then,

$$\begin{aligned} u \ge l \text{ in } \Omega, \quad & u(0) = l(0), \\ u = 0 \text{ on } \partial\Omega, \quad & \max_{\partial\Omega} l = 0. \end{aligned}$$

Set $v = u - l$ and write $\varphi = -l$. Then, $v(0) = 0$ and

$$\begin{aligned} v \ge 0 \quad & \text{in } \Omega, \\ v = \varphi \quad & \text{on } \partial\Omega. \end{aligned}$$

Here, φ is an affine function with

$$\min_{\partial\Omega} \varphi = 0.$$

We assume $\text{dist}(0, \partial\Omega) \ge \delta$ for some constant $\delta > 0$.

By a rotation, we assume, for some $a_0 \ge 0$ and some a_1,

$$\varphi(x) = a_0 x_n + a_1.$$

We first note that $a_0 \ne 0$. Otherwise, $a_0 = 0$ and then $\varphi|_{\partial\Omega} = a_1 = 0$. Since v is convex and $v \ge 0$ in Ω, then v is identically zero, contradicting $\nu(\Omega) > 0$. With $a_0 \ne 0$, we have $a_0 > 0$. Then, $a_1 = \varphi(0) \ge 0$. This is because $0 = v(0) = u(0) + \varphi(0) \le \varphi(0)$. Hence,

$$\{x \in \partial\Omega : v(x) = 0\} \subset \left\{x_n = -\frac{a_1}{a_0}\right\}.$$

Set $(0, \bar{x}_n) \in \partial\Omega \cap \{x' = 0\}$ with $\bar{x}_n < 0$. Then, $\varphi(0, \bar{x}_n) \ge 0$ and hence $\frac{a_1}{a_0} \ge -\bar{x}_n$. Since $\text{dist}(0, \partial\Omega) \ge \delta$, then $-\bar{x}_n \ge \delta$ and hence

$$\frac{a_1}{a_0} \ge \delta.$$

We point out that $(0, \bar{x}_n)$ is not necessarily the lowest point of $\partial\Omega$ and that v may not be zero at $(0, \bar{x}_n)$. Take a point $\widetilde{x} = (\widetilde{x}_1, \dots, \widetilde{x}_n) \in \partial\Omega$ with $v(\widetilde{x}) = 0$, and some point $x^0 = (x_1^0, \dots, x_n^0) \in \Omega$ with $v(x^0) = 0$ and $x_n^0 = \sup\{x_n : v(x) = 0\}$. In other words, $x_n \le x_n^0$, for any $x \in \Omega$ with $v(x) = 0$. Note that $x_n^0 \ge 0$ since $v(0) = 0$. Then,

$$x_n^0 - \widetilde{x}_n \ge -\widetilde{x}_n = \frac{a_1}{a_0} \ge \delta.$$

Set

$$G_\varepsilon = \{x \in \Omega : v(x) < \varepsilon(x_n - \widetilde{x}_n)\}.$$

It is easy to check that G_ε is a convex domain, with $\widetilde{x} \in \partial G_\varepsilon$ and $x^0 \in G_\varepsilon$. Take $\widehat{x}_\varepsilon \in \partial G_\varepsilon$ such that $\widetilde{x}, x^0, \widehat{x}_\varepsilon$ lie in the same straight line. As $\varepsilon \to 0$, $G_\varepsilon \to \{v = 0\}$ and then,

$$\widehat{x}_\varepsilon \to x^0. \tag{1}$$

Set

$$w_\varepsilon(x) = v(x) - \varepsilon(x_n - \widetilde{x}_n).$$

With $v = \varphi$ on $\partial\Omega$ and $\widetilde{x}_n = -a_1/a_0$, we have

$$w_\varepsilon(x) = (a_0 - \varepsilon)\left(x_n + \frac{a_1}{a_0}\right) \quad \text{on } \partial\Omega.$$

Hence, $w_\varepsilon > 0$ on $\partial\Omega \cap \{x_n > -a_1/a_0\}$, for any $\varepsilon > 0$ sufficiently small. By the definition of G_ε, we have

$$\partial G_\varepsilon \cap \partial\Omega \subset \left\{x_n = -\frac{a_1}{a_0}\right\}.$$

Since w_ε differs from v by an affine function, then $w_\varepsilon \in C(\bar{G}_\varepsilon)$ is a generalized solution of

$$\begin{aligned} \det \nabla^2 w_\varepsilon &= \nu \quad \text{in } G_\varepsilon, \\ w_\varepsilon &= 0 \quad \text{on } \partial G_\varepsilon. \end{aligned}$$

Next, we note that

$$\begin{aligned} &w_\varepsilon(x^0) = -\varepsilon(x_n^0 - \widetilde{x}_n), \\ &\inf_{G_\varepsilon} w_\varepsilon \geq \inf_{G_\varepsilon} \big(- \varepsilon(x_n - \widetilde{x}_n)\big). \end{aligned}$$

Since $w_\varepsilon < 0$ in G_ε, then

$$\begin{aligned} 1 &\leq \frac{\inf_{G_\varepsilon} w_\varepsilon}{w_\varepsilon(x^0)} = \frac{\inf_{G_\varepsilon} w_\varepsilon}{-\varepsilon(x_n^0 - \widetilde{x}_n)} \\ &\leq \frac{\inf_{G_\varepsilon}\big(-\varepsilon(x_n - \widetilde{x}_n)\big)}{-\varepsilon(x_n^0 - \widetilde{x}_n)} = \sup_{G_\varepsilon} \frac{x_n - \widetilde{x}_n}{x_n^0 - \widetilde{x}_n} \to 1 \quad \text{as } \varepsilon \to 0. \end{aligned}$$

This implies, as $\varepsilon \to 0$,

$$\frac{\inf_{G_\varepsilon} w_\varepsilon}{w_\varepsilon(x^0)} \to 1.$$

Hence, for ε sufficiently small,

$$w_\varepsilon(x^0) < \frac{1}{2} \inf_{G_\varepsilon} w_\varepsilon.$$

By Corollary 8.3.6, we obtain

$$|\widehat{x}_\varepsilon - x^0| \geq c|\widehat{x}_\varepsilon - \widetilde{x}| \geq c|\widetilde{x} - x^0|,$$

where c is a positive constant depending only on n and b. This contradicts (1). We point out that we need the doubling condition of ν in G_ε in order to apply Corollary 8.3.6. □

Lemma 8.4.2 asserts that the set $\{x \in \Omega : u(x) = l_{x_0}(x)\}$ does not extend to the boundary $\partial\Omega$. We now improve Lemma 8.4.2 to get an estimate.

Lemma 8.4.3. *Let Ω be a convex domain in $\mathbb{R}^n$ with $B_{1/n} \subset \Omega \subset B_1$ and ν be a nontrivial finite measure in Ω satisfying the doubling condition with the doubling constant b. Suppose that $u \in C(\bar{\Omega})$ is a convex generalized solution of*

$$\begin{aligned} \det \nabla^2 u &= \nu \quad \text{in } \Omega, \\ u &= 0 \quad \text{on } \partial\Omega. \end{aligned}$$

Then, for any constant $\delta > 0$, there exists a positive constant h_0, depending only on n, b, δ, and $\nu(\Omega)$, such that, for any point $x_0 \in \Omega$ with $\operatorname{dist}(x_0, \partial\Omega) > \delta$ and any supporting function l_{x_0} of u at x_0,

$$\sup_{\partial\Omega} l_{x_0} \leq -h_0.$$

Proof. We prove the desired result by a contradiction argument. Suppose there exist sequences of convex domains Ω_i in $\mathbb{R}^n$ with $B_{1/n} \subset \Omega_i \subset B_1$, convex functions $u_i \in C(\bar{\Omega}_i)$ with $u_i = 0$ on $\partial\Omega_i$, points $x_i \in \Omega_i$ with $\operatorname{dist}(x_i, \partial\Omega_i) > \delta$, and affine functions l_i such that, for some positive constants b and C_0,

$$Mu_i(\omega) \leq bMu_i(2^{-1}\omega) \quad \text{for any convex subset } \omega \Subset \Omega_i$$

and

$$C_0^{-1} \leq Mu_i(\Omega_i) \leq C_0.$$

Moreover, l_i is a supporting function of u_i at x_i satisfying

$$\sup_{\partial\Omega_i} l_i \to 0 \quad \text{as } i \to \infty.$$

By Lemma 8.4.1, there exist a convex domain Ω in $\mathbb{R}^n$ with $B_{1/n} \subset \Omega \subset B_1$, a convex function $u \in C(\bar{\Omega})$ with $u = 0$ on $\partial\Omega$, a point $x_0 \in \Omega$ with $\operatorname{dist}(x_0, \partial\Omega) \geq \delta$, and a supporting function l of u at x_0 such that, up to a subsequence,

$$\begin{aligned} \partial\Omega_i &\to \partial\Omega \quad \text{in the Hausdorff metric}, \\ u_i &\to u \quad \text{locally uniformly in } \Omega, \end{aligned}$$

and

$$x_i \to x_0, \quad l_i \to l.$$

By Theorem 8.1.22, Mu_i converges to Mu weakly in Ω and, for any convex subset $\omega \Subset \Omega$,

$$Mu(\omega) \leq bMu(2^{-1}\omega).$$

For each i, take $\bar{x}_i \in \partial\Omega_i \subset B_1$ such that $l_i(\bar{x}_i) = \max_{\partial\Omega_i} l_i$. We assume, by passing to a subsequence, $\bar{x}_i \to \bar{x}$. Lemma 8.4.1(iii) implies $\bar{x} \in \partial\Omega$. By the assumption $l_i(\bar{x}_i) = \max_{\partial\Omega_i} l_i \to 0$, we have

$$l(\bar{x}) = \max_{\partial\Omega} l = 0.$$

In summary, we conclude that there exist a convex domain $\Omega \subset \mathbb{R}^n$ with $B_{1/n} \subset \Omega \subset B_1$ and a convex generalized solution $u \in C(\bar{\Omega})$ of

$$\begin{aligned} \det \nabla^2 u &= \nu \quad \text{in } \Omega, \\ u &= 0 \quad \text{on } \partial\Omega, \end{aligned}$$

for some nontrivial finite measure ν in Ω satisfying the doubling condition with the doubling constant b, such that, for some $x_0 \in \Omega$ with $\operatorname{dist}(x_0, \partial\Omega) \geq \delta$ and some supporting function l of u at x_0,

$$\max_{\partial\Omega} l = 0.$$

Lemma 8.4.2 asserts that such a function u does not exist. □

In the next result, we estimate the difference of generalized solutions and their supporting functions from below.

Theorem 8.4.4. *Let Ω be a convex domain in $\mathbb{R}^n$ with $B_{1/n} \subset \Omega \subset B_1$ and ν be a nontrivial finite measure in Ω satisfying the doubling condition with the doubling constant b. Suppose that $u \in C(\bar{\Omega})$ is a convex generalized solution of*

$$\begin{aligned} \det \nabla^2 u &= \nu \quad \text{in } \Omega, \\ u &= 0 \quad \text{on } \partial\Omega. \end{aligned}$$

Then, there exists a positive constant β, depending only on n and b, such that, for any constant $\delta > 0$, any point $x_0 \in \Omega$ with $\operatorname{dist}(x_0, \partial\Omega) > \delta$, and any supporting function l_{x_0} of u at x_0,

$$u(x) \geq l_{x_0}(x) + C|x - x_0|^{1+\beta} \quad \text{for any } x \in \Omega,$$

where C is a positive constant depending only on n, b, δ, and $\nu(\Omega)$.

Proof. Let $\delta > 0$ be a positive constant and take an arbitrarily fixed $x_0 \in \Omega$ with $\operatorname{dist}(x_0, \partial\Omega) > \delta$. First, Theorem 8.3.1 implies

$$\sup_{\Omega} |u| \leq C,$$

where C is a positive constant depending only on n and $\nu(\Omega)$. Then, by Lemma 8.1.7, we have, for any $p \in \partial u(x_0)$,

$$|p| \leq \frac{C}{\delta}.$$

Let l_{x_0} be a supporting function of u at x_0. Then, $l_{x_0} \leq 0$ on $\partial\Omega$ since $u = 0$ on $\partial\Omega$, and $|\nabla l_{x_0}| \leq C/\delta$. Hence, with $u(x_0) = l_{x_0}(x_0)$, we obtain

$|l_{x_0}| \le C/\delta$ in Ω. Moreover, Lemma 8.4.3 implies $S_h(x_0) \Subset \Omega$ for any $h < h_0$, for a positive constant h_0 depending only on n, b, δ, and $\nu(\Omega)$.

The proof of the desired result is divided into two steps. In the first step, we prove that there exists a positive constant $\theta \in (1/2, 1)$, depending only on n and b, such that, for any point $x_0 + y \in S_{h_0}(x_0)$,

$$u(x_0+\theta y) - l_{x_0}(x_0+\theta y) \ge \frac{1}{2}\big[u(x_0+y) - l_{x_0}(x_0+y)\big]. \tag{1}$$

In the second step, we prove the desired result.

For convenience, we assume $x_0 = 0$. Hence, $0 \in \Omega$, $\operatorname{dist}(0, \partial\Omega) > \delta$, and $B_{1/n}(x_*) \subset \Omega \subset B_1(x_*)$ for some $x_* \in B_1$. Let l be a supporting function of u at 0 and set $v = u - l$. Then, $v(0) = 0$ and

$$0 \le v \le C \quad \text{in } \Omega.$$

In the following, all sections are with respect to v. Set, for any $h \le h_0$,

$$S_h = \{x \in \Omega : \ v(x) < h\}.$$

Then, $S_h \Subset \Omega$ for any $h < h_0$.

Step 1. We claim, for any $x \in S_{h_0}$,

$$v(\theta x) \ge \frac{1}{2} v(x), \tag{2}$$

where $\theta \in (1/2, 1)$ is a constant depending only on n and b. This is (1) for v at 0.

To prove (2), we take an arbitrary constant $h < h_0$ and restrict v to S_h. Then,

$$\begin{aligned} \det \nabla^2 v &= \nu \quad \text{in } S_h, \\ v &= h \quad \text{on } \partial S_h. \end{aligned}$$

By Theorem 8.3.5 and $h = \sup_{S_h}(v - h)$, we have

$$C_1^{-1}|S_h|\nu(S_h) \le h^n \le C_1|S_h|\nu(S_h).$$

Consider an affine transform $y = Tx$ such that $B_{1/n} \subset T(S_h) \subset B_1$. We write $Tx = Ax + y_0$, for some point $y_0 \in \mathbb{R}^n$ and some invertible matrix A. Then,

$$c^{-1} \le |T(S_h)| = |\det A||S_h| \le c.$$

Set, for any $y = Tx \in T(S_h)$,

$$w(y) = \frac{1}{h} v(x).$$

Then,

$$\begin{aligned} \det \nabla^2 w &= \widetilde{\nu} \quad \text{in } T(S_h), \\ w &= 1 \quad \text{on } \partial T(S_h), \end{aligned}$$

where $\widetilde{\nu}$ is a measure in $T(S_h)$ given by, for any $\omega \subset S_h$,

$$\widetilde{\nu}(T(\omega)) = \frac{1}{h^n}|\det A|^{-1}\nu(\omega).$$

Then,

$$\widetilde{\nu}\big(T(S_h)\big) = \frac{1}{h^n}|\det A|^{-1}\nu(S_h) \le C.$$

Next, by applying Theorem 8.3.1 to w in $T(S_h)$, we get, for any $y_1, y_2 \in T(S_h)$,

$$|w(y_1) - w(y_2)| \le C|y_1 - y_2|^{\frac{1}{n}}.$$

By taking $y_1 \in T(\partial S_h)$ and $y_2 \in T(S_{h/2})$, we have $w(y_1) = 1$, $w(y_2) \le 1/2$, and hence

$$|y_1 - y_2|^{\frac{1}{n}} \ge c.$$

Therefore,

$$\text{(3)} \qquad \operatorname{dist}\big(T(S_{h/2}), T(\partial S_h)\big) \ge c_0,$$

where c_0 is a positive constant depending only on n and b. We may assume $c_0 \in (0,1)$ and then set $\theta = 1 - c_0/2$. Hence, $\theta \in (1/2, 1)$. Take an arbitrary $x \in \partial S_h$ and recall that $Tx = Ax + y_0$. Since $0 \in S_h$, then $y_0 = T0 \in B_1$, and hence

$$|Ax| = |Tx - y_0| \le |Tx| + |y_0| < 2.$$

Therefore,

$$|Tx - T(\theta x)| = (1-\theta)|Ax| = \frac{1}{2}c_0|Ax| < c_0.$$

With (3) and $x \in \partial S_h$, we have $T(\theta x) \in T(S_h) \setminus T(S_{h/2})$, or $\theta x \in S_h \setminus S_{h/2}$. This implies

$$v(\theta x) \ge \frac{1}{2}h = \frac{1}{2}v(x),$$

and hence (2).

Step 2. With $\theta \in (1/2, 1)$, we can take a constant $\beta > 0$ such that $\theta^{1+\beta} = 1/2$. Then, β is a positive constant depending only on n and b. We claim, for any $x \in \Omega$,

$$\text{(4)} \qquad v(x) \ge C|x|^{1+\beta},$$

where C is a positive constant depending only on n, b, δ, and $\nu(\Omega)$.

First, for any $x \in S_{h_0}$, let the ray from 0 to x intersect ∂S_{h_0} at $\widehat{x}$. Then, there exists a nonnegative integer k such that $\theta^{k+1}|\widehat{x}| \le |x| < \theta^k|\widehat{x}|$. A simple iteration of (2) implies

$$v(x) \ge v(\theta^{k+1}\widehat{x}) \ge \frac{1}{2}v(\theta^k\widehat{x}) \ge \cdots \ge \frac{1}{2^{k+1}}v(\widehat{x}) = \theta^{(k+1)(\beta+1)}h_0.$$

With $S_{h_0} \subset \Omega \subset B_2$, we have $|\widehat{x}| \leq 2$ and

$$v(x) \geq \frac{\theta^{1+\beta}}{|\widehat{x}|^{1+\beta}}(\theta^k|\widehat{x}|)^{1+\beta}h_0 \geq \frac{\theta^{1+\beta}}{2^{1+\beta}}h_0|x|^{1+\beta}.$$

Next, for any $x \in \Omega \setminus S_{h_0}$, we have $|x| \leq 2$ and hence

$$v(x) \geq h_0 \geq \frac{1}{2^{1+\beta}}h_0|x|^{1+\beta}.$$

By combining both cases, we have (4). □

Theorem 8.4.4 implies that u is strictly convex at x_0. Then, l_{x_0} is the unique supporting function by Lemma 8.3.8. As a consequence of Lemma 8.1.10, u is differentiable at x_0. Next, we prove that u is in fact $C^{1,\alpha}$ at x_0.

Theorem 8.4.5. *Let $\Omega \subset \mathbb{R}^n$ be a convex domain with $B_{1/n} \subset \Omega \subset B_1$ and ν be a nontrivial finite measure in Ω satisfying the doubling condition with the doubling constant b. Suppose that $u \in C(\bar{\Omega})$ is a convex generalized solution of*

$$\det \nabla^2 u = \nu \quad \text{in } \Omega,$$
$$u = 0 \quad \text{on } \partial\Omega.$$

Then, there exists a positive constant $\alpha \in (0,1)$, depending only on n and b, such that, for any constant $\delta > 0$, any point $x_0 \in \Omega$ with $\operatorname{dist}(x_0, \partial\Omega) > \delta$, and any supporting function l_{x_0} of u at x_0,

$$u(x) \leq l_{x_0}(x) + C|x - x_0|^{1+\alpha} \quad \text{for any } x \in \Omega,$$

where C is a positive constant depending only on n, b, δ, and $\nu(\Omega)$.

Proof. By Theorem 8.3.1, we have

$$|u| \leq C \quad \text{in } \Omega,$$

and, for any $x_1, x_2 \in \Omega$,

$$|u(x_1) - u(x_0)| \leq C|x_1 - x_2|^{\frac{1}{n}},$$

where C is a positive constant depending only on n and $\nu(\Omega)$.

As in the proof of Theorem 8.4.4, we assume $x_0 = 0$. Then, $0 \in \Omega$, $\operatorname{dist}(0, \partial\Omega) > \delta$, and $B_{1/n}(x_*) \subset \Omega \subset B_1(x_*)$ for some $x_* \in B_1$. Let l be a supporting function of u at 0 and set $v = u - l$. Then, $v(0) = 0$,

$$0 \leq v \leq C \quad \text{in } \Omega, \tag{1}$$

and, for any $x_1, x_2 \in \Omega$,

$$|v(x_1) - v(x_2)| \leq C|x_1 - x_2|^{\frac{1}{n}}. \tag{2}$$

In the following, all sections are with respect to v. Set, for any $h \leq h_0$,

$$S_h = \{x \in \Omega : v(x) < h\}.$$

By Lemma 8.4.3, $S_h \Subset \Omega$ for any $h < h_0$, where h_0 is a positive constant depending only on n, b, δ, and $\nu(\Omega)$. Since v is convex, then, for any $y \in \Omega$,

$$v\left(\frac{1}{2}y\right) \leq \frac{1}{2}v(0) + \frac{1}{2}v(y) = \frac{1}{2}v(y).$$

Now we claim that there exists a constant $h_* \in (0, h_0)$ such that, for any $y \in S_{h_*}$,

$$v\left(\frac{1}{2}y\right) \leq \sigma v(y), \tag{3}$$

where $\sigma \in (0, 1/2)$ is a positive constant depending only on n and b and h_* is a positive constant depending only on n, b, δ, and $\nu(\Omega)$.

Take any $h < h_0$ and an arbitrarily fixed $y \in \partial S_h$. Theorem 8.4.4 implies

$$v(y) \geq c|y|^{1+\beta}.$$

With $v(y) = h$, we have

$$|y| \leq \left(\frac{h}{c}\right)^{\frac{1}{1+\beta}}.$$

Hence, $|y| < \delta/2$ if

$$h < c\left(\frac{\delta}{2}\right)^{1+\beta} \equiv C_1\delta^{1+\beta}.$$

In particular, $\text{dist}(y, \partial\Omega) > \delta/2$.

Set

$$\bar{x}_0 = \frac{1}{2}y. \tag{4}$$

Then, $|\bar{x}_0| = |y|/2 < \delta/4$ and hence, $\text{dist}(\bar{x}_0, \partial\Omega) > 3\delta/4$. Let $l_{\bar{x}_0}$ be a supporting function of v at $\bar{x}_0$. Then, for some $p \in \partial v(\bar{x}_0)$,

$$l_{\bar{x}_0}(z) = v(\bar{x}_0) + p \cdot (z - \bar{x}_0).$$

Note that $|p| \leq C$. We also set

$$\bar{y} = -\frac{1}{2\theta}y, \tag{5}$$

where θ is the positive constant depending only on n and b, as for (1) in the proof of Theorem 8.4.4. Then, $\bar{x}_0 + \theta\bar{y} = 0$ and

$$v(\bar{x}_0 + \bar{y}) - l_{\bar{x}_0}(\bar{x}_0 + \bar{y}) = v(\bar{x}_0 + \bar{y}) - v(\bar{x}_0) - p \cdot \bar{y}.$$

Set

$$\bar{h} = v(\bar{x}_0 + \bar{y}) - l_{\bar{x}_0}(\bar{x}_0 + \bar{y}).$$

By (2), we have

$$\bar{h} \leq C|\bar{y}|^{\frac{1}{n}} + C|\bar{y}| \leq C|\bar{y}|^{\frac{1}{n}} \leq C|y|^{\frac{1}{n}} \leq Ch^{\frac{1}{n(1+\beta)}}.$$

Since $\text{dist}(\bar{x}_0, \partial\Omega) > 3\delta/4$, Lemma 8.4.3 implies $\partial S_{\bar{h}}(\bar{x}_0) \Subset \Omega$ if $\bar{h} < h_1$, where h_1 is the positive constant, depending only on n, b, $3\delta/4$, and $\nu(\Omega)$, for the point $\bar{x}_0$ as in Lemma 8.4.3. To have $\bar{h} < h_1$, we require

$$Ch^{\frac{1}{n(1+\beta)}} \le h_1,$$

or

$$h \le C_2 h_1^{n(1+\beta)}.$$

In summary, we take

$$h < h_* = \min\left\{h_0, C_1\delta^{1+\beta}, C_2 h_1^{n(1+\beta)}\right\}.$$

With such a choice of h, we obtain $\bar{x}_0 + \bar{y} \in \partial S_{\bar{h}}(\bar{x}_0) \Subset \Omega$.

By applying (1) in the proof of Theorem 8.4.4 to v at $\bar{x}_0$, we obtain

$$v(\bar{x}_0 + \theta\bar{y}) - l_{\bar{x}_0}(\bar{x}_0 + \theta\bar{y}) \ge \frac{1}{2}\big[v(\bar{x}_0 + \bar{y}) - l_{\bar{x}_0}(\bar{x}_0 + \bar{y})\big].$$

Since $v(\bar{x}_0 + \theta\bar{y}) = v(0) = 0$, we have

$$v(\bar{x}_0 + \bar{y}) \le l_{\bar{x}_0}(\bar{x}_0 + \bar{y}) - 2l_{\bar{x}_0}(\bar{x}_0 + \theta\bar{y}). \tag{6}$$

Recall that $\bar{x}_0$ and $\bar{y}$ are defined in (4) and (5), respectively. Set, for any $t \in \left[-\frac{1}{2\theta}, \frac{1}{2}\right]$,

$$\psi(t) = v(\bar{x}_0 + ty)$$

and

$$\chi(t) = l_{\bar{x}_0}(\bar{x}_0 + ty).$$

Then, for some $a \in \partial\psi(0)$,

$$\chi(t) = \psi(0) + at.$$

With $v \ge 0$ and the expression of χ, (6) implies

$$0 \le \chi\left(-\frac{1}{2\theta}\right) - 2\chi\left(-\frac{1}{2}\right) = -\psi(0) + \frac{2\theta - 1}{2\theta}a.$$

Since ψ is a convex function, then

$$a \le \frac{\psi(1/2) - \psi(0)}{1/2}.$$

With $\theta \in (1/2, 1)$, we have

$$0 \le -\psi(0) + \frac{2\theta - 1}{2\theta} \cdot 2\left(\psi\left(\frac{1}{2}\right) - \psi(0)\right),$$

and hence

$$\psi(0) \le \frac{2\theta - 1}{3\theta - 1}\psi\left(\frac{1}{2}\right).$$

We note that

$$0 < \frac{2\theta - 1}{3\theta - 1} < \frac{1}{2}.$$

This proves (3).

Since $\sigma \in (0, 1/2)$, we take a constant $\alpha > 0$ such that $\sigma = 2^{-(1+\alpha)}$. We now prove, for any $y \in \Omega$,

$$v(y) \le C|y|^{1+\alpha}. \tag{7}$$

We first consider $y \in \bar{S}_{h_*}$. For such a y, there exists a nonnegative integer k such that $2^k y \in \bar{S}_{h_*}$ and $2^{k+1}y \notin \bar{S}_{h_*}$. Assume the ray from 0 to y intersects ∂S_{h_*} at some point z. Then,

$$\begin{aligned} v(y) &\le \sigma v(2y) \le \cdots \le \sigma^k v(2^k y) \\ &\le 2^{-k(1+\alpha)} v(z) = \frac{2^{1+\alpha}}{2^{(k+1)(1+\alpha)}} v(z). \end{aligned}$$

Since $2^{k+1}|y| \ge |z|$ and $z \in \partial S_{h_*}$, we have

$$v(y) \le \frac{2^{1+\alpha}|y|^{1+\alpha}}{|z|^{1+\alpha}} v(z) = \frac{2^{1+\alpha}}{|z|^{1+\alpha}} h_* |y|^{1+\alpha}.$$

By (2), we obtain

$$|z| \ge c|v(z) - v(0)|^n \ge c h_*^n.$$

Hence, for any $y \in \bar{S}_{h_*}$,

$$v(y) \le C|y|^{1+\alpha}.$$

Next, we consider $y \in \Omega \setminus \bar{S}_{h_*}$. Again, we assume the ray from 0 to y intersects ∂S_{h_*} at some z. Then, $|y| \ge |z| \ge c h_*^n$. Since $v(y) \le C$ by (1), then

$$v(y) \le C|y|^{1+\alpha}.$$

This ends the proof of (7). □

As a consequence of Theorem 8.4.5, we have the following $C^{1,\alpha}$-regularity and estimates.

Theorem 8.4.6. *Let $\Omega \subset \mathbb{R}^n$ be a bounded convex domain with $B_{1/n} \subset \Omega \subset B_1$ and ν be a nontrivial finite measure in Ω satisfying the doubling condition with the doubling constant b. Suppose that $u \in C(\bar{\Omega})$ is a convex generalized solution of*

$$\begin{aligned} \det \nabla^2 u &= \nu \quad \text{in } \Omega, \\ u &= 0 \quad \text{on } \partial\Omega. \end{aligned}$$

Then, there exists a positive constant $\alpha \in (0,1)$, depending only on n and b, such that $u \in C^{1,\alpha}(\Omega)$. Moreover, for any subdomain $\Omega' \Subset \Omega$,

$$|\nabla u|_{C^\alpha(\Omega')} \le C,$$

where C is a positive constant depending only on n, b, $\mathrm{dist}(\Omega', \partial\Omega)$, and $\nu(\Omega)$.

We leave the proof as an exercise.

8.5. Interior $C^{2,\alpha}$-Regularity

In this section, we discuss the interior $C^{2,\alpha}$-regularity, due to Caffarelli [**15**], for solutions of the Monge-Ampère equation under the condition that Monge-Ampère measures are induced by Hölder continuous functions. We start with the following existence result.

Theorem 8.5.1. *Let Ω be a convex domain in $\mathbb{R}^n$ with $B_{1/n} \subset \Omega \subset B_1$. Then, there exists a unique convex solution $u \in C(\bar{\Omega}) \cap C^\infty(\Omega)$ of*

$$\begin{aligned} \det \nabla^2 u &= 1 \quad \text{in } \Omega, \\ u &= 0 \quad \text{on } \partial\Omega. \end{aligned}$$

Moreover, for any subdomain $\Omega' \Subset \Omega$ and any nonnegative integer k,

$$|u|_{C^k(\Omega')} \le C,$$

where C is a positive constant depending only on n, k, Ω', and $\mathrm{dist}(\Omega', \partial\Omega)$.

Proof. By Lemma 8.2.6, there exists a unique generalized convex solution $u \in C(\bar{\Omega})$ of

$$\begin{aligned} \det \nabla^2 u &= 1 \quad \text{in } \Omega, \\ u &= 0 \quad \text{on } \partial\Omega. \end{aligned}$$

We will prove $u \in C^\infty(\Omega)$ and the desired estimate.

To this end, we take a sequence of decreasing uniformly convex C^∞-domains $\{\Omega_i\}$ such that $\partial\Omega_i$ converges to $\partial\Omega$ in the Hausdorff metric. By Theorem 6.2.6 and Proposition 6.1.4, let $u_i \in C^\infty(\bar{\Omega})$ be the convex solution of

$$\begin{aligned} \det \nabla^2 u_i &= 1 \quad \text{in } \Omega_i, \\ u_i &= 0 \quad \text{on } \partial\Omega_i. \end{aligned}$$

By the comparison principle and $\Omega_{i+1} \subset \Omega_i$, we have $u_i \le u_{i+1}$ in Ω_{i+1}. By Lemma 8.4.1(ii), we conclude easily that u_i converges to u locally uniformly in Ω. Then, $u_i \le u$ in Ω for any i.

Define, for any $t < 0$,

$$\begin{aligned} S_{u,t} &= \{x \in \Omega : \ u(x) < t\}, \\ S_{u_i,t} &= \{x \in \Omega_i : \ u_i(x) < t\}. \end{aligned}$$

Now, we take an arbitrarily fixed $h < 0$. Then, $S_{u,h} \subset S_{u_i,h}$ and there exists a positive constant $\varepsilon = \varepsilon(h)$ such that, for any i sufficiently large and $l = 1, 2, 3$,

$$\text{dist}\left(S_{u_i,lh/3}, \partial S_{u_i,(l-1)h/3}\right) \geq \varepsilon.$$

We note that $S_{u_i,0} = \Omega_i$. Without loss of generality, we assume, for some $R > 0$, $\Omega_i \subset B_R$ for any i. By Theorem 8.3.1, we get, for any i,

$$\sup_{\Omega_i} |u_i| \leq C_1,$$

where C_1 is a positive constant depending only on n and R, independent of i. By applying Lemma 8.1.7 to u_i in Ω_i for sufficiently large i, we have

$$\sup_{S_{u_i,h/3}} |\nabla u_i| \leq C_1 \varepsilon^{-1}.$$

Now we apply Theorem 6.3.1, Pogorelov's interior estimates, to $u_i - h/3$ in $S_{u_i,h/3}$ and obtain

$$\sup_{S_{u_i,h/3}} \left[-(u_i - h/3)|\nabla^2 u_i| \right] \leq C_2,$$

where C_2 is a positive constant depending only on n, R, and h. This implies

$$\sup_{S_{u_i,2h/3}} |\nabla^2 u_i| \leq C_3,$$

where C_3 is a positive constant depending only on n, R, and h, independent of i. Therefore, the equation $\det \nabla^2 u_i = 1$ is uniformly elliptic in $S_{u_i,2h/3}$. By Theorem 5.4.1 and Proposition 5.1.9, the interior estimates for fully nonlinear elliptic equations, we obtain, for any integer k and constant $\alpha \in (0, 1)$,

$$|u_i|_{C^{k,\alpha}(S_{u_i,h})} \leq C_4,$$

where C_4 is a positive constant depending only on n, k, α, R, and h, independent of i. In particular,

$$|u_i|_{C^{k,\alpha}(S_{u,h})} \leq C_4.$$

This holds for any $h < 0$, with C_4 depending on h. Therefore, u_i converges to u in $C^k(S_{u,h})$, for any $h < 0$ and any integer $k \geq 0$. Then, we conclude the desired interior regularity and estimates for u. □

Now, we are ready to prove the pointwise $C^{2,\alpha}$-regularity. We first discuss this at the minimum point.

Theorem 8.5.2. *Let $\alpha \in (0, 1)$ be a constant, Ω be a convex domain in $\mathbb{R}^n$ with $B_{1/n} \subset \Omega \subset B_1$, $x_0 \in \Omega$ be a point, and f be a bounded function in Ω satisfying, for some positive constants λ, Λ, and M,*

$$\lambda \leq f \leq \Lambda \quad \text{in } \Omega$$

and, for any $x \in \Omega$,

$$|f(x) - f(x_0)| \leq M|x - x_0|^\alpha.$$

Suppose that $u \in C(\bar{\Omega})$ *is a convex generalized solution of*

$$\det \nabla^2 u = f \quad \textit{in } \Omega,$$
$$u = 0 \quad \textit{on } \partial\Omega,$$

and that u *attains its minimum at* x_0. *Then, there exists a quadratic polynomial* P *such that*

$$\det \nabla^2 P = f(x_0),$$
$$|P(x_0)| + |\nabla P(x_0)| + \|\nabla^2 P(x_0)\| \leq C,$$

and, for any $x \in \Omega$,

$$|u(x) - P(x)| \leq C|x - x_0|^{2+\alpha},$$

where C *is a positive constant depending only on* n, α, λ, Λ, $f(x_0)$, *and* M.

Proof. Without loss of generality, we assume $x_0 = 0$ and $B_{1/n}(x_*) \subset \Omega \subset B_1(x_*)$ for some $x_* \in \mathbb{R}^n$. Furthermore, we assume $f(0) = 1$. Otherwise, we consider $u/[f(0)]^{1/n}$ instead of u. Hence, for any $x \in \Omega$,

$$|f(x) - 1| \leq M|x|^\alpha. \tag{1}$$

By subtracting a constant from u, we assume $u(0) = 0$ and $u = h_0$ on $\partial\Omega$. By Theorem 8.3.5, $c_1 \leq h_0 \leq c_2$ for some positive constants c_1 and c_2 depending only on n, λ, and Λ. Moreover, by Theorem 8.3.1, we have, for any $x \in \partial\Omega$,

$$h_0 = |u(x) - u(0)| \leq C|x|^\alpha,$$

and hence, $\text{dist}(0, \partial\Omega) \geq c_0$, for some positive constant c_0 depending only on n, λ, and Λ. We divide the rest of the proof into three parts.

Step 1. First, we prove that M *in* (1) *can be chosen as a small constant.* Set, for any constant $h \in (0, h_0)$,

$$S_h = \{x \in \Omega : u(x) < h\}.$$

Take an arbitrarily fixed $h < h_0$ and consider

$$\det \nabla^2 u = f \quad \text{in } S_h,$$
$$u = h \quad \text{on } \partial S_h.$$

By Theorem 8.3.5, we have

$$C^{-1}|S_h|^{\frac{2}{n}} \leq h \leq C|S_h|^{\frac{2}{n}},$$

or

$$C^{-1}h^{\frac{n}{2}} \leq |S_h| \leq Ch^{\frac{n}{2}}, \tag{2}$$

where C is a positive constant depending only on n, λ, and Λ. By Theorem 8.4.4, we have, for any $x \in \Omega$,

$$u(x) \geq C^{-1}|x|^{1+\beta},$$

where β is a positive constant depending only on n, λ, and Λ. Hence,

$$S_h \subset B_{Ch^{\frac{1}{1+\beta}}}. \tag{3}$$

Let T_h be a linear transform such that, for some $y_h \in \mathbb{R}^n$,

$$B_{1/n}(y_h) \subset T_h(S_h) \subset B_1(y_h).$$

We view T_h as an invertible matrix. Then,

$$C^{-1} \leq |\det T_h||S_h| \leq C.$$

By combining with (2), we have

$$C^{-1} \leq h^{\frac{n}{2}}|\det T_h| \leq C. \tag{4}$$

Since $0 = T_h 0 \in T_h(S_h) \subset B_1(y_h)$, then $y_h \in B_1$ and hence $B_1(y_h) \subset B_2$. Therefore, $T_h(S_h) \subset B_2$.

Set $y = T_h x$ for any $x \in S_h$ and, for any $y \in T_h(S_h)$,

$$u_h(y) = |\det T_h|^{\frac{2}{n}} u(x).$$

Then, $u_h \in C(\overline{T_h(S_h)})$ is a generalized solution of

$$\begin{aligned} \det \nabla^2 u_h &= f_h \quad \text{in } T_h(S_h), \\ u_h &= a_h \quad \text{on } \partial(T_h(S_h)), \end{aligned}$$

where a_h and f_h are given by

$$a_h = h|\det T_h|^{\frac{2}{n}}$$

and, for any $y \in T_h(S_h)$,

$$f_h(y) = f(T_h^{-1}y).$$

By (4), we get

$$C^{-1} \leq a_h \leq C.$$

Next, by Theorem 8.3.1, we have, for any $y \in \partial(T_h(S_h))$,

$$a_h = |u_h(y) - u_h(0)| \leq C|y|^{\frac{1}{n}}.$$

Therefore,

$$B_{2r_0} \subset T_h(S_h) \subset B_2, \tag{5}$$

where r_0 is a positive constant depending only on n, λ, and Λ. Combining (3) and (5), we have

$$T_h^{-1}(B_{2r_0}) \subset B_{Ch^{\frac{1}{1+\beta}}},$$

which implies

$$\|T_h^{-1}\| \leq Ch^{\frac{1}{1+\beta}}.$$

With the help of (5) and (1), we obtain, for any $y \in T_h(S_h)$,

$$|f_h(y) - 1| = |f(T_h^{-1}y) - 1| \le M|T_h^{-1}y|^\alpha \le CMh^{\frac{\alpha}{1+\beta}}|y|^\alpha.$$

For any given positive constant σ, we take h such that

$$CMh^{\frac{\alpha}{1+\beta}} \le \sigma.$$

We note that h is a positive constant depending only on n, α, λ, Λ, M, and σ. Then, for any $y \in T_h(S_h)$,

$$|f_h(y) - 1| \le \sigma|y|^\alpha. \tag{6}$$

In particular,

$$1 - 2\sigma \le f_h \le 1 + 2\sigma \quad \text{in } T_h(S_h).$$

Step 2. Next, we prove that u_h is close to a quadratic polynomial in a fixed ball. We consider

$$\begin{aligned} \det \nabla^2 w &= 1 \quad \text{in } T_h(S_h), \\ w &= a_h \quad \text{on } \partial(T_h(S_h)). \end{aligned}$$

By Theorem 8.5.1, there exists a solution $w \in C(\overline{T_h(S_h)}) \cap C^\infty(T_h(S_h))$. By applying Theorem 8.3.1 to $w - a_h$, we get

$$a_h - w \le C_1 \quad \text{in } T_h(S_h),$$

where C_1 is a positive constant depending only on n. By Theorem 8.2.2, we have

$$(1 + 2\sigma)(w - a_h) \le u_h - a_h \le (1 - 2\sigma)(w - a_h) \quad \text{in } T_h(S_h),$$

and hence,

$$|u_h - w| \le 2\sigma(a_h - w) \le 2C_1\sigma \quad \text{in } T_h(S_h).$$

By evaluating at $y = 0$, we get $|w(0)| \le 2C_1\sigma$, and hence

$$|u_h - (w - w(0))| \le 4C_1\sigma \quad \text{in } T_h(S_h).$$

Note that $\text{dist}(B_{r_0}, \partial(T_h(S_h))) \ge r_0$ by (5). Hence, by Theorem 8.5.1, we have

$$|w|_{C^3(B_{r_0})} \le C_2.$$

Then,

$$|\nabla w(0)| \le C_2, \quad \|\nabla^2 w(0)\| \le C_2. \tag{7}$$

Moreover, for any $y \in B_{r_0}$,

$$\left| w(y) - w(0) - \nabla w(0) \cdot y - \frac{1}{2} y^T \nabla^2 w(0) y \right| \le C_2 |y|^3,$$

and hence,

$$\left| u_h(y) - \nabla w(0) \cdot y - \frac{1}{2} y^T \nabla^2 w(0) y \right| \le 4C_1\sigma + C_2|y|^3. \tag{8}$$

By a rotation, we assume that $\nabla^2 w(0)$ is a diagonal matrix given by

$$\nabla^2 w(0) = \operatorname{diag}(\lambda_1, \ldots, \lambda_n),$$

for some positive constants $\lambda_1, \ldots, \lambda_n$. By $\det \nabla^2 w(0) = \lambda_1 \cdots \lambda_n = 1$ and (7), we obtain, for each $i = 1, \ldots, n$,

$$C_2^{1-n} \leq \lambda_i \leq C_2.$$

For a positive constant μ to be determined, we consider the change of coordinates

$$\mu z_i = \sqrt{\lambda_i} y_i \quad \text{for } i = 1, \ldots, n.$$

Note that

$$|y| \leq \mu C_2^{\frac{n-1}{2}} |z|.$$

By taking μ with $C_2^{(n-1)/2}\mu \leq r_0$, $|z| < 1$ implies $|y| < r_0$. In the following, we set, for any $z \in B_1$,

$$u_{h,\mu}(z) = \frac{1}{\mu^2}\big[u_h(y) - \nabla w(0) \cdot y\big].$$

Then, $u_{h,\mu} \in C(\bar{B}_1)$ is a convex generalized solution of

$$\det \nabla^2 u_{h,\mu} = f_{h,\mu} \quad \text{in } B_1,$$

where $f_{h,\mu}(z) = f_h(y)$. Hence, by (8) and (6), we have, for any $z \in B_1$,

$$\left| u_{h,\mu}(z) - \frac{1}{2}|z|^2 \right| \leq 4C_1\sigma\mu^{-2} + C_2^{\frac{3n-1}{2}}\mu$$

and

$$|f_{h,\mu}(z) - 1| \leq C_2^{\frac{\alpha(n-1)}{2}} \sigma\mu^\alpha |z|^\alpha.$$

For any given positive constants $\varepsilon, \delta \in (0,1)$, we first take μ such that

$$C_2^{\frac{n-1}{2}}\mu \leq r_0, \quad C_2^{\frac{3n-1}{2}}\mu \leq \frac{1}{2}\varepsilon,$$

and then take σ such that

$$4C_1\sigma\mu^{-2} \leq \frac{1}{2}\varepsilon, \quad C_2^{\frac{\alpha(n-1)}{2}}\sigma\mu^\alpha \leq \varepsilon\delta.$$

Hence, for any $z \in B_1$,

$$\left| u_{h,\mu}(z) - \frac{1}{2}|z|^2 \right| \leq \varepsilon, \tag{9}$$

and

$$|f_{h,\mu}(z) - 1| \leq \varepsilon\delta|z|^\alpha. \tag{10}$$

Step 3. Last, we prove that $u_{h,\mu}$ *is close to a quadratic polynomial at the desired rate.* We formulate this in a separate result.

Theorem 8.5.3. *Let* $\alpha \in (0,1)$ *be a constant, and let* $f \in L^\infty(B_1)$*. Suppose that* $u \in C(B_1)$ *is a convex generalized solution of*

$$\det \nabla^2 u = f \quad \text{in } B_1,$$

with $u(0) = 0$*. Then, there exist constants* $\varepsilon, \delta \in (0, 1/2)$*, depending only on* n *and* α*, such that, if, for any* $x \in B_1$*,*

$$|f(x) - 1| \le \varepsilon\delta|x|^\alpha,$$
$$\left|u(x) - \frac{1}{2}|x|^2\right| \le \varepsilon,$$

then there exist a vector $b \in \mathbb{R}^n$ *and a positive definite matrix* A*, with* $|b| \le 1$*,* $\det A = 1$*, and* $\|A\| \le 2$*, such that, for any* $x \in B_1$*,*

$$\left|u(x) - b\cdot x - \frac{1}{2}x^T Ax\right| \le C|x|^{2+\alpha},$$

where C *is a positive constant depending only on* n *and* α*.*

Proof. We will prove by induction that, for each nonnegative integer m, there exist a constant c_m, a linear function l_m, and a homogeneous quadratic polynomial Q_m, with $\det \nabla^2 Q_m = 1$, $|c_m| < 1$, $|\nabla l_m| < 1$, and $\|\nabla^2 Q_m\| < 2$, such that

$$|u - c_m - l_m - Q_m| \le \varepsilon r^{(2+\alpha)m} \quad \text{in } B_{r^m}, \tag{11}$$

where $r \in (0, 1/2)$ is a constant depending only on n and α. Moreover, for each m,

$$\begin{aligned} |c_{m+1} - c_m| &\le C\varepsilon r^{(2+\alpha)m}, \\ |\nabla(l_{m+1} - l_m)| &\le C\varepsilon r^{(1+\alpha)m}, \\ \|\nabla^2(Q_{m+1} - Q_m)\| &\le C\varepsilon r^{\alpha m}, \end{aligned} \tag{12}$$

where C is a positive constant depending only on n and α.

We note that (11) holds for $m = 0$ by the assumption, with $c_0 = 0$, $l_0 = 0$, and $Q_0(x) = |x|^2/2$. Now we assume (11) holds for $0, 1, \ldots, m$ and we consider the case $m + 1$. We restrict u to B_{r^m} and set, for any $y \in B_1$,

$$v(y) = \frac{1}{r^{2m}}\big[u(r^m y) - c_m - l_m(r^m y)\big].$$

Then, (11) implies

$$|v - Q_m| \le \varepsilon r^{m\alpha} \quad \text{in } B_1.$$

By $\|\nabla^2 Q_m\| < 2$ and $\det \nabla^2 Q_m = 1$, we have, for any $y \in \mathbb{R}^n$,

$$\frac{1}{2^n}|y|^2 \le Q_m(y) \le |y|^2.$$

Then, for any $y \in B_1$,

$$\frac{1}{2^n}|y|^2 - \varepsilon \le v(y) \le |y|^2 + \varepsilon.$$

This implies, in particular, $-\varepsilon \le v \le \varepsilon + 1$ in B_1. We now take positive constants a and r_0, depending only on n, such that, for any ε sufficiently small,

$$B_{4r_0} \subset \{y \in B_1 : v(y) < a\} \subset B_{1/2}. \tag{13}$$

In fact, we can take

$$a = \frac{1}{2^{n+3}}, \quad r_0 = \frac{1}{2^{\frac{n}{2}+4}}.$$

Then, (13) holds if

$$\varepsilon < \frac{1}{2^{n+4}}.$$

In the following, we set $S = \{y \in B_1 : v(y) < a\}$.

Note that $v \in C(\bar{B}_1)$ is a convex generalized solution of

$$\det \nabla^2 v = \widetilde{f} \quad \text{in } B_1,$$

where $\widetilde{f}(y) = f(r^m y)$ satisfies, for any $y \in B_1$,

$$|\widetilde{f}(y) - 1| \le \varepsilon\delta r^{\alpha m}|y|^\alpha$$

and, in particular,

$$1 - \varepsilon\delta r^{\alpha m} \le \widetilde{f} \le 1 + \varepsilon\delta r^{\alpha m} \quad \text{in } B_1.$$

Consider the Dirichlet problem

$$\det \nabla^2 w = 1 \quad \text{in } S,$$
$$w = a \quad \text{on } \partial S.$$

Theorem 8.5.1 implies the existence of a convex solution $w \in C(\bar{S}) \cap C^\infty(S)$. By applying Theorem 8.3.1 to $w - a$, we get

$$a - w \le C_1 \quad \text{in } S,$$

where C_1 is a positive constant depending only on n. By Theorem 8.2.2, we have

$$(1 + \delta\varepsilon r^{\alpha m})(w - a) \le v - a \le (1 - \delta\varepsilon r^{\alpha m})(w - a) \quad \text{in } S,$$

and then,

$$|v - w| \le \delta\varepsilon r^{\alpha m}(a - w) \le C_1\delta\varepsilon r^{\alpha m} \quad \text{in } S.$$

Hence, if $\delta < 1/C_1$,

$$|w - Q_m| \le |w - v| + |v - Q_m| \le C_1\delta\varepsilon r^{\alpha m} + \varepsilon r^{\alpha m} \le 2\varepsilon r^{\alpha m} \quad \text{in } B_1.$$

Note that $\operatorname{dist}(B_{2r_0}, \partial S) \ge 2r_0$ by (13). Then, by Theorem 8.5.1, we obtain

$$|\nabla^2 w|_{C^{1,1}(B_{2r_0})} \le C.$$

Since $\det \nabla^2 w = \det \nabla^2 Q_m = 1$, by taking a difference, we get

$$a_{ij}\partial_{ij}(w - Q_m) = 0 \quad \text{in } B_{2r_0},$$

where $a_{ij} \in C^{1,1}$ is uniformly elliptic in B_{2r_0}, with the ellipticity constants depending only on n. By the interior Schauder estimates, we have

$$|w - Q_m|_{C^3(B_{r_0})} \leq C|w - Q_m|_{L^\infty(B_{2r_0})} \leq C\varepsilon r^{\alpha m}.$$

By expanding $w - Q_m$ up to degree 2, we get, for any $y \in B_{r_0}$,

$$|w(y) - Q_m(y) - \widetilde{c}_m - \widetilde{l}_m(y) - \widetilde{Q}_m(y)| \leq C_2\varepsilon r^{\alpha m}|y|^3,$$

where $\widetilde{c}_m$, $\widetilde{l}_m$, and $\widetilde{Q}_m$ are the constant, linear, and homogeneous quadratic parts of $w - Q_m$, respectively; namely, $\widetilde{c}_m = w(0)$, $\nabla\widetilde{l}_m = \nabla w(0)$, and $\nabla^2\widetilde{Q}_m = \nabla^2(w - Q_m)(0)$. Then,

$$|\widetilde{c}_m| \leq C_2\varepsilon r^{\alpha m}, \quad |\nabla\widetilde{l}_m| \leq C_2\varepsilon r^{\alpha m}, \quad \|\nabla^2\widetilde{Q}_m\| \leq C_2\varepsilon r^{\alpha m}.$$

Note that $Q_m + \widetilde{c}_m + \widetilde{l}_m + \widetilde{Q}_m$ is the degree 2 expansion of w. Hence,

$$\det \nabla^2(Q_m + \widetilde{Q}_m) = 1.$$

With $r \leq r_0$, we have, for any $y \in B_r$,

$$\begin{aligned}
&|v(y) - Q_m(y) - \widetilde{c}_m - \widetilde{l}_m(y) - \widetilde{Q}_m(y)| \\
&\quad \leq |v(y) - w(y)| + |w(y) - Q_m(y) - \widetilde{c}_m - \widetilde{l}_m(y) - \widetilde{Q}_m(y)| \\
&\quad \leq C_1\delta\varepsilon r^{\alpha m} + C_2\varepsilon r^{\alpha m} r^3.
\end{aligned}$$

We first take r such that

$$C_2 r^{1-\alpha} \leq \frac{1}{2}$$

and then δ such that

$$C_1\delta \leq \frac{1}{2} r^{2+\alpha}.$$

Hence, for any $y \in B_r$,

$$|v(y) - Q_m(y) - \widetilde{c}_m - \widetilde{l}_m(y) - \widetilde{Q}_m(y)| \leq \varepsilon r^{\alpha m} r^{2+\alpha}.$$

With the definition of v, we obtain, for any $y \in B_r$,

$$\begin{aligned}
&\left|\frac{1}{r^{2m}}\left[u(r^m y) - c_m - l_m(r^m y)\right] - Q_m(y) - \widetilde{c}_m - \widetilde{l}_m(y) - \widetilde{Q}_m(y)\right| \\
&\quad \leq \varepsilon r^{\alpha m} r^{2+\alpha},
\end{aligned}$$

and hence, for any $x \in B_{r^{m+1}}$,

$$\begin{aligned}
&\left|u(x) - c_m - l_m(x) - Q_m(x) - r^{2m}\widetilde{c}_m - r^m\widetilde{l}_m(x) - \widetilde{Q}_m(x)\right| \\
&\quad \leq \varepsilon r^{(m+1)(2+\alpha)}.
\end{aligned}$$

Set

$$\begin{aligned} c_{m+1} &= c_m + r^{2m}\widetilde{c}_m, \\ l_{m+1} &= l_m + r^m \widetilde{l}_m, \\ Q_{m+1} &= Q_m + \widetilde{Q}_m. \end{aligned}$$

Then,

$$|u - c_{m+1} - l_{m+1} - Q_{m+1}| \leq \varepsilon r^{(m+1)(2+\alpha)} \quad \text{in } B_{r^{m+1}},$$

and

$$\begin{aligned} |c_{m+1} - c_m| &= r^{2m}|\widetilde{c}_m| \leq C_2 \varepsilon r^{(2+\alpha)m}, \\ |\nabla(l_{m+1} - l_m)| &= r^m |\nabla \widetilde{l}_m| \leq C_2 \varepsilon r^{(1+\alpha)m}, \\ \|\nabla^2(Q_{m+1} - Q_m)\| &= \|\nabla^2 \widetilde{Q}_m\| \leq C_2 \varepsilon r^{\alpha m}. \end{aligned} \tag{14}$$

We note that (14) holds for any $k \leq m$ in place of m by the induction hypothesis. Then, a simple iteration yields

$$\begin{aligned} |c_{m+1}| &\leq \frac{C_2\varepsilon}{1 - r^{2+\alpha}}, \\ |\nabla l_{m+1}| &\leq \frac{C_2\varepsilon}{1 - r^{1+\alpha}}, \\ \|\nabla^2(Q_{m+1} - Q_0)\| &\leq \frac{C_2\varepsilon}{1 - r^{\alpha}}. \end{aligned}$$

By requiring

$$r^\alpha \leq \frac{1}{2},$$

we have

$$\begin{aligned} |c_{m+1}| &< 2C_2\varepsilon, \\ |\nabla l_{m+1}| &< 2C_2\varepsilon, \\ \|\nabla^2(Q_{m+1} - Q_0)\| &< 2C_2\varepsilon. \end{aligned}$$

Next, we require

$$2C_2\varepsilon \leq 1.$$

Then,

$$\begin{aligned} |c_{m+1}| &< 1, \\ |\nabla l_{m+1}| &< 1, \\ \|\nabla^2(Q_{m+1} - Q_0)\| &< 1. \end{aligned}$$

In particular, $\|\nabla^2 Q_{m+1}\| < 2$. This finishes the proof of (11) for $m+1$ and the proof of (12).

By another iteration of (14), we obtain, for any integers $k \geq m \geq 0$,

$$|c_k - c_m| \leq \frac{C_2 \varepsilon r^{(2+\alpha)m}}{1 - r^{2+\alpha}},$$
$$|\nabla(l_k - l_m)| \leq \frac{C_2 \varepsilon r^{(1+\alpha)m}}{1 - r^{1+\alpha}},$$
$$\|\nabla^2(Q_k - Q_m)\| \leq \frac{C_2 \varepsilon r^{\alpha m}}{1 - r^{\alpha}}.$$

Hence, $c_k \to c$, $l_k \to l$, and $Q_k \to Q$ for some constant c, some linear function l, and some homogeneous quadratic polynomial Q, and, with $r^\alpha \leq 1/2$,

$$|c_m - c| \leq 2C_2 \varepsilon r^{(2+\alpha)m},$$
$$|\nabla(l_m - l)| \leq 2C_2 \varepsilon r^{(1+\alpha)m},$$
$$\|\nabla^2(Q_m - Q)\| \leq 2C_2 \varepsilon r^{\alpha m}.$$

Moreover, we obtain, for any $x \in \mathbb{R}^n$,

$$|l_m(x) - l(x)| \leq 2C_2 \varepsilon r^{(1+\alpha)m}|x|,$$
$$|Q_m(x) - Q(x)| \leq 2C_2 \varepsilon r^{\alpha m}|x|^2.$$

For each $x \in B_1$, we take a nonnegative integer m such that $r^{m+1} \leq |x| < r^m$. Then, by (11),

$$\begin{aligned}
&|u(x) - c - l(x) - Q(x)| \\
&\quad \leq |u(x) - c_m - l_m(x) - Q_m(x)| \\
&\qquad + |c - c_m| + |l(x) - l_m(x)| + |Q(x) - Q_m(x)| \\
&\quad \leq \varepsilon r^{(2+\alpha)m} + C\varepsilon r^{(2+\alpha)m} + C\varepsilon r^{(1+\alpha)m} r^m + C\varepsilon r^{\alpha m} r^{2m} \\
&\quad \leq C\varepsilon r^{(2+\alpha)m} \leq \frac{C\varepsilon}{r^{2+\alpha}}|x|^{2+\alpha}.
\end{aligned}$$

This implies $c = 0$ since $u(0) = 0$. In conclusion, we take

$$\varepsilon = (2C_2)^{-1}, \quad r = \min\left\{r_0, 2^{-\frac{1}{\alpha}}, (2C_2)^{-\frac{1}{1-\alpha}}\right\},$$

and then,

$$\delta = \frac{r^{2+\alpha}}{2C_1}.$$

We then have the desired estimate. □

We now continue the proof of Theorem 8.5.2. Let ε and δ be as in Theorem 8.5.3, μ and σ be determined as in Step 2, and h be as in Step 1. These are fixed positive constants depending only on n, α, λ, Λ, and M. Recall that, from Steps 1 and 2,

$$y = T_h x, \quad u_h(y) = |\det T_h|^{\frac{2}{n}} u(x),$$

and, for $i = 1, \ldots, n$,

$$z_i = \frac{\sqrt{\lambda_i}}{\mu} y_i, \quad u_{h,\mu}(z) = \frac{1}{\mu^2}\big[u_h(y) - l_\mu(y)\big],$$

where T_h is a linear transform and l_μ is a linear function satisfying

$$|\det T_h| \leq C, \quad |\nabla l_\mu| \leq C.$$

Note that (9) and (10) imply that $u_{h,\mu}$ satisfies the assumptions in Theorem 8.5.3. By applying Theorem 8.5.3 to $u_{h,\mu}$, we obtain, for any $z \in B_1$,

$$|u_{h,\mu}(z) - l(z) - Q(z)| \leq C|z|^{2+\alpha},$$

where l and Q are a linear function and a homogeneous quadratic polynomial, respectively, satisfying

$$|\nabla l| + \|\nabla^2 Q\| \leq C.$$

By a simple substitution, we obtain, for any $x \in B_{r_*}$,

$$|u(x) - P(x)| \leq C|x|^{2+\alpha}, \tag{15}$$

where r_* is a positive constant depending only on n, λ, and Λ and P is a quadratic polynomial given by

$$P(x) = \frac{1}{|\det T_h|^{\frac{2}{n}}}\big[l_\mu(y) + \mu^2 l(z)\big] + \frac{\mu^2}{|\det T_h|^{\frac{2}{n}}} Q(z).$$

By $\det \nabla_z^2 Q = 1$ and $\lambda_1 \cdots \lambda_n = 1$, we have $\det \nabla_x^2 P = 1$. We also point out that (15) holds easily for any $x \in \Omega \setminus B_{r_*}$ by the bounds of u, $\nabla P(0)$, and $\nabla^2 P(0)$. □

We now generalize Theorem 8.5.2 to arbitrary interior points.

Theorem 8.5.4. *Let $\alpha \in (0,1)$ and $\delta > 0$ be constants, Ω be a convex domain in $\mathbb{R}^n$ with $B_{1/n} \subset \Omega \subset B_1$, $x_0 \in \Omega$ be a point with* $\operatorname{dist}(x_0, \partial\Omega) \geq \delta$, *and f be a bounded function in Ω satisfying, for some positive constants λ, Λ, and M,*

$$\lambda \leq f \leq \Lambda \quad \text{in } \Omega$$

and, for any $x \in \Omega$,

$$|f(x) - f(x_0)| \leq M|x - x_0|^\alpha.$$

Suppose that $u \in C(\bar{\Omega})$ is a convex generalized solution of

$$\begin{aligned} \det \nabla^2 u &= f \quad \text{in } \Omega,\\ u &= 0 \quad \text{on } \partial\Omega. \end{aligned}$$

Then, there exists a quadratic polynomial P such that

$$\begin{aligned} &\det \nabla^2 P = f(x_0),\\ &|P(x_0)| + |\nabla P(x_0)| + \|\nabla^2 P(x_0)\| \leq C \end{aligned}$$

and, for any $x \in \Omega$,

$$|u(x) - P(x)| \le C|x - x_0|^{2+\alpha},$$

where C is a positive constant depending only on n, α, δ, λ, Λ, $f(x_0)$, and M.

Proof. By Lemma 8.4.3, we have $S_{h_0}(x_0) \Subset \Omega$ for some constant h_0 depending only on n, λ, Λ, and δ. For brevity, we write $S = S_{h_0}(x_0)$. By Theorem 8.4.4 and Theorem 8.4.5, we obtain, for any $x \in S$,

$$c_1^{-1}|x - x_0|^{1+\beta_1} \le u(x) - l_{x_0}(x) \le c_2^{-1}|x - x_0|^{1+\beta_2},$$

where l_{x_0} is a supporting function of u at x_0, β_1 and β_2 are positive constants depending only on n, λ, and Λ, and c_1 and c_2 are positive constants depending only on n, λ, Λ, and δ. Therefore,

$$B_{r_2}(x_0) \subset S \subset B_{r_1}(x_0),$$

where

$$r_1 = (c_1 h_0)^{\frac{1}{1+\beta_1}} \quad \text{and} \quad r_2 = (c_2 h_0)^{\frac{1}{1+\beta_2}}.$$

Consider an affine transform $y = Tx$ such that $B_{1/n} \subset T(S) \subset B_1$. Then,

$$T\left(B_{r_2}(x_0)\right) \subset B_1 \quad \text{and} \quad B_{1/n} \subset T\left(B_{r_1}(x_0)\right).$$

With $Tx_0 \in B_1$, we have $T\left(B_{r_2}(x_0)\right) \subset B_2(Tx_0)$. Next, we note that $T\left(B_{r_1}(x_0)\right)$ is an ellipsoid with its center Tx_0. By the symmetry with respect to the center and the convexity of the ellipsoid, we have $B_{1/n}(Tx_0) \subset T\left(B_{r_1}(x_0)\right)$. Therefore, for any $x_1, x_2 \in \mathbb{R}^n$,

$$c_0^{-1}|x_1 - x_2| \le |Tx_1 - Tx_2| \le c_0|x_1 - x_2|, \tag{1}$$

where c_0 is a positive constant depending only on n, λ, Λ, and δ. In the following, we write $y = Tx = Ax + y_*$, for some point $y_* \in \mathbb{R}^n$ and some invertible matrix A. Then, (1) implies

$$c^{-1} \le |\det A| \le c,$$

where c is a positive constant depending only on n, λ, Λ, and δ.

Set, for any $y \in T(S)$,

$$v(y) = |\det A|^{\frac{2}{n}}\left[u(x) - l_{x_0}(x) - h_0\right],$$

where l_{x_0} is a supporting function of u at x_0 as above. Then,

$$\begin{aligned} \det \nabla^2 v &= \widetilde{f} \quad \text{in } T(S), \\ v &= 0 \quad \text{on } \partial(T(S)), \end{aligned}$$

where $\widetilde{f}$ is given by $\widetilde{f}(y) = f(x)$. Set $y_0 = Tx_0$. Then, v attains at y_0 its minimum in $T(S)$. By (1), we have, for any $y \in T(S)$,

$$|\widetilde{f}(y) - \widetilde{f}(y_0)| \le M c_0^{\alpha}|y - y_0|^{\alpha}.$$

By Theorem 8.5.2, there exists a quadratic polynomial $\widetilde{P}$ such that

$$\det \nabla^2 \widetilde{P} = \widetilde{f}(y_0),$$
$$|\widetilde{P}(y_0)| + |\nabla \widetilde{P}(y_0)| + \|\nabla^2 \widetilde{P}(y_0)\| \le C$$

and, for any $y \in T(S)$,

$$|v(y) - \widetilde{P}(y)| \le C|y - y_0|^{2+\alpha},$$

where C is a positive constant depending only on n, α, δ, λ, Λ, $f(x_0)$, and M. Set

$$P(x) = |\det A|^{-\frac{2}{n}} \widetilde{P}(y) + l_{x_0}(x) + h_0.$$

Then,

$$\det \nabla^2 P = f(x_0),$$
$$|P(x_0)| + |\nabla P(x_0)| + \|\nabla^2 P(x_0)\| \le C,$$

and, for any $x \in S$,

$$|u(x) - P(x)| \le C|x - x_0|^{2+\alpha}. \tag{2}$$

We point out that we used (1) again in the verification of (2). Note that (2) holds also for any $x \in \Omega \setminus S$. □

As a consequence of Theorem 8.5.4, we have the following $C^{2,\alpha}$-regularity and estimates.

Theorem 8.5.5. *Let $\alpha \in (0,1)$ be a constant, $\Omega \subset \mathbb{R}^n$ be a bounded convex domain with $B_{1/n} \subset \Omega \subset B_1$, and $f \in C^\alpha(\Omega)$ with a bounded $C^\alpha(\Omega)$-norm and a positive lower bound. Suppose that $u \in C(\bar{\Omega})$ is a convex generalized solution of*

$$\det \nabla^2 u = f \quad \text{in } \Omega,$$
$$u = 0 \quad \text{on } \partial\Omega.$$

Then, $u \in C^{2,\alpha}(\Omega)$ and, for any $\Omega' \Subset \Omega$,

$$|\nabla^2 u|_{C^\alpha(\Omega')} \le C,$$

where C is a positive constant depending only on n, α, $\operatorname{dist}(\Omega', \partial\Omega)$, $\inf_\Omega f$, *and $|f|_{C^\alpha(\Omega)}$.*

We leave the proof as an exercise.

Bibliography

[1] Aleksandrov, A. D., *Certain estimates for the Dirichlet problem*, Dokl. Akad. Nauk. SSSR, 134(1960), 1001–1004 [Russian]; English translation in Soviet Math. Dokl., 1(1960), 1151–1154.

[2] Aleksandrov, A. D., *Uniqueness conditions and estimates for the solution of the Dirichlet problem*, Vestnik Leningrad Univ., 18(1963), 5–29; English translation in Amer. Math. Soc. Transl., 68(1968), 89–119.

[3] Aubin, T., *Real Monge-Ampère equations*, J. Funct. Anal., 41(1981), 354–377.

[4] Aubin, T., *Some Nonlinear Problems in Riemannian Geometry*, Springer, Berlin, 1998.

[5] Bakelman, I. Ya., *Generalized solutions of Monge-Ampère equations*, Dokl. Akad. Nauk SSSR, 114(1957) 1143–1145.

[6] Bakelman, I. Ya., *Mean curvature and quasilinear elliptic equations*, Sibirsk. Mat. Ž., 9(1968), 1014–1040 [Russian]; English translation in Siberian Math. J., 9(1968), 752–771.

[7] Bakelman, I. J., *The Dirichlet problem for the elliptic N-dimensional Monge-Ampère equations and related problems in the theory of quasilinear elliptic equations*, Monge-Ampère equations and related topics (Florence, 1980), Ist. Naz. Alta Mat. Francesco Severi, Rome, 1–78, 1982.

[8] Bakelman, I. J., *Variational problems and elliptic Monge-Ampère equations*, J. Diff. Geom., 18(1983), 669–699.

[9] Bakelman, I. J., *Convex Analysis and Nonlinear Geometric Elliptic Equations*, Springer-Verlag, Berlin, 1994.

[10] Bernstein, S., *Sur les équations du calcul des variations*, Ann. Sci. École Norm. Sup., 29(1912), 431–485.

[11] Bombieri, E., de Giorgi, E., Giusti, E., *Minimal cones and the Bernstein problem*, Invent. Math., 7(1969), 243–268.

[12] Bombieri, E., de Giorgi, E., Miranda, M., *Una maggioriazione a priori relativa alle ipersurfici minimali non parametriche*, Arch. Ration. Mech. Anal., 32(1969), 255–267.

[13] Caffarelli, L., *Interior a priori estimates for solutions of fully nonlinear equations*, Ann. of Math., 130(1989), 189–213.

[14] Caffarelli, L., *A localization property of viscosity solutions to the Monge-Ampère equation and their strict convexity*, Ann. of Math., 131(1990), 129–134.

[15] Caffarelli, L., *Interior $W^{2,p}$ estimates for solutions of Monge-Ampère equations*, Ann. of Math., 131(1990), 135–150.

[16] Caffarelli, L., *Some regularity properties of solutions of Monge-Ampère equation*, Comm. Pure Appl. Math., 44(1991), 965–969.

[17] Caffarelli, L., *The regularity of mappings with a convex potential*, J. Amer. Math. Soc., 5(1992), 99–104.

[18] Caffarelli, L., *Boundary regularity of maps with convex potentials*, Comm. Pure Appl. Math., 45(1992), 1141–1151.

[19] Caffarelli, L., *Boundary regularity of maps with convex potentials II*, Ann. of Math., 144(1996), 453–496.

[20] Caffarelli, L., Cabré, X., *Fully Nonlinear Elliptic Equations*, AMS Colloquium Publications, Vol. 43, Amer. Math. Soc., Providence, 1993.

[21] Caffarelli, L., Gutiérrez, C., *Real analysis related to the Monge-Ampère equation*, Trans. Amer. Math. Soc., 348(1996), 1075–1092.

[22] Caffarelli, L., Gutiérrez, C., *Properties of the solutions of the linearized Monge-Ampère equations*, Amer. J. Math., 119(1997), 423–465.

[23] Caffarelli, L., Kohn, J. J., Nirenberg, L., Spruck, J., *The Dirichlet problem for nonlinear second-order elliptic equations. II. Complex Monge-Ampère, and uniformly elliptic, equations*, Comm. Pure Appl. Math., 38(1985), 209–252.

[24] Caffarelli, L., Li, Y., *An extension to a theorem of Jörgens, Calabi, and Pogorelov*, Comm. Pure Appl. Math., 56(2003), 549–583.

[25] Caffarelli, L., Li, Y., *A Liouville theorem for solutions of the Monge-Ampère equation with periodic data*, Ann. Inst. H. Poincaré Anal. Non Linéaire, 21(2004), 97–120.

[26] Caffarelli, L., Li, Y., *Some multi-valued solutions to Monge-Ampère equations*, Comm. Anal. Geom., 14(2006), 411–441.

[27] Caffarelli, L., Nirenberg, L., Spruck, J., *The Dirichlet problem for nonlinear second-order elliptic equations I. Monge-Ampère equation*, Comm. Pure Appl. Math., 37(1984), 369–402.

[28] Caffarelli, L., Nirenberg, L., Spruck, J., *The Dirichlet problem for nonlinear second-order elliptic equations III. Functions of the eigenvalues of the Hessian*, Acta Math., 155(1985), 261–301.

[29] Caffarelli, L., Nirenberg, L., Spruck, J., *Nonlinear second-order elliptic equations IV. Starshaped compact Weingarten hypersurfaces*, Current topics in partial differential equations, ed. by Y. Ohya, K. Kasahara, N. Shimakura, Kinokunize Co., Tokyo, 1–26, 1986.

[30] Caffarelli, L., Nirenberg, L., Spruck, J., *The Dirichlet problem for nonlinear second-order elliptic equations V. The Dirichlet problem for Weingarten surfaces*, Comm. Pure Appl. Math., 41(1988), 47–70.

[31] Calabi, E., *Improper affine hyperspheres of convex type and a generalization of a theorem by K. Jörgens*, Michigan Math. J., 5(1958), 105–126.

[32] Calderon, A. P., Zymnund, A., *On the existence of certain singular integrals*, Acta Math., 88(1952), 85–139.

[33] Cheng, S.-Y., Yau, S.-T., *On the regularity of the solution of the n-dimensional Minkowski problem*, Comm. Pure Appl. Math., 29(1976), 495–516.

[34] Cheng, S.-Y., Yau, S.-T., *On the regularity of the Monge-Ampère equation* $\det(\partial^2/\partial x_i \partial x_j) = F(x,u)$, Comm. Pure Appl. Math., 30(1977), 41–68.

[35] Cheng, S.-Y., Yau, S.-T., *On the existence of a complete Kähler metric on non-compact complex manifolds and the regularity of Fefferman's equation*, Comm. Pure Appl. Math., 33(1980), 507–544.

[36] Cheng, S.-Y., Yau, S.-T., *The real Monge-Ampère equation and affine flat structures*, Proc. 1980 Beijing Symposium on Differential Geometry and Differential Equations, Vol. I, ed. by S.-S. Chern, W.-T. Wu, 339–370, 1982.

[37] Cheng, S.-Y., Yau, S.-T., *Complete affine hypersurfaces I, The completeness of affine metrics*, Comm. Pure Appl. Math., 39(1986), 839–866.

[38] Chou, K.-S., *On a real Monge Ampère functional*, Invent. Math., 101(1990), 425 448.

[39] Chou, K.-S., *Convex hypersurfaces with prescribed Gauss-Kronecker curvature*, J. Diff. Geom., 34(1991), 389–410.

[40] Chou, K.-S., Wang, X.-J., *Entire solutions of the Monge-Ampère equation*, Comm. Pure Appl. Math., 49(1996), 529–539.

[41] Chou, K.-S., Wang, X.-J., *A variational theory of the Hessian equation*, Comm. Pure Appl. Math., 54(2001), 1029–1064.

[42] Dai, Q., Trudinger, N. S., Wang, X.-J., *The mean curvature measure*, J. Eur. Math. Soc., 14(2012), 779–800.

[43] Daskalopoulos, P., Savin, O., *On Monge-Ampère equations with homogeneous right-hand sides*, Comm. Pure Appl. Math., 62(2009), 639–676.

[44] de Giorgi, E., *Sulla differenziabilità e l'analiticità delle estremali degli integrali multipli regolari*, Mem. Accad. Sci. Torino Cl. Sci. Fis. Mat. Natur., 3(1957), 25–43.

[45] Delanoë, P., *Equations du type de Monge-Ampère sur les varietes riemanniennes coompactes I, II, III*, J. Funct. Anal., 40(1981), 358–386; 41(1981), 341–353; 45(1982), 403–430.

[46] De Philippis, G., Figalli, A., Savin, O., *A note on interior $W^{2,1+\varepsilon}$ estimates for the Monge-Ampère equation*, Math. Ann., 357(2013), 11–22.

[47] Dierkes, U., Hildebrandt, S., Sauvigny, F., *Minimal Surfaces*, 2nd ed., Springer, Berlin, 2010.

[48] Dierkes, U., Hildebrandt, S., Tromba, A., *Regularity of Minimal Surfaces*, 2nd ed., Springer, Berlin, 2010.

[49] Dierkes, U., Hildebrandt, S., Tromba, A., *Global Analysis of Minimal Surfaces*, 2nd ed., Springer, Berlin, 2010.

[50] Donaldson, S. K., *Interior estimates for solutions of Abreu's equation*, Collect. Math., 56(2005), 103–142.

[51] Evans, L. C., *Classical solutions of fully nonlinear, convex, second order elliptic equations*, Comm. Pure Appl. Math., 25(1982), 333–363.

[52] Evans, L. C., *Classical solutions of the Hamilton-Jacobi Bellman equation for uniformly elliptic equations*, Trans. Amer. Math. Soc., 275(1983), 245–255.

[53] Evans, L. C., Gariepy, R., *Measure Theory and Fine Properties of Functions*, CRC Press, Boca Raton, 1992.

[54] Evans, L. C., Lions, P.-L., *Optimal stochastic switching and the Dirichlet problem for the Bellman equations*, Trans. Amer. Math. Soc., 253(1979), 365–389.

[55] Evans, L. C., Lions, P.-L., *Résolution des équations de Hamilton-Jacobi-Bellman pour des opérateurs uniformèment elliptiques*, C. R. Acad. Sci. Paris, 290(1980), 1049–1052.

[56] Evans, L. C., Lions, P.-L., *Fully non-linear second order elliptic equations with large zeroth order coefficient*, Ann. Inst. Fourier (Grenoble), 31(1981), 175–191.

[57] Federer, H., *Geometric Measure Theory*, Springer, Berlin, 1969.

[58] Finn, R., *On equations of minimal surface type*, Ann. Math., 60(1954), 397–416.

[59] Gilbarg, D., Trudinger, N., *Elliptic Partial Differential Equations of Second Order*, 3rd ed., Springer, Berlin, 2001.

[60] Giusti, E., *Minimal Surfaces and Functions of Bounded Variation*, Birkhäuser, Boston, 1984.

[61] Guan, B., *The Dirichlet problem for Monge-Ampère equations in non-convex domains and spacelike hypersurfaces of constant Gauss curvature*, Trans. Amer. Math. Soc., 350(1998), 4955–4971.

[62] Guan, B., *The Dirichlet problem for complex Monge-Ampère equations and regularity of the pluri-complex Green function*, Comm. Anal. Geom., 6(1998), 687–703.

[63] Guan, B., *The Dirichlet problem for complex Monge-Ampère equations and applications*, Trends in Partial Differential Equations, Higher Ed. Press, 53–97, 2009.

[64] Guan, B., Guan, P., *Convex hypersurfaces of prescribed curvatures*, Ann. Math., 156(2002), 655–673.

[65] Guan, B., Li, Y., *Monge-Ampère equations on Riemannian manifolds*, J. Diff. Equations, 132(1996), 126–139.

[66] Guan, B., Spruck, J., *Boundary value problem on $\mathbb{S}^n$ for surfaces of constant Gauss curvature*, Ann. Math., 138(1993), 601–624.

[67] Guan, B., Spruck, J., *The existence of hypersurfaces of constant Gauss curvature with prescribed boundary*, J. Diff. Geom., 62(2002), 259–287.

[68] Guan, P., *C^2 a priori estimates for degenerate Monge-Ampère equations*, Duke Math. J., 86(1997), 323–346.

[69] Guan, P., Li, J., Li, Y., *Hypersurfaces of prescribed curvature measure*, Duke Math. J., 161(2012), 1927–1942.

[70] Guan, P., Li, Y., *The Weyl problem with nonnegative Gauss curvature*, J. Diff. Geom., 39(1994), 331–342.

[71] Guan, P., Li, Y., *$C^{1,1}$ estimates for solutions of a problem of Alexandrov*, Comm. Pure Appl. Math., 50(1997), 789–811.

[72] Guan, P., Ma, X., *The Christoffel-Minkowski problem, I. Convexity of solutions of a Hessian equation*, Invent. Math., 151(2003), 553–577.

[73] Guan, P., Trudinger, N., Wang, X.-J., *On the Dirichlet problem for degenerate Monge-Ampère equations*, Acta Math., 182(1999), 87–104.

[74] Gutiérrez, C. E., *The Monge-Ampère Equation*, Birkhäuser, Boston, 2001.

[75] Han, Q., Hong, J.-X., *Isometric Embedding of Riemannian Manifolds in Euclidean Spaces*, Mathematical Surveys and Monographs, 130, Amer. Math. Soc., Providence, 2006.

[76] Han, Q., Lin, F.-H., *Elliptic Partial Differential Equations*, 2nd ed., Amer. Math. Soc., Providence, 2011.

[77] Heinz, E., *On elliptic Monge-Ampère equations and Weyl's embedding problem*, J. Analyse Math., 7(1959), 1–52.

[78] Heinz, E., *Über die Differentialungleichung* $0 < \alpha \leq rt - s^2\beta < \infty$, Math. Z., 72(1959/1960), 107–126.

[79] Heinz, E., *Interior estimates for solutions of elliptic Monge-Ampère equations*, Proc. Symp. Pure Math., Vol. IV, Amer. Math. Soc., Providence, 149–155, 1996.

[80] Hong, J.-X, *Darboux equations and isometric embedding of Riemannian manifolds with nonnegative curvature in* $\mathbb{R}^3$, Chin. Ann. of Math., 20B(1999), 123–136.

[81] Hong, J.-X., Huang, G., L^p *and Hölder estimates for a class of degenerate elliptic partial differential equations and its applications*, Inter. Math. Res. Notices, 13(2012), 2889–2941.

[82] Hong, J.-X., Huang, G., Wang, W., *Existence of global smooth solutions to Dirichlet problem for degenerate elliptic Monge-Ampère equations*, Comm. P. D. E., 36(2011), 635–656.

[83] Hong, J.-X., Zuily, C., *Isometric embedding of the 2-sphere with nonnegative curvature in* $\mathbb{R}^3$, Math. Z., 219(1995), 323–334.

[84] Hopf, E., *Elementare Bemerkungen über die Lösungen partieller Differentialgleichungen zweiter Ordnung vom elliptischen Typus*, Sitz. Ber. Preuss. Akad. Wissensch. Berlin, Math.-Phys. Kl., 19(1927), 147–152.

[85] Huang, Q., *On the mean oscillation of the Hessian of solutions to the Monge-Ampère equation*, Adv. Math., 207(2006), 599–616.

[86] Ivochkina, N. M., *A priori estimate of* $\|u\|_{C^2(\bar{\Omega})}$ *of convex solutions of the Dirichlet problem for the Monge-Ampère equation*, Zap. Nauchn. Sem. Leningrad. Otdel. Mat. Inst. Steklov. (LOMI), 96(1980), 69–79 (Russian); English translation in J. Soviet Math., 21(1983), 689–697.

[87] Ivochkina, N. M., *Classical solvability of the Dirichlet problem for the Monge-Ampère equation*, Zap. Nauchn. Sem. Leningrad. Otdel. Mat. Inst. Steklov. (LOMI), 131(1983), 72–79.

[88] Ivochkina, N., Trudinger, N. S., Wang, X.-J., *The Dirichlet problem for degenerate Hessian equations*, Comm. P. D. E., 29(2004), 219–235.

[89] Jian, H., Wang, X.-J., *Continuity estimates for the Monge-Ampère equation*, SIAM J. Math. Anal., 39(2007), 608–626.

[90] Jian, H., Wang, X.-J., *Bernstein theorem and regularity for a class of Monge-Ampère equations*, J. Diff. Geom., 93(2013), 431–469.

[91] John, F., *Extremum problems with inequalities as subsidiary conditions*, Studies and Essays Presented to R. Courant on His 60th Birthday, Interscience Publishers, New York, 187–204, 1948.

[92] John, F., Nirenberg, L., *On functions of bounded mean oscillation*, Comm. Pure Appl. Math., 14(1961), 415–426.

[93] Jörgens, K., *Über die Lösungen der Differentialgleichung* $rt - s^2 = 1$, Math. Ann., 127(1954), 130–134.

[94] Kellogg, O. D., *On the derivatives of harmonic functions on the boundary*, Trans. Amer. Math. Soc., 33(1931), 486–510.

[95] Korevaar, N., *An easy proof of the interior gradient bound for solutions to the prescribed mean curvature equation*, Nonlinear Functional Analysis and Its Applications, Proc. Symp. Pure Math., 45, (2), Amer. Math. Soc., Providence, 81–90, 1986.

[96] Krylov, N. V., *Boundedly inhomogeneous elliptic and parabolic equations*, Izvestia Akad. Nauk. SSSR, 46(1982), 487–523 [Russian]; English translation in Math. USSR Izv., 20(1983), 459–492.

[97] Krylov, N. V., *Boundedly inhomogeneous elliptic and parabolic equations in a domain*, Izvestia Akad. Nauk. SSSR, 47(1983), 75–108 [Russian]; English translation in Math. USSR Izv., 22(1984), 67–97.

[98] Krylov, N. V., *Nonlinear Elliptic and Parabolic Equations of the Second Order*, Reidel, Dordrecht-Boston, 1987.

[99] Krylov, N. V., Safonov, M. V., *An estimate of the probablitily that a diffusion process hits a set of positive measure*, Dokl. Akad. Nauk. SSSR, 245(1979), 253–255 [Russian]; English translation in Soviet Math. Dokl., 20(1979), 253–255.

[100] Krylov, N. V., Safonov, M. V., *Certain properties of solutions of parabolic equations with measurable coefficients*, Izvestia Akad. Nauk. SSSR, 40(1980), 161–175 [Russian]; English translation in Math. USSR Izv., 161(1981), 151–164.

[101] Ladyzhenskaya, O. A., Ural'tseva, N. N., *Quasilinear elliptic equations and variational problems with several independent variables*, Uspehi Mat. Nauk, 16(1961), 19–90 [Russian]; English translation in Russian Math Survey, 16(1961), 17–92.

[102] Ladyzhenskaya, O. A., Ural'tseva, N. N., *On Hölder continuity of solutions and their derivatives of linear and quasilinear elliptic and parabolic equations*, Trudy Mat. Inst. Steklov., 73(1964), 172–220 [Russian]; English translation in Amer. Math. Soc. Transl., 61(1967), 207–269.

[103] Ladyzhenskaya, O. A., Ural'tseva, N. N., *Linear and Quasilinear Elliptic Equations*, Academic Press, New York, 1968.

[104] Li, A.-M., Xu, R., Simon, U., Jia, F., *Affine Bernstein Problems and Monge-Ampère Equations*, World Scientific, Singapore, 2010.

[105] Li, Y., *Some existence results for fully nonlinear elliptic equations of Monge-Ampère type*, Comm. Pure Appl. Math., 43(1990), 233–271.

[106] Li, Y., *Interior gradient estimates for solutions of certain fully nonlinear elliptic equations*, J. Diff. Equations, 90(1991), 172–185.

[107] Lieberman, G., Trudinger, N. S., *Nonlinear oblique boundary value problems for nonlinear elliptic equations*, Trans. Amer. Math. Soc., 295(1986), 509–546.

[108] Lions, P.-L., *Résolution analytique des problèmes de Bellman-Dirichlet*, Acta Math., 146(1981), 151–166.

[109] Lions, P.-L., *Sur les équations de Monge-Ampère I, II*, Manu. Math., 41(1983), 1–44; Arch. Ration. Mech. Anal., 89(1985), 93–122.

[110] Lions, P.-L., *Two remarks on Monge-Ampère equations*, Ann. Mat. Pura Appl. 142(1985), 263–275.

[111] Lions, P.-L., Trudinger, N. S., Urbas, J. I. E., *The Neumann problem for equations of Monge-Ampère type*, Comm. Pure Appl. Math., 39(1986), 539–563.

[112] Liu J., Trudinger, N. S., *On Pogorelov estimates for Monge-Ampère type equations*, Discrete Contin. Dyn. Syst., 28(2010), 1121–1135.

[113] Michael, J. H., Simon, L. M., *Sobolev and mean value inequalities on generalized submanifolds of* $\mathbb{R}^n$, Comm. Pure Appl. Math., 26(1973), 361–379.

[114] Miranda, C., *Partial Differential Equations of Elliptic Type*, 2nd ed., Springer, Berlin, 1970.

[115] Morrey, C. B., Jr., *Multiple Integrals in the Calculus of Variations*, Springer, Berlin, 1966.

[116] Morrey, C. B., Jr., *Second order elliptic equations in several variables and Hölder continuity*, Math. Z., 72(1959), 146–164.

[117] Moser, J., *A new proof of de Giorgi's theorem concerning the regularity problem for elliptic differential equations*, Comm. Pure Appl. Math., 13(1960), 457–468.

[118] Moser, J., *On Harnack's theorem for elliptic differential equations*, Comm. Pure Appl. Math., 14(1961), 577–591.

[119] Nash, J., *Continuity of solutions of parabolic and elliptic equations*, Amer. J. Math., 80(1958), 931–954.

[120] Pogorelov, A. V., *Monge-Ampère Equations of Elliptic Type*, Noordhoff, Groningen, 1964.

[121] Pogorelov, A. V., *On the improper affine hyperspheres*, Geom. Dedicata, 1(1972), 33–46.

[122] Pogorelov, A. V., *The Multidimensional Minkowski Problems*, J. Wiley & Sons, New York, 1978.

[123] Protter, M. H., Weinberger, H. F., *Maximum Principles in Differential Equations*, Prentice-Hall, Englewood Cliffs, 1978.

[124] Pucci, C., *Limitazioni per soluzioni di equazioni ellittiche*, Ann. Mat. Pura Appl., 74(1966), 15–30.

[125] Safonov, M. V., *Harnack inequalities for elliptic equations and Hölder continuity of their solutions*, Zap. Naučn. Sem. Leningrad. Otdel. Mat. Inst. Steklov., 96(1980), 272–287 [Russian]; English translation in J. Soviet Math., 21(1983), 851–863.

[126] Safonov, M. V., *Classical solution of second-order nonlinear elliptic equations*, Izv. Akad. Nauk SSSR Ser. Mat., 52(1988), 1272–1287 (Russian); English translation in Math. USSR-Izv., 33(1989), 597–612.

[127] Savin, O., *A Liouville theorem for solutions to the linearized Monge-Ampère equation*, Discrete Contin. Dyn. Syst., 28(2010), 865–873.

[128] Savin, O., *Pointwise $C^{2,\alpha}$ estimates at the boundary for the Monge-Ampère equation*, J. Amer. Math. Soc., 26(2013), 63–99.

[129] Savin, O., *Global $W^{2,p}$ estimates for the Monge-Ampère equation*, Proc. Amer. Math. Soc., 141(2013), 3573–3578.

[130] Schauder, J., *Über lineare elliptische Differentialgleichungen zweiter Ordnung*, Math. Z., 38(1934), 257–282.

[131] Schauder, J., *Numerische Abschätzungen in elliptische linearen Differentialgleichungen*, Studia Math., 5(1935), 34–42.

[132] Schoen, R., Simon, L., Yau, S.-T., *Curvature estimates for minimal hypersurfaces*, Acta Math., 134(1974), 275–288.

[133] Serrin, J., *Local behavior of solutions of quasi-linear elliptic equations*, Acta Math., 111(1964), 247–302.

[134] Serrin, J., *The problem of Dirichlet for quasilinear elliptic differential equations with many independent variables*, Philos. Trans. Roy. Soc. London Ser. A, 264(1969), 413–496.

[135] Sheng, W., Trudinger, N., Wang, X.-J., *Prescribed Weingarten curvature equations*, Recent Developments in Geometry and Analysis, 359–386, International Press, Somerville, 2012.

[136] Sheng, W., Urbas, J., Wang, X.-J., *Interior curvature bounds for a class of curvature equations*, Duke Math. J., 123(2004), 235–264.

[137] Simon, L., *Remarks on curvature estimates for minimal hypersurfaces*, Duke Math. J., 43(1976), 545–553.

[138] Simon, L., *Lectures on geometric measure theory*, Proc. Centre Math. Anal. 3, Australian National University, Centre for Mathematical Analysis, Canberra, 1984.

[139] Simon, L., *The minimal surface equation*, Geometry V, Minimal Surfaces, ed. by R. Osserman, Encycl. Math. Sci, 90, Springer, Berlin, 239–266, 1997.

[140] Simons, J., *Minimal varieties in Riemannian manifolds*, Ann. Math., 88(1968), 62–105.

[141] Stampacchia, G., *Problemi al contorno ellittici, con dati discontinui, dotati di soluzioni hölderiane*, Ann. Mat. Pura Appl. 51(1960), 1–37.

[142] Trudinger, N. S., *On Harnack type inequalities and their application to quasilinear elliptic equations*, Comm. Pure Appl. Math., 20(1967), 721–747.

[143] Trudinger, N. S., *The boundary gradient estimate for quasilinear elliptic and parabolic differential equations*, Indiana Univ. Math. J., 21(1972), 657–670.

[144] Trudinger, N. S., *Gradient estimates and mean curvature*, Math. Z., 131(1973), 165–175.

[145] Trudinger, N. S., *Local estimates for subsolutions and supersolutions of general second order elliptic quasilinear equations*, Invent. Math., 61(1980), 67–79.

[146] Trudinger, N. S., *Fully nonlinear, uniformly elliptic equations under natural structure conditions*, Trans. Amer. Math. Soc., 28(1983), 217–231.

[147] Trudinger, N. S., *Boundary value problems for fully nonlinear elliptic equations*, Proc. Center for Math. Anal. Aust. Nat. Univ., 8(1984), 65–83.

[148] Trudinger, N. S., *On the Dirichlet problem for Hessian equations*, Acta Math., 175(1995), 151–164.

[149] Trudinger, N. S., Urbas, J. I. E., *The Dirichlet problem for the equation of prescribed Gauss curvature*, Bull. Austral. Math. Soc., 28(1983), 217–231.

[150] Trudinger, N. S., Urbas, J. I. E., *On second derivative estimates for equations of Monge-Ampère type*, Bull. Austral. Math. Soc., 30(1984), 321–334.

[151] Trudinger, N. S., Wang, X.-J., *Hessian measures. I, II, III*, Dedicated to Olga Ladyzhenskaya., Topol. Methods Nonlinear Anal., 10(1997), 225–239; Ann. of Math., 150(1999), 579–604; J. Funct. Anal., 193(2002), 1–23.

[152] Trudinger, N. S., Wang, X.-J., *The Bernstein problem for affine maximal hypersurfaces*, Invent. Math., 140(2000), 399–422.

[153] Trudinger, N. S., Wang, X.-J., *Bernstein-Jörgens theorem for a fourth order partial differential equation*, J. Partial Diff. Equations, 15(2002), 78–88.

[154] Trudinger, N. S., Wang, X.-J., *The Monge-Ampère equation and its geometric applications*, Handbook of geometric analysis, No. 1, 467–524, Adv. Lect. Math. (ALM), 7, International Press, Somerville, 2008.

[155] Trudinger, N. S., Wang, X.J., *Boundary regularity for the Monge-Ampère and the affine maximal surface equations*, Ann. Math., 167(2008), 993–1028.

[156] Trudinger, N. S., Wang, X.-J., *On the second boundary value problem for Monge-Ampère type equations and optimal transportation*, Ann. Sc. Norm. Super. Pisa Cl. Sci., 8(2009), 143–174.

[157] Urbas, J. I. E., *The generalized Dirichlet problem for equations of Monge-Ampère type*, Ann. Inst. H. Poincaré Anal. Non Linéaire, 3(1986), 209–228.

[158] Urbas, J. I. E., *Global Hölder estimates for equations of Monge-Ampère type*, Invent. Math., 91(1988), 1–29.

[159] Urbas, J. I. E., *Regularity of generalized solutions of Monge-Ampère equations*, Math. Z., 197(1988), 365–393.

[160] Urbas, J. I. E., *On the second boundary value problem for equations of Monge-Ampère type*, J. Reine Angew. Math., 487(1997), 115–124.

[161] Wang, X.-J., *Regularity for Monge-Ampère equation near the boundary*, Analysis, 16(1996), 101–107.

[162] Wang, X.-J., *Interior gradient estimates for mean curvature equations*, Math. Z., 228(1998), 73–81.

[163] Wang, X.-J., *Schauder estimates for elliptic and parabolic equations*, Chinese Ann. Math. Ser. B, 27(2006), 637–642.

[164] Wang, X.-J., *The k-Hessian equation*, Geometric analysis and PDEs, 177–252, Lecture Notes in Math., 1977, Springer, Dordrecht, 2009.

[165] Yau, S.-T., *Open problems in geometry*, Proc. Symposia Pure Math., Vol. 54, 1–28, 1993.

Index